T0224127

Lecture Notes in Computer Science 7311

Commenced Publication in 1973
Founding and Former Series Editors:
Gerhard Goos, Juris Hartmanis, and Jan van Leeuwen

Editorial Board

Jing Liu Cesare Alippi
Bernadette Bouchon-Meunier
Garrison W. Greenwood
Hussein A. Abbass (Eds.)

Advances
in Computational
Intelligence

IEEE World Congress
on Computational Intelligence, WCCI 2012
Brisbane, Australia, June 10-15, 2012
Plenary/Invited Lectures

 Springer

Volume Editors

Jing Liu
Xidian University
710071 Xi'an, Shaanxi, China
E-mail: neouma@mail.xidian.edu.cn

Cesare Alippi
Politecnico di Milano
20133 Milano, Italy
E-mail: cesare.alippi@polimi.it

Bernadette Bouchon-Meunier
Université Pierre et Marie Curie - Paris 6
75005 Paris, France
E-mail: bernadette.bouchon-meunier@lip6.fr

Garrison W. Greenwood
Portland State University
Portland, OR 97207, USA
E-mail: greenwd@ece.pdx.edu

Hussein A. Abbass
University of New South Wales
Canberra, ACT 2600, Australia
E-mail: h.abbass@adfa.edu.au

ISSN 0302-9743   e-ISSN 1611-3349
ISBN 978-3-642-30686-0   e-ISBN 978-3-642-30687-7
DOI 10.1007/978-3-642-30687-7
Springer Heidelberg Dordrecht London New York

Library of Congress Control Number: Applied for

CR Subject Classification (1998): H.3, H.2.8, H.2, I.2.3, I.2.6, H.4, I.5.3

LNCS Sublibrary: SL 3 – Information Systems and Application, incl. Internet/Web and HCI

Typesetting: Camera-ready by author, data conversion by Scientific Publishing Services, Chennai, India

Printed on acid-free paper

Springer is part of Springer Science+Business Media (www.springer.com)

Preface

The 2012 IEEE World Congress on Computational Intelligence (IEEE WCCI) is a premier event organized by the IEEE Computational Intelligence Society. On even years it hosts three large conferences: The IEEE Congress on Evolutionary Computation (CEC), the International Joint Conference on Neural Networks (IJCNN), and the IEEE International Conference on Fuzzy Systems (FUZZ-IEEE).

It has been a tradition for IEEE WCCI to issue a book composed of chapters written by the Plenary (Congress Level) and Invited Lecture (Conference Level) speakers. This volume is the culmination of this activity in 2012.

The Organizing Committee of IEEE WCCI 2012 wish to take this opportunity to acknowledge the contributions made by these speakers. Each of them presents a chapter capturing the state of the art in their own domain. Together, these chapters form a snapshot of recent research, trends and future directions in computational intelligence. We sincerely hope that the readers of this volume find useful information relevant to their own interest in computational intelligence.

June 2012

Jing Liu
Cesare Alippi
Bernadette Bouchon-Meunier
Garrison W. Greenwood
Hussein A. Abbass

Table of Contents

Lazy Meta-Learning: Creating Customized Model Ensembles on Demand

Piero P. Bonissone

GE Global Research Center, Niskayuna, NY 12309, USA
bonissone@ge.com

Abstract. In the not so distant future, we expect analytic models to become a commodity. We envision having access to a large number of data-driven models, obtained by a combination of crowdsourcing, crowdservicing, cloud-based evolutionary algorithms, outsourcing, in-house development, and legacy models. In this new context, the critical question will be model ensemble selection and fusion, rather than model generation. We address this issue by proposing *customized model ensembles on demand*, inspired by *Lazy Learning*. In our approach, referred to as *Lazy Meta-Learning*, for a given query we find the most relevant models from a DB of models, using their meta-information. After retrieving the relevant models, we select a subset of models with highly uncorrelated errors. With these models we create an ensemble and use their meta-information for dynamic bias compensation and relevance weighting. The output is a weighted interpolation or extrapolation of the outputs of the models ensemble. Furthermore, the confidence interval around the output is reduced as we increase the number of uncorrelated models in the ensemble. We have successfully tested this approach in a power plant management application.

Keywords: Machine learning, lazy learning, meta-learning, computational intelligence, fusion, ensemble, entropy, Pareto set, neural networks, coal-fired power plant management.

1 Analytic Model Building in the Near Future

Until recently, analytic model building has been a specialized craft. Usually, such models were handcrafted by specialized researchers and manually maintained. Typically, the model builder selected a data set for training, test, and validation, extracted a run-time model from the training data set using machine learning (ML) techniques as a compiler, validated the model using a validation set, and finally used the model to handle new queries. When a model deteriorated, the model builder created a new model by following a similar build cycle. As noted by the author in earlier papers [1-2], this lack of automation in model creation has led to bottlenecks in the models lifecycle, preventing their scalability and eventually leading to their obsolescence.

To address the lack of automation in model building, we have proposed the use of meta-heuristics, specifically the use of evolutionary algorithms, to generate runtime

J. Liu et al. (Eds.): WCCI 2012 Plenary/Invited Lectures, LNCS 7311, pp. 1–23, 2012.

analytical models, while limiting the amount of human intervention in the process [3-5]. Although this proposal was a step in the right direction, it was still aimed at generating single models or, in the best case, static ensembles of models [6-9]. While the model builder was no longer the bottleneck in the loop defining and running the experiments needed to create and test the model, s/he was still an integral part of the process.

We believe that the advent of cloud computing has changed dramatically the process of developing analytic models. Cloud computing has lowered the barriers to entry, enabling a vast number of analytics-skilled people to access in a flexible manner large amounts of computing cycles for relatively low operational cost (and without any capital expenses). The literature on cloud computing is growing exponentially, making it difficult to provide the interested reader with a single reference. However, reference [10] provides a great explanation of the opportunities created by cloud computing for machine learning and crowdsourcing.

1.1   Multiple Sources of Analytic Models Enabled by Cloud Computing

To validate our assumption that analytic models are trending to become a commodity, let us explore some of the ways used to generate or obtain analytic models:

1. Crowdsourcing analytics

   (a) Using traditional crowdsourcing markets
   (b) Using competitions and prizes
   (c) Using games or puzzles
2. Evolving populations of models using evolutionary algorithms (GA's, GP's)
3. Outsourcing analytics
4. Traditional model development (including legacy models)

We will focus on the first two sources, which are enabled or accelerated by cloud computing. Crowdsourcing is a relatively new phenomenon that started about a decade ago [11]. It is becoming increasingly popular for outsourcing micro-tasks to a virtual crowd, creating new marketplaces - see for instance *Amazon's Mechanical Turk* (*mturk.com*). Recently, however these tasks have become more complex and knowledge intensive, requiring a more specialized crowd. Web portals such as *Kaggle* (*kaggle.com*), *TunedIT* (*TunedIT.com*), and *CrowdANALYTICX* (*crowdanalytix.com*) allow organizations to post problems, training data sets, performance metrics, and timelines for ML competitions. The portals register the potential competitors and manage the competition during its various stages. At the end, the models are scored against an unpublished validation set, the winners receive their prizes, and the sponsoring organizations have access to new analytic models [12]. Alternative ways to incentivize a crowd to solve a given problem is by transforming the problem into a game or a puzzle. A successful example of this approach can be found at *Foldit* (*foldit.it*), in which the gaming community was able to solve a molecular folding problem that had baffled biologists for over a decade.

A second possible source of models is the use of evolutionary algorithms to search the model space. The author has explored this approach over a decade ago [13-14] inspired by many efforts in this area. In the early days, researchers resorted to clusters of computers, such as the Beowulf [15] to distribute the computational load of GP-driven search. Recently, the MIT CSAIL Department developed a *Flexible Genetic Programming* (*groups.csail.mit.edu/EVO-DesignOpt/evo.php?n=Site.FlexGP*) framework, leveraging cloud computing to evolve a population of models, whose outputs are ultimately averaged to produce an answer [16].

The last two approaches are more traditional ways of generating analytical models by outsourcing them to universities, developing them internally, or using legacy models. Our goal is to create *a touch-free, domain agnostic* process that can use any subset of models, regardless of their sources, and determine at run-time the most suitable and diverse subset that should be used to construct the answer to a given query.

To ensure a more focused discussion, we will limit the scope of this paper to the use of analytics to support Prognostics and Health Management (PHM) capabilities. However, the approach illustrated in this paper is application domain agnostic.

1.2 Prognostics and Health Management (PHM) Motivation for Analytics

The main goal of Prognostics and Health Management (PHM) for assets such as locomotives, medical scanners, aircraft engines, and turbines, is to maintain these assets' performance over time, improving their utilization while minimizing their maintenance cost. This tradeoff is typical of contractual service agreements offered by OEMs to their valued customers.

PHM analytic models are used to provide *anomaly detection and identification* (leveraging unsupervised learning techniques, such as clustering), *diagnostic analysis* (leveraging supervised learning techniques, such as classification), *prognostics* (leveraging prediction techniques to produce estimates of remaining useful life), *fault accommodation* (leveraging intelligent control techniques), and *logistics and maintenance optimization* (leveraging optimization techniques). A more detailed description of PHM functionalities and how they can be addressed individually by Computational Intelligence techniques can be found in [17]. In this paper, we will take a more holistic view on how to address PHM needs, while at the same time we will remain agnostics on the specific technologies used to build each model.

Since analytics play such a critical role in the development of PHM services, it is necessary to ensure that the underlying analytic models are accurate, up-to-date, robust, and reliable. There are at least two PHM applications for which such accuracy is critical: (1) *anomaly detection* (1-class classification), in which high volume of *false positives* might decrease the usefulness of the PHM system; (2) *prognostics* (prediction), in which high prediction variance might prevent us from acting on the results.

We will focus on a PHM prediction application in which prediction accuracy is a stringent requirement for production optimization. We will show how to leverage computational intelligence and ML techniques, combined with the elasticity of cloud computing, to address these accuracy requirements.

1.3 The Novel Idea

In this paper we are shifting our focus from model creation to model ensemble assembly. Rather than creating and optimizing models based on expected queries, we want to build a vast library of robust, local or global models, and compile relevant meta-information about each model. At run-time, for a specific query, we will select an ensemble of the most appropriate models from the library and determine their weights in the model fusion schema, based on their local performance around the query. The model ensemble will be constructed dynamically, on the basis of the models' meta-information. The model fusion will use the meta-information to determine bias compensation and relevance weight for each model's output. Finally, the models run-time versions will be executed via a function call at the end of the fusion stage. This concept is illustrated in Figure 1.

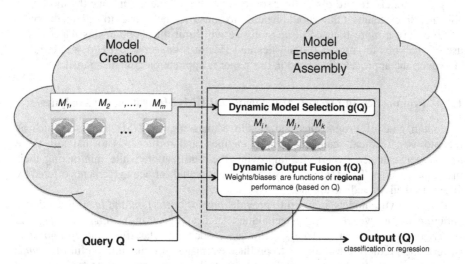

Fig. 1. The two modeling stages: Model Creation and Model Assembly

1.4   Paper Organization

In section 2, we will examine the relevant state of the art for this problem, while in section 3 we will describe a summary of our approach. In section 4, we will provide an in-depth analysis of the proposed approach, while in section 5 we will show some preliminary results using an ensemble of thirty neural networks to predict load, efficiency, and emissions in a power plant. In section 6, we will draw some conclusions from our experiments and highlight future work and extensions.

2   Related Work

We are proposing an approach for creating the best model ensemble on demand, based on the query information (as in Lazy-Learning), and performing the selection and dynamic fusion of the ensemble based on several performance criteria. In the literature we find vast amount of work covering model ensembles, meta-learning, lazy-learning, and multi-criteria decision making, but to the best of our knowledge there is no related work covering the intersection of these topics.

2.1   Model Ensembles

Individual models (classifiers or predictors) have a performance ceiling, which limits their performance, regardless of the amount of training or tuning.   One way to raise this ceiling is by creating an ensemble of highly *diverse* models and performing a fusion of their outputs.   There is currently an entire scientific community, *Multi-Classifier Systems* (MCS), devoted to this area.

The design of a successful classifier fusion system consists of three parts: design of the individual classifiers [18], selection of a set of classifiers [19-20], and design of the classifier fusion mechanism.   The most critical factor for an effective model fusion, however, is the *diversity* of the individual classifiers, where model diversity is defined in terms of the orthogonality of their errors [21].   Strategies for boosting such diversity include: 1) using different types of classifiers; 2) training individual classifiers with different data set (bagging and boosting); and 3) using different subsets of features.

2.2   Meta- Learning and Lazy Learning

Meta-learning literally means *learning how to learn*, but in our context it means *learning how to use ML models*.   Most meta-learning approaches deal with topics such as: Discovering meta-knowledge   (e.g. rule induction of rules from data to create a rule-based system that will solve the object-level problem); Stacked generalization [22] (e.g. combining a number of different learning algorithms); Boosting (e.g. combining the same learning algorithm trained in different ways); Dynamic bias selection (e.g. modifying the ML algorithm bias to match a given problem); and Inductive transfer (e.g. trying to improve the learning process over time) [23].

An interesting approach is one proposed by Duch [24], in which he creates a framework of Similarity-Based Methods to represent many algorithms, such as k-NN, MLP, RBF, etc. A variant of the Best First Search is used to perform a local search for optimal parameters.

In [25] Schaul attributes to Utgoff [26] the development of the first meta-learning system that learns parameters and to Schmidhuber [27] the first learning algorithm to learn other ML algorithms using evolutionary search in the model space (using GP for improving GP). According to Schaul: *"...meta-learning can be used for automating human design decisions (e.g. parameter tuning) and then automatically revisiting and optimizing those decisions dynamically in the light of new experience obtained during learning. Another application is transfer of knowledge between multiple related tasks..."* [25].

We endorse this goal, but our proposal is not limited to parameter tuning. Our key idea is not to focus on the optimization and tuning of pre-computed models. Rather, we aim to create model ensembles *on demand*, guided by the location of the query in the feature space.   This approach can be traced back to memory-based approaches [27-29], instance-based learning, and lazy-learning [30].   Figure 2 shows the Lazy Learning approach for locally weighted interpolation.

For a query Q, defined as a point $\bar{X}_Q$ in the state space $\mathbf{X}$, and a set of data points defined in the cross-product $\mathbf{X}$ x $\mathbf{Y}$, we find the points $u_j = (\bar{X}_j, y_j)$ that are close to

Q in their projection on **X**. We compute a matching score between the query and each of these data points, as a function of its proximity to the query. Such proximity is measured by a distance $d(\bar{X}_Q, \bar{X}_{Qj})$ that is interpreted as the dissimilarity between the two points in **X**. The distance is smoothed by a parameter h that defines the scale or range of the model. Usually h is computed by minimizing a cross-validation error. Each output y_j (corresponding to each of the points close to Q) is weighted by applying a kernel function K to this smoothed distance. A convex sum is used for the aggregation, which is identical to the Nadaraya-Watson estimator for non-parametric regressions using locally weighted averages. When dealing with extrapolation, weighted linear models are instead used of convex sums in the aggregation box, to provide for a better generalization [4].

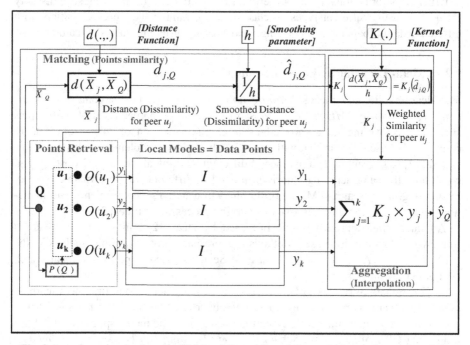

Fig. 2. Lazy Learning (or Locally Weighted Kernel-based) Interpolation - adapted from [4]

2.3   MCDM: Selection of Best Model Ensemble Based on Constrained Multi-Criteria

We plan to create a dynamic ensemble that best satisfies a set of performance metrics, leveraging techniques in Multi Criteria Decision Making (MCDM). There is a vast amount of literature on MCDM, ranging from pioneering books [31-32], to many conference proceedings and articles focused on improving search methods, preference aggregations, interactive solution visualization, etc.

The author's perspective on MCDM is described in [33]. It can be summarized as defining a proper problem representation followed by the intersection of search, preference aggregation, and (when needed) interactive visualization of their solutions.

- *Representation.* A typical MCDM process starts with a set of constraints over the solution space, defining the feasible solution set. Each point from this set is mapped into a performance space, whose dimensions are the criteria used to evaluate the MCDM solutions.
- *Multi Objective Search.* We search for the set of non-dominated solutions, forming the Pareto set. This step induces a partial ordering on the set of feasible solutions, and defines the concept of Multi-Objective Optimization (MOO).
- *Preferences.* MCDM requires an additional step over MOO, which is the selection of one or more non-dominated solutions, to maximize our aggregated preferences, thus creating a complete order over the feasible solutions sets.
- *Interactive Visualization.* In cases when the decision-maker is part of the solution refinement and selection loop, we need a process to enhance our cognitive view of the problem and enable us to perform interim decisions. We need to understand and present the impacts that intermediate tradeoffs in one sub-space could have in the other ones, while allowing him/her to retract or modify any intermediate decision steps to strike appropriate tradeoff balances. This last step will not be needed in our proposed MCDM approach.

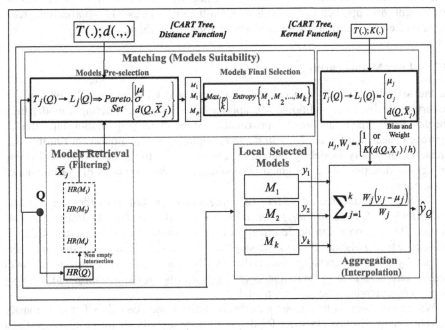

Fig. 3. Meta-Lazy Learning

In our case, model creation is equivalent to creating the solutions set, while model filtering generates the *feasible* solution set. The model pre-selection, based on the meta-information attached to each model, maps the solutions in the performance space and the extraction of the points in the Pareto set represents the MOO step. The model final selection, i.e. the selection of the k-tuple of local models with the least amount of error correlation (via entropy maximization) reflects our preferences. The aggregation is the final ensemble with the parameters defining dynamic fusion represents the outcome of the MCDM process, i.e. our customized ensemble on demand. Our approach is summarized in Fig. 3 and will be further explained in sections 3 and 4.

We have briefly covered the state of the art in multi classifier systems, lazy learning, meta-learning, and MCDM. Our goal is to address their intersection in a way that has never been investigated before.

3   Summary of the Approach

3.1   Cloud Computing, the Enabler

Although the concept of model ensemble has been proposed and analyzed since the early 2000 [18], the idea of creating dynamic ensembles at run-time has not been proposed yet. Such idea would not have been feasible or practical, had it not been for the advent of Grid- and Cloud-computing. After constructing an offline library of models with their associate meta-information, we can now leverage the cloud environment, with its parallel computation and automated provisioning of processors, to implement this approach and provide fast run-time responses. Our approach is predicated on a cloud-based Software as a Service (*SaaS*) paradigm.

3.2   Lazy Meta-Learning

In the ML literature, the concept of Lazy learning (LL) or memory-based learning departs radically from traditional ML approaches. Instead of training a model from the data to become a functional approximation of the underlying relationships, the LL approach states that the *model is in the data.* Upon receiving a query, at run-time LL creates a temporary model by finding the closest points to the query and performing an aggregation (usually a weighted interpolation or extrapolation) of the outputs of those points.

Our approach, labeled *Lazy Meta-Learning,* can be described by the analogy: *Meta-learning is Lazy Learning for models like learning is Lazy-Learning for data points.* This approach can be described as a multi-criteria decision-making (MCDM) process, whose structure follows the steps defined in [5]. The model design is performed by an offline meta-heuristics (the MCDM process), while the run-time model architecture is formed by an online meta-heuristics (the fusion module), and a collection of object models (the analytic models.)

Within the scope of this paper, a model could be a one-class classifier for anomaly detection, a multi-class classifier for diagnostics, or a predictor for prognostics. Each model will have associated meta-information, such as its region of competence and applicability (based on its training set statistics), a summary of its (local) performance during validation, an assessment of its remaining useful life (based on estimate of its obsolescence), etc.

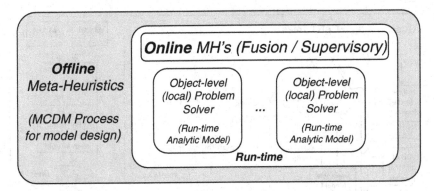

Fig. 4. Model Design: Offline Meta-Heuristics to design, tune, optimize, adapt to changes, and maintain the runtime models over time. Model Architecture: (a) Online Meta-Heuristics to integrate or interpolate among multiple local object-models, manage their complexity, and improve their overall performance; (b) Multiple object-models, either in parallel configuration (ensemble) or sequential configuration (cascade, loop), to integrate functional approximation with optimization and reasoning with imperfect data (imprecise and uncertain).

3.3   MCDM Process for Model Creation and Dynamic Model Assembly

We decompose the MCDM process into two stages:

- *Model Creation*, an off-line stage in which we create the initial building blocks for the assembly and we compile their meta-information
- *Dynamic Model Assembly*, an on-line stage in which, for a given query we select the best subset of models

This process is followed by the execution stage, *Dynamic Model Fusion*, in which we evaluate the selected models and dynamically fuse them to solve the query. We will use different metrics to evaluate each stage, looking for coverage and diversity in the creation stage, while looking for accuracy and precision in the assembly and fusion stages.

Model Creation: The Building Blocks. We assume the availability of an initial training set that samples an underlying mapping from a feature space X to an output y. In the case of supervised learning, we also know the ground truth-value t for each record in the training set. We create a library of diverse, local or global models. We increase model diversity by using competing ML techniques trained on the same local regions. We assume that any of the sources mentioned in section 1.1 could be used to create these models.

Dynamic Model Assembly: Query Driven Model Selection and Ensemble. This stage is divided into three steps:

- *Model Filtering*, in which we retrieve the applicable models from the DB for the given query;
- *Model Pre-Selection*, in which we reduce the number of models based on their local performance characteristics (bias, variability, and distance from the query);
- *Model Final Selection*, in which we define the final model subset.

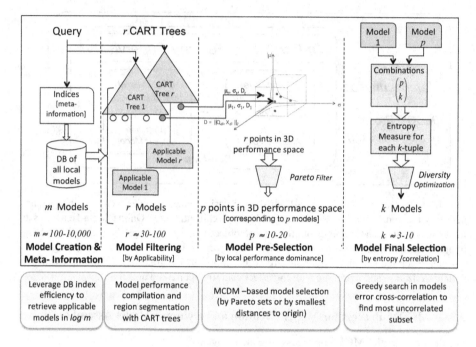

Fig. 5. Dynamic Model Assembly on Demand (Filtering, Selection)

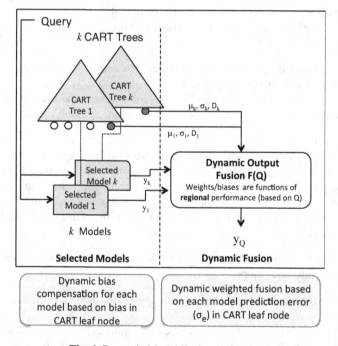

Fig. 6. Dynamic Model Fusion on Demand

Dynamic Model Fusion: Generating the Answer. Finally, we evaluate the selected models and we aggregate their outputs after compensating for their biases and weighting them by their proximity to the query (or by their estimated error) to generate the solution to the query.

4 (The Devil Is in the) Details of Our Approach

4.1 Model Creation

This step is common to both regression and classification problems. The premise is that local models, trained on regions of the feature space, usually will have smaller variances than global models, trained on the entire space, providing that the local models are used *only* within their region of competence. Typically this requires a supervision mechanism (at the meta-level) to determine their degree of applicability and contribution to the final output. We rely upon the model meta-information to make this assessment. There are many ways to segment the original training set to define regions of competence for local models, but this topic is outside the scope of this paper.

Library of m Local Models. As stated above, for each prediction or classification problem we want to generate a large number of global and local models. We will assume that m is the total number of available models for each problem.

Prediction Problems. Each regression model M_i will define a mapping:
$$M_i: X \to Y, where\ i = 1, \dots, m;\ |X| = n;\ |Y| = 1;\ X \epsilon \mathbb{R}^n;\ Y \epsilon \mathbb{R}$$
In a more general case, we might want to predict multiple i.e., g variables, i.e.:
$$M_i: X \to Y, where\ i = 1, \dots, m;\ |X| = n;\ |Y| = 1;\ X \epsilon \mathbb{R}^n;\ Y \epsilon \mathbb{R}^g$$
Within the scope of this article we will limit ourselves to a single output.

Classification Problems. Each classification model M_i will define a mapping: $M_i: X \to Y, where\ i = 1, \dots, m;\ |X| = n;\ |Y| = (C + 1)$, and C is the number of classes. Within the scope of this article, we will assume that the classifier output is a probability density function *(pdf)* over C classes. The first C components of the *pdf* are the probabilities of the corresponding classes. The $(C+1)^{th}$ element of the *pdf* allows the classifier to represent the choice *"none of the above"* (i.e., it permits to deal with the Open World Assumption). The $(C+1)^{th}$ element of the *pdf* is computed as the complement to 1 of the sum of the first C components. The final decision of classifier M_i is the *argmax* of the *pdf.*

Meta-information. Every time that we create a model, we need to capture its associated meta-information, i.e., information about the model itself, its training set, and its local/global performance in the validation set. Over time, we will define a standard API for this purpose. Table 1 summarizes our current preliminary thoughts on meta-information. The essence is to capture information that we can use later on to reason about the applicability and suitability of a model for a given situation (e.g., a set of queries).

Table 1. Meta-Information (Preliminary Version)

Model	Training Set	Validation Set
Label: M_i	Label: TS_i	Label: VS_i
Model Mapping:	Model Applicability:	Local Model Perfor-
$\quad M_i : X_i \to Y_i$	$\quad$ Hyper-rectangle: HR_i	mance:
$\quad X_i \in R^n; Y_i$	$\quad HR_i \in R^{2,n}$ is the range	$\quad$ CART Tree: $\quad T_i : X_i \to$
$\quad \in R$ (regressions)	$\quad$ of values of each of the	$\quad e_i$
$\quad Y_i \in [0,1]^{(C+1)}(classif.)$	$\quad n$	$\quad X_i \in R^n; e_i$
	$\quad$ features over TS_i	$\quad \in R$ (regressions)
		$\quad e_i \in [0,1]^{(C+1)}(classif.)$

Model Applicability: Hyper-rectangle HR_i. The Hyper-rectangle in the feature space defines each model's region of competence[1]. Each model M_i has a training set TS_i, which is a region of the feature space X. We define the Hyper-rectangle of each model M_i, $HR(M_i)$, to be the smallest hyper-rectangle that encloses all the training points in training set TS_i. If a query point q is contained in $HR(M_i)$, we consider model M_i applicable for such query. For a set of query points Q, we consider the model applicable if $HR(Q)$ is not disjoint with $HR(M_i)$.

Local Model Performance. We want to capture the local performance of the model by answering the question: *"For this type of query, how reliable is this model?"* For regression problems, we have used continuous case-based reasoning and fuzzy constraints [35], and lazy learning [36] to estimate the local prediction error. Within the same context of regression problems, the authors replaced the run-time use of lazy learning with the compilation of local performance via CART trees [34] for the purpose of correcting the prediction via bias compensation [7]. For classification problems, we find a similar lazy learning approach [37] to estimate the local classification error.

After many experiments to find the most efficient summary of the model performance, we opted for using a CART Tree T_i that maps the feature space to the **signed error** computed in the validation set, i.e. $T_i : X \to e_i$.

Each CART tree T_i will have a depth d_i, such that there will be up to 2^{di} paths from the roots to the leaf nodes (for a fully balanced tree). For each tree, we store each path from the root node to each leaf node. The path is as a conjunct of constraint rules that need to be satisfied to reach the leaf node. The leaf node is a pointer to a table containing the leaf node statistics:

- N_i $\quad$ – Number points in the leaf node (from the training/testing set)
- $\mu_i(e)$ – *Bias* (Average error computed over N_i points)
- $\sigma_i(e)$ – *Standard Deviation* of the error computed over N_i points
- X_{di} $\quad$ – Normalized centroid as percentage of its average $(1,...,1)$ of the N_i points

In the future, we will extend the meta-information to capture temporal and usage information, such as model creation date, last usage date, and usage frequencies, which will be used by the model lifecycle management to select the models to maintain and update.

---

[1] By defining the applicability of a model as its hyper-rectangle in the feature space, we are limiting the use of the model to interpolation. Should we choose to use it for extrapolation, we would consider queries outside the hyper-rectangle (within some fuzzy tolerance).

4.2   Dynamic Model Assembly on Demand

Query Formulation: q or Q. To simplify the description of our approach we have used the case of the query being a single point query q. However, our approach can be easily generalized to the case when the query is a data set (time-series or time independent). Let's refer to such query as $Q=[q_1, \ldots, q_z]$. In such case, the model filtering described in section 4.3.1 will be modified. Instead of retrieving for each point q those models M_i whose associated hyper-rectangles $HR(M_i)$ contain q, i.e., $q \in HR(M_i)$, we retrieve all models M_i whose associated hyper-rectangles $HR(M_i)$ are not disjoint with the Hyper-rectangle of Q, i.e. $HR(Q) \cap HR(M_i) \neq \emptyset$. This will avoid the overhead of multiple accesses to the DB[2]. The model assembly, composed by the *model pre-selection* and *model final selection* steps, will be performed iteratively for every point query q in Q. So these steps, as well as final model evaluation and fusion, will be linear in the number of query points z.   The most efficient way to generate the set Q is to cluster similar query points using a *k-d tree* [38-39].

Model Filtering: From m to r Models. After creating m models, trained on their corresponding regions of the training space, we will organize the models in a database DB, whose indices will be derived from the models meta-information (see section 4.1.2). For a given query q, the MCDM process starts with a set of constraints to define the feasibility set.   In this case the constraints are:

- Model soundness, i.e., there are sufficient points in the training/testing set to develop a reliable model, competent in its region of applicability
- Model vitality, i.e., the model is up-to-date, not obsolete
- Model applicability to the query, i.e., the query is in the model's competence region

For each of the r retrieved models M_i, we will use a compiled summary of its performance, represented by a CART tree T_i, of depth d_i, trained on the model error vector obtained during the validation of the model.

Model Pre-selection: From r to p Models. We will classify the query using the same CART tree T_i, reaching leaf node $L_i(q)$. Each leaf will be defined by its path to the root of the tree and will contain d_i constraints over (at most) d_i features. Leaf $L_i(q)$ will provide estimates of model M_i performance in the region of the query.   Following the MCDM process described in section 3, we map each feasible solution (the r models) into the performance space. This space is defined by the following criteria:

- Model bias: $|\mu_i(e)|$
- Model variability $\sigma_i(e)$
- Model suitability for the query: $\|q_{di}, X_{di}\|_2$ (Distance of q_{di} to the normalized centroid (X_{di}), computed in a reduced, normalized feature space d

---

[2] This step might allow models that are irrelevant for a subset of points in Q. These models will be rejected in the model pre-selection (due to their high distances from the query point).

In this three dimensional space, we want to minimize all three objectives. We may further impose limits to the largest value that each dimension can have.

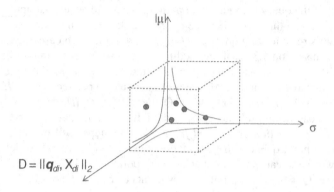

Fig. 7. Local Model Performance Space – r points represent r model performances

We can use a Pareto Filter to extract all the non-dominated points in this space. Should the result be too limited, we can resort to additional Pareto set (after removing the previous tier), until we have p models.

Model Final Selection: From p to k Models. During the previous steps, we relied on meta-information to reduce the number of models from the original m models available in the DB, to r applicable models, to p suitable models. Before performing the fusion, we need to *make sure that the models are diverse*, i.e. we need to explore the error correlation among smaller subsetS of k models that we will use for generating the answer to the query. This step would require generating all possible k-tuples chosen from p models to evaluate their error correlations, i.e. $\binom{p}{k}$. Although, this is a much smaller number than $\binom{m}{k}$, it is still an onerous step to take for each point. So we will use a *greedy search* to further decrease this complexity.

Models in an ensemble should be different from each other for the ensemble's output to be better than the individual models outputs. The goal is to use an ensemble whose elements have the most uncorrelated errors. In reference [21], Kuncheva and Whitaker propose six different *non-pairwise diversity measures* to determine the models difference. We will use the *Entropy Measure E*, proposed in the same reference, as the way to find the k most diverse models to form the ensemble. Let's create an N by k matrix M, such that N is the number of records evaluated by k models.

Classification Problems. When the models are classifiers, cell $M[i,j]$ contains a binary value $Z[i,j]$ (1 if classifier j classified record i correctly, 0 otherwise). This metric assumes that we already obtained each classifier decision on the training/validation records, by applying the *argmax* function to the *pdf* generated by the classifier. Then, we compute diversity of the k classifiers by using the Entropy measure E (modified from reference [21]):

$$E = \frac{1}{N} \sum_{i=1}^{N} \left[\frac{1}{k - floor\left(\frac{k+1}{2}\right)} min\left(\sum_{j=1}^{k} M[i,j], k - \sum_{j-1}^{k} M[i,j]\right) \right] \quad (1)$$

E takes values in $[0,1]$.

Prediction Problems. When the models are predictors, cell $M[i,j]$ contains the error value $e[i,j]$, which is the prediction error made by model i on record j. In such case, we will follow the following process:

- *Histogram of Record Error:* Compute a histogram of the errors for each record $M[i,.]$. We need to define a reasonable bin size for the histogram, thus defining the total number of bins, *nmax*. Let $H(i,r)$ be the histogram for record i, where r defines the bin number $(r=1, nmax)$.
- *Normalized Histogram of Record Error:* Normalize histogram $H(i,r)$, so that its area is equal to one (becoming a *pdf*). Let $H_N(i,r)$ be the normalized histogram, i.e.:

$$H_N(i,r) = \frac{H(i,r)}{\sum_{r=1}^{nmax} H(i,r)} \quad (2)$$

- *Normalized Record Entropy:* Compute the normalized record entropy of the *pdf* (so that its value is in $[0,1]$), i.e.:

$$ent(i) = -\left(\frac{1}{\ln nmax}\right) \sum_{r=1}^{nmax} H_N(i,r) \times \ln H_N(i,r) \quad (3)$$

where $\frac{1}{\ln nmax}$ is a normalizing factor so that $ent(i)$ takes values in $[0,1]$

$$H_N(i,r) = \frac{H(i,r)}{\sum_{r=1}^{nmax} H(i,r)} \quad (4)$$

- *Overall Normalized Entropy:* Average the normalized entropy over all N records:

$$E = \frac{1}{N} \sum_{i}^{N} ent(i) \quad (5)$$

E takes values in $[0,1]$.

For both classifiers and prediction problems, *higher overall normalized entropy values indicate higher models diversity.*

Greedy Search in Combinatorial Space. In both cases we will use a greedy search, starting from $k=2$ and compute the normalized entropy for each 2-tuple, to find the one(s) with the highest entropy. We will then increase the value of k to explore all 3-tuples. If the maximum normalized entropy for the explored 3-tuples is lower than the maximum value obtained for the 2-tuples, we will stop and use the 2-tuple with the highest entropy. Otherwise we will keep the 3-tuple with the highest entropy and explore the next level ($k=4$) and so on, until no further improvement can be found. With all the caveats of local search, this greedy approach will substantially reduce search complexity, as we will not need to explore all the combinations of ensembles.

4.3   Dynamic Model Fusion

The last step in our approach is to perform the dynamic fusion of the selected k models. When probed with the query q, each model M_i will produce and output $y_i(q)$. Each model also has a corresponding CART tree T_i, which will be used to classify the query q. In that case, the query will reach a leaf node $L_i(q)$ in the corresponding tree T_i. The leaf node will contain the local statistics for the points similar to the query, such as the bias (average of the error) $\mu_i(e|q)$, and the standard deviation of the error, $\sigma_i(e|q)$.

Fusion for Regression Problems. After many design experiments, which will be described in details in a follow-up paper, we concluded that the best fusion schema is accomplished by using dynamic bias compensation and by weighting the compensated output using a kernel function of the standard deviation of the error, i.e.:

$$\hat{y}(q) = \frac{\sum_{s=1}^k w_s(y_s(q) - \mu_s(e|q))}{\sum_{s=1}^k w_s} \tag{6}$$

where:

$$w_s = K\left(\frac{\sigma_s(e|q)}{h}\right) \tag{7}$$

and h is the usual smoothing factor for the kernel function $K(.)$ obtained by minimizing the cross-validation error.

Fusion for Classification Problems.   As shown in reference [37], the classification problem can be cast in a fashion similar to the regression problem. In reference [37] we used a local weighting fusion (similar to [7] but for classifications), and used dynamic bias compensation with equal weight contributions for the selected models.   In section 6, we will discuss how to extend our approach, based on these preliminary results, to cover classification problems.

5   Customized Analytics Applied to Power Plant Management

5.1   Problem Definition

In references [40-41] we described the optimization problem for a power plant management system, in which a coal-fired boiler drives a steam turbine to generate electric power. For given environmental conditions, the problem was to determine the control variable set points that could generate the *load* (equality constraint), without exceeding *CO* and *SO* limits (inequality constraints), while minimizing both *Heat Rate* and *NOx* emissions. After using first-principles-based methods and domain-knowledge to identify the relevant model inputs, we built a nonlinear neural-network to map the inputs space (control variable set-points) and time variable, ambient uncontrollable variables, to each of the outputs of interest, which represented our objectives and constraints. As shown in the above references, we used an evolutionary multi-objective optimizer to evolve the set points and identify the Pareto-optimal set of input-output vector tuples that satisfy operational constraints.

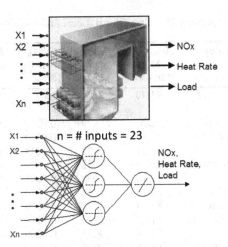

Fig. 8. Power plant input-output relationships. Each output (*NOx, Heat Rate, Load*) is modeled by a committee of predictive neural networks (adapted from reference [7]).

We noted the importance of reducing the neural networks uncertainty, as they generated the fitness function that drove the evolutionary search. In reference [41] we showed how we managed the model extrapolation error by using the equivalent of continuous case-based reasoning. We also need to address the intrinsic model uncertainty of the NNs. So, we performed preliminary experiments using the same data set, and created a *fixed committee* of neural networks, injecting diversity by bootstrapping their training set. We noted that model performance was significantly improved by fusing the outputs from the ensemble of models.

In references [36] we presented a method called *locally weighted fusion*, which aggregated the results of multiple predictive models based on local accuracy measures of these models in the neighborhood of the probe point for which we want to make a prediction. This neighborhood was computed on demand, following a Lazy Learning technique. In reference [7], we extended our experimentation by compiling the error information and avoiding the run-time search for the closest points. Figure 8, adapted from reference [7] shows the (23-3-1) structure of the NN's used in the ensemble.

This paper is an extension of the work initiated in reference [7]. We have proposed a complete architecture for selecting a *dynamic committee* of neural networks, and a *dynamic fusion*, based on the query and the meta- information. We extended the experiments with the same data set to compare our results with the ones previously obtained in references [7, 36].

5.2 Preliminary Experimental Results

The following tables show a sample of early experiments performed in [36] and [7], followed by the current experiments, in which we used part of the proposed architecture.

Table 2. Experimental Results (Baseline, Global, Local fusion without bias compensation)

Exp. #	Fusion Strategy	Heat Rate MAE [Btu/KwHr]	Heat Rate pg [%]	NOx MAE [lb/MBtu]	NOx pg [%]	Load MAE [MW]	Load pg [%]
1	Baseline: average of 30 predictors	91.79	0.00	0.0228	0.00	1.05	0.00
2	Best of 30 predictors	85.1	7.29	0.0213	6.58	0.987	6.00
3	Global Average	87.15	5.06	0.0214	6.14	1.042	0.78
4	Global GWF	86.91	5.32	0.0214	6.14	1.04	0.95
5	Global Least Square	83.05	9.52	0.02	**12.28**	0.984	**6.29**
6	H-Rect+W1 No bias	82.19	10.46	0.0202	11.40	1.024	2.48
7	H-Rect No bias	87.15	5.06	0.0214	6.14	1.042	0.78
8	H-Rect+W2 No bias	83.93	8.56	0.0208	8.77	1.03	1.93
9	1-nn No bias	81.19	**11.55**	0.0214	6.14	1.008	4.02
10	5-nn No bias	84.31	8.15	0.0206	9.65	1.029	1.97
11	CART EW No bias	87.15	5.06	0.0213	614	1.042	0.78
12	CART UW0 No bias	87.15	5.06	0.0213	614	1.042	0.78

The data set was comprised by 8,000+ records of daily operations, sampling the mapping between operational set points under given environmental conditions and power plant outputs, such as *Heat Rate (HR)*, *NOx* emissions, and generated *load*. Roughly 25% of these records were used as the validation set.

In table 2, the baseline labeled as experiment 1, is the simple average of the outputs of the thirty neural networks. As such, the column labeled *pg (percentage gain)* is 0%. We will use *percentage gain* to show the percentage improvement over the baseline for each experiment. Experiment 2 reports the output of the best predictor (a posteriori). The remaining ten experiments in Table 2, show the results of various fusion combinations, *without performing bias compensation*, i.e., weighing the outputs of the models using different fusion schemes: variations of global fusion (experiments 3-5), local fusion based on Lazy learning (experiments 6-10), and CART trees to selected the local points (experiments 11-12). Most of these results were reported in reference [36]. The conclusion from this table is that *without bias compensation* the improvements are marginal: *HR pg: 11%, NOx pg: 12%; Load pg: 6%.*

In Table 3, we *perform bias compensation for all experiments*. In a separate set of experiments (not shown for sake of brevity) we concluded that the best CART trees are the one mapping the feature space to the **signed error** (as opposite to the unsigned error or the output of the model). Thus, with the exception of experiments 13-17, which show the results for lazy leaning type of local fusion (i.e., without using CART tree as part of the meta-information), all other experiments (18-33) use the signed error derived CART tree to compile their local performance in the validation set.

Experiments 18-27 show the result of using all thirty models with different weighting schemas: from equal weights (exp. 18), to unequal weights derived from linear or exponential kernel functions using as arguments various elements of the performance space in the pre-selection stage, e.g., distance of the query from the centroid, standard deviation of the error, etc. Overall the results are comparable (with very similar standard deviations of the error). The best result in this set of the experiments is from Unequal Weights-7 (UW7), in which we used a linear kernel function and the standard deviation of the error, i.e.: $K\left(\frac{\sigma_s(e|q)}{h}\right)$.

Table 3. Experimental Results (All with bias compensation, CART trees)

Exp. #	Fusion Strategy	Heat Rate MAE [Btu/KwHr]	Heat Rate pg [%]	NOx MAE [lb/MBtu]	NOx pg [%]	Load MAE [MW]	Load pg [%]
13	H-Rect+W1 bias	69.20	24.61	0.0140	38.60	0.855	18.57
14	H-Rect bias	69.23	24.58	0.0140	38.60	0.855	18.56
15	H-Rect+W2 bias	69.16	24.66	0.0140	38.60	0.854	18.63
16	1-nn bias	72.99	20.48	0.0143	37.28	0.861	17.98
17	5-nn bias	76.34	16.83	0.0169	25.88	0.903	14.04
18	CART EW bias	60.62	33.96	0.0117	48.68	0.718	31.62
19	CART UW0 bias (nf)	68.56	25.31	0.0148	35.09	0.817	22.19
20	CART UW1 bias (nf)	60.57	34.01	0.0111	**51.32**	0.725	30.92
21	CART UW2 bias (nf)	60.45	34.14	0.0116	49.12	0.719	31.50
22	CART UW3 bias (nf)	64.62	29.60	0.0130	42.98	0.733	30.20
23	CART UW4 bias (nf)	62.31	32.12	0.0125	45.18	0.721	31.35
24	CART UW5 bias (nf)	60.10	34.53	0.0145	36.40	0.792	24.59
25	CART UW6 bias (nf)	59.75	34.90	0.0129	43.42	0.745	29.04
26	CART UW7 bias (nf)	59.77	34.88	0.0115	49.56	0.713	**32.10**
27	CART UW8 bias (nf)	61.87	32.60	0.0122	46.49	0.743	29.25
28	CART UW7 bias (f1)	59.75	34.91	0.0117	48.68	0.721	31.35
29	CART UW7 bias (f2)	65.04	29.14	0.0135	40.79	0.776	26.10
30	CART UW7 bias (f3)	64.82	29.38	0.0130	42.98	0.772	26.46
31	CART UW7 bias (f4)	62.18	32.26	0.0122	46.49	0.744	29.12
32	CART UW7 bias (f5)	60.54	34.04	0.0117	48.68	0.726	30.86
33	CART UW7 bias (f6)	59.69	**34.98**	0.0116	49.12	0.716	31.86

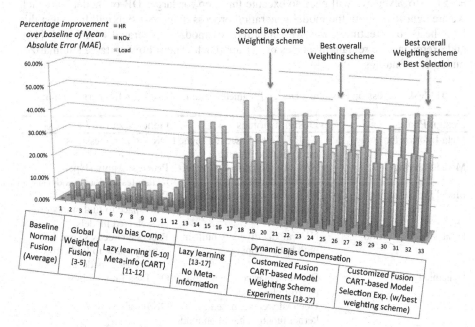

Fig. 9. Summary of experimental results

Finally, in experiments 28-33, we fixed this weighting scheme and searched for the best value of p, the number of models to use in the ensemble. The results, only valid for this specific application and for the initial DB of models, indicate that ~80% of the models provide the best results. All experiments results are illustrated in figure 9.

6   Conclusions

6.1   Analysis for Regression Problems

The experiments presented in this paper are preliminary in nature and have not completely exercised our approach. By using thirty neural networks trained on the entire feature space, we did not compare global and local models. As a result, we did not need to perform the filtering stage, i.e., $r = m = 30$. The selection process indicated that the best performance was obtained with about 80% of our models (i.e., $p = 25$). This was the result of performing a local search in the number of models. This process should be repeated for different applications and for a different DB of models. From the same experiments, we learned that there are statistically significant differences in performing dynamic bias compensation, and in using the local bias compiled in the CART trees. After experimenting with unsigned error, model output, and signed error, we concluded that the latter was the best way to generate the CART trees as part of the models meta-information. Finally, given that we controlled the original model generation and used bootstrapped training sets to inject diversity in the models, we did not need to perform an entropy-based final selection stage (i.e., $p = k = 25$). However, we will need to execute this step for larger DB of models in which we are agnostics about the model generation process. Our next step will be to test this hypothesis, by creating a DB with hundreds of models, generated by *FlexGP* [16], and exercising the remaining stages of our approach. The following table summarizes our current findings.

Table 4. Regression Problems: Design Parameters and Processes (Preliminary Version)

Design Phase	Design Choices	Generalization / Process
Meta-Information	CART Trees based on signed error; limited pruning	OK for all regressions
Model Filtering	N/A: All global models applicable by construction	**Process**: Apply Hyper-rectangles for local models
Model Pre-Selection	80% of applicable models (for this application)	**Process**: Greedy search in error space to find number of pre-selected models
Final Model Selection	N/A: model diversity built-in with bootstrapping	**Process**: Greedy search in entropy space to find error-uncorrelated models
Dynamic Fusion	Dynamic bias compensation (μ_e from leaf node)	OK for all regressions
	Unequal weights based from kernel function based on model prediction error (σ_e)	OK for all regressions

6.2   Future work for Classification Problems

In reference [37], we used lazy learning to perform local fusion for classification problems. We want to extend our approach to these types of problems and design a set of experiments to validate it. Specifically, let us assume that a classifier maps a n-dimensional feature space X into a C-dimensional space Y, where C is the number of classes. Then, the output of each model k for the training record j, is a normalized probability density $\varpi_k(j)$, where $\varpi_k(j)$ is a $(C+1)$ dimensional vector. The error of each model k for the training record j is computed as $= e_k(j) = \varpi_k(j) - t_k(j)$, where $t_k(j)$ is a binary $C+1$ dimensional vector in which only one element is 1, indicating the correct classification for training record j. Since $\varpi_k(j)$ is a normalized probability density, then the sum of all the elements of $e_k(j)$ equals 0.

By following a process similar to the one for the regression problems, we will train a CART Tree T_k for *each* class value. The tree will map the feature space X to the error vector e (for that specific class value), such that in each leaf node we will cluster the subset of the training records that have similar classification errors. We will compute the mean error of the kth classifier over the points in each leaf node. In a fashion similar to the regression, we will refer to the average error of the leaf node $L_s(q)$, in which query q was classified, as $\mu_k(e|q)$. For a given query we will use the same model assembly steps (filtering, preference, and final selection based on entropy maximization). For the selected k models, we will perform a similar *bias compensation*. For the case when all k models are equally weighted, we have:

$$\hat{y}(q) = argmax\left\{\tfrac{1}{k}\textstyle\sum_{s=1}^{k}(y_s(e) - \mu_s(e|q))\right\} \tag{8}$$

We also plan to perform a design of experiments to determine the most appropriate weighting scheme for each model in a manner similar to our regression experiments.

Acknowledgements. I would like to thank my colleagues Ya Xue, Fred Xue, Dustin Garvey, Steve Gustafson, Dongrui Wu, Scott Evans, Jianhui Chen, Weizhong Yan, and many others for their feedback and constructive criticism that helped shaping this idea.

References

1. Bonissone, P.: The life cycle of a fuzzy knowledge-based classifier. In: Proc. North American Fuzzy Information Processing Society (NAFIPS 2003), Chicago, IL, pp. 488–494 (2003)
2. Patterson, A., Bonissone, P., Pavese, M.: Six Sigma Quality Applied Throughout the Life-cycle of and Automated Decision System. Journal of Quality and Reliability Engineering International 21(3), 275–292 (2005)
3. Bonissone, P., Varma, A., Aggour, K.: An Evolutionary Process for Designing and Maintaining a Fuzzy Instance-based Model (FIM). In: Proc. First Workshop of Genetic Fuzzy Systems (GFS 2005), Granada, Spain (2005)
4. Bonissone, P., Varma, A., Aggour, K., Xue, F.: Design of local fuzzy models using evolutionary algorithms. Computational Statistics and Data Analysis 51, 398–416 (2006)

5. Bonissone, P.: Soft Computing: A Continuously Evolving Concept. Int. J. Computational Intelligence Systems 3(2), 237–248 (2010)
6. Bonissone, P., Cadenas, J.M., Garrido, M.C., Diaz, R.A.: A Fuzzy Random Forest. International Journal of Approximate Reasoning 51(7), 729–747 (2010), doi:10.1016/j.ijar.2010.02.003
7. Bonissone, P., Xue, F., Subbu, R.: Fast Meta-models for Local Fusion of Multiple Predictive Models. Applied Soft Computing Journal 11(2), 1529–1539 (2008), doi:10.1016/j.asoc.2008.03.006
8. Bonissone, P., Eklund, N., Goebel, K.: Using an Ensemble of Classifiers to Audit a Production Classifier. In: Oza, N.C., Polikar, R., Kittler, J., Roli, F. (eds.) MCS 2005. LNCS, vol. 3541, pp. 376–386. Springer, Heidelberg (2005)
9. Evangelista, P., Embrechts, M., Bonissone, P., Szymanski, B.: Fuzzy ROC Curves for Unsupervised Nonparametric Ensemble Techniques. In: IJCNN 2005, Montreal, Canada, pp. 3040–3045 (2005)
10. Armbrust, M., Fox, A., Griffith, R., Joseph, A.D., Katz, R., Konwinski, A., Lee, G., Patterson, D., Rabkin, A., Stoica, I., Zaharia, M.: Above the Clouds: A Berkeley View of Cloud Computing. Technical Report EECS-2009-28, EECS Department, University of California, Berkeley (2009)
11. Howe, J.: Crowdsourcing- Why the Power of the Crowd Is Driving the Future of Business. Random House, New York (2008)
12. Vance, A.: Kaggle's Contests: Crunching Numbers for Fame and Glory. Businessweek, January 04 (2012)
13. Bonissone, P., Subbu, R., Aggour, K.: Evolutionary Optimization of Fuzzy Decision Systems for Automated Insurance Underwriting. In: Proc. FUZZ-IEEE 2002, Honolulu, HI, pp. 1003–1008 (2002)
14. Aggour, K., Bonissone, P., Cheetham, W., Messmer, R.: Automating the Underwriting of Insurance Applications. AI Magazine 27(3), 36–50 (2006)
15. Bennett III, F.H., Koza, J.R., Shipman, J., Stiffelman, O.: Building a parallel computer system for $18,000 that performs a half peta-flop per day. In: Banzhaf, W., Daida, J., Eiben, A.E., Garzon, M.H., Honavar, V., Jakiela, M., Smith, R.E. (eds.) GECCO 1999: Proceedings of the Genetic and Evolutionary Computation Conference, Orlando, FL, pp. 1484–1490. Morgan Kaufmann, San Francisco (1999)
16. Sherry, D., Veeramachaneni, K., McDermott, J., O'Reilly, U.-M.: FlexGP: Genetic Programming on the Cloud. To Appear in Parallel Implementation of Evolutionary Algorithms, EvoStar 2012, Malaga, Spain (2012)
17. Bonissone, P.P., Iyer, N.: Soft Computing Applications to Prognostics and Health Management (PHM): Leveraging Field Data and Domain Knowledge. In: Sandoval, F., Prieto, A.G., Cabestany, J., Graña, M. (eds.) IWANN 2007. LNCS, vol. 4507, pp. 928–939. Springer, Heidelberg (2007)
18. Roli, F., Giacinto, G., Vernazza, G.: Methods for Designing Multiple Classifier Systems. In: Kittler, J., Roli, F. (eds.) MCS 2001. LNCS, vol. 2096, pp. 78–87. Springer, Heidelberg (2001)
19. Kuncheva, L.: Switching between selection and fusion in combining classifiers: An experiment. IEEE Transactions on Systems, Man, and Cybernetics, Part B 32(2), 146–156 (2002)
20. Tumer, K., Ghosh, J.: Error correlation and error reduction in ensemble classifiers. Connection Science 8, 385–404 (1996)

21. Kuncheva, L., Whitaker, C.: Ten measures of diversity in classifier ensembles: Limits for two classifiers. In: Proceedings of IEE Workshop on Intelligent Sensor Processing, Birmingham, p. 10/1-6 (2001)
22. Wolpert, D.H.: Stacked generalization. Neural Networks 5, 241–259 (1992)
23. http://en.wikipedia.org/wiki/Meta_learning_(computer_science)
24. Duch, W., Grudzinski, K.: Meta-learning: searching in the model space. In: Proc. of the Int. Conf. on Neural Information Processing (ICONIP), Shanghai, China (2001)
25. Schaul, T., Schmidhuber, J.: Meta-learning. Scholarpedia 5(6), 4650 (2010)
26. Utgoff, P.: Shift of bias for inductive concept learning. In: Michalski, R., Carbonell, J., Mitchell, T. (eds.) Machine Learning, pp. 163–190 (1986)
27. Schmidhuber, J.: Evolutionary principles in self-referential learning. Diploma thesis, Institut für Informatik, Technische Universität München (1987)
28. Atkeson, C.G.: Memory-based approaches to approximating continuous functions. In: Casdagli, M., Eubank, S. (eds.) Nonlinear Modeling and Forecasting, pp. 503–521. Addison Wesley, Harlow (1992)
29. Atkeson, C.G., Moore, A., Schaal, S.: Locally Weighted Learning. Artificial Intelligence Review 11(1-5), 11–73 (1997)
30. Bersini, H., Bontempi, G., Birattari, M.: Is readability compatible with accuracy? From neuro-fuzzy to lazy learning. In: Freksa, C. (ed.) Proceedings in Artificial Intelligence 7, pp. 10–25. Infix/Aka, Berlin (1998)
31. Deb, K.: Multi-objective optimization using evolutionary algorithms. J. Wiley (2001)
32. Coello Coello, C.A., Van Veldhuizen, D.A., Lamont, G.B.: Evolutionary Algorithm MOP Approaches, Evolutionary Algorithms for Solving Multi-Objective Problems. Kluwer Academic (2002)
33. Bonissone, P., Subbu, R., Lizzi, J.: Multi Criteria Decision Making (MCDM): A Framework for Research and Applications. IEEE Computational Intelligence Magazine 4(3), 48–61 (2009)
34. Breiman, L., Friedman, J., Olshen, R.A., Stone, C.J.: Classification and regression trees. Wadsworth (1984)
35. Bonissone, P., Cheetham, W.: Fuzzy Case-Based Reasoning for Decision Making. In: Proc. FUZZ-IEEE 2001, Melbourne, Australia, vol. 3, pp. 995–998 (2001)
36. Xue, F., Subbu, R., Bonissone, P.: Locally Weighted Fusion of Multiple Predictive Models. In: IEEE International Joint Conference on Neural Networks (IJCNN 2006), Vancouver, BC, Canada, pp. 2137–2143 (2006), doi:10.1109/IJCNN.2006.246985
37. Yan, W., Xue, F.: Jet Engine Gas Path Fault Diagnosis Using Dynamic Fusion of Multiple Classifiers. In: IJCNN 2008, Hong Kong, pp. 1585–1591 (2008)
38. Bentley, J.L.: Multidimensional binary search trees used for associative searching. Communications of the ACM 18(9), 509–517 (1975)
39. Gray, A., Moore, A.: N-Body Problems in Statistical Learning. In: Proc. Advances in Neural Information Processing Systems, NIPS (2001)
40. Subbu, R., Bonissone, P., Eklund, N., Yan, W., Iyer, N., Xue, F., Shah, R.: Management of Complex Dynamic Systems based on Model-Predictive Multi-objective Optimization. In: CIMSA 2006, La Coruña, Spain, pp. 64–69 (2006)
41. Subbu, R., Bonissone, P., Bollapragada, S., Chalermkraivuth, K., Eklund, N., Iyer, N., Shah, R., Xue, F., Yan, W.: A review of two industrial deployments of multi-criteria decision-making systems at General Electric. In: First IEEE Symposium on Computational Intelligence in Multi-Criteria Decision-Making (MCDM 2007), Honolulu, Hawaii (2007), doi:10.1109/MCDM.2007.369428

Multiagent Learning through Neuroevolution

Risto Miikkulainen, Eliana Feasley, Leif Johnson, Igor Karpov,
Padmini Rajagopalan, Aditya Rawal, and Wesley Tansey

Department of Computer Science
The University of Texas at Austin, Austin, TX 78712, USA
{risto,elie,leif,ikarpov,padmini,aditya,tansey}@cs.utexas.edu

Abstract. Neuroevolution is a promising approach for constructing intelligent agents in many complex tasks such as games, robotics, and decision making. It is also well suited for evolving team behavior for many
multiagent tasks. However, new challenges and opportunities emerge in
such tasks, including facilitating cooperation through reward sharing and
communication, accelerating evolution through social learning, and measuring how good the resulting solutions are. This paper reviews recent
progress in these three areas, and suggests avenues for future work.

Keywords: Neuroevolution, neural networks, intelligent agents, games.

1 Introduction

Neuroevolution, i.e. evolution of artificial neural networks, has recently emerged
as a powerful approach to constructing complex behaviors for intelligent agents
[11,24]. Such networks can take a number of simulated or real sensor values
as input, and perform a nonlinear mapping to outputs that represent actions
in the world such as moving around, picking up objects, using a tool or firing a weapon. Recurrency in neural networks allow then to integrate information over time, and make decisions robustly even in partially observable
domains where traditional value-function based reinforcement learning techniques [42] have difficulty. Neuroevolution has thus been useful in building intelligent agents for e.g. video games, board games, mobile robots, and autonomous
vehicles [40,13,45,12,21,16,23].

Much of the work so far has focused on developing intelligent behaviors for
single agents in a complex environment. As such behaviors have become more
successful, a need for principled multiagent interactions has also risen. In many
domains such as video games and robotics, there are actually several agents that
work together to achieve a goal. A major part of being effective in such domains
is to evolve principled mechanisms for interacting with other agents. Neuroevolution is a natural approach to multiagent systems as well: The evolving population
provides a natural team setting, and neural networks allow implementing team
sensing and interactions in a natural manner.

It turns out the multiagent perspective brings entirely new challenges and
opportunities to neuroevolution research. This paper reviews recent progress

J. Liu et al. (Eds.): WCCI 2012 Plenary/Invited Lectures, LNCS 7311, pp. 24–46, 2012.

in three of them: Setting up evolution so that effective collaboration emerges, combining evolution with learning within the team, and evaluating the team behaviors quantitatively.

First, how should evolution be set up to promote effective team behaviors. That is, when the team is successful, should the rewards be distributed among team members equally, or should individuals be rewarded for their own performance? Should the team members communicate explicitly to coordinate their behavior, or is it sufficient to rely on changes in the environment (i.e. stigmergy)? How much should collaboration be rewarded for it to emerge over simpler individual behaviors? Experiments illustrating these issues will be reviewed in section 2.1.

Second, being part of a team provides an opportunity not only for coordinating actions of several team members, but also of learning from one another in the team. How should just learning be best established? Should the population champion be used as a teacher, or is it better to learn from any successful behavior in the population, in an egalitarian fashion? If everyone is learning based on the same successful behaviors, how can diversity be maintained in the population? Is learning useful in driving evolution through the Baldwin effect, or is it more effective to encode the learned behaviors directly to the genome through Lamarckian evolution? Section 3 evaluates possible solutions to these issues.

Third, given that multiagent behaviors can be particularly complex, depending on interactions between the team members, the environment, and opponents, how can they be best characterized and evaluated? For instance in a competitive environment, can a tournament be set up to evaluate the strengths of teams quantitatively? Is there a single best behavior or are multiple roughly equally good different solutions possible? Are best behaviors shared by everyone on the team, or is it better to have different specialties, or even on-line adaptation? These issues are discussed in the context of a comprehensive NERO tournament in section 4.

2   Setting up Multiagent Neuroevolution

As described below in separate sections, prey capture by a team of predators is used as the experimental domain to study how reward structure and amount and coordination mechanism affect multiagent evolution. An advanced neuroevolution method of multi-component-ESP will be used to evolve the controller neural networks.

2.1   Predator-Prey Environment

A significant body of work exists on computational modeling of cooperation in nature. For instance, flocking behaviors of birds and schooling of fish have been modeled extensively using rule-based approaches [6,31,37]. Cooperative behavior of micro-organisms like bacteria and viruses has been modeled with genetic algorithms [22,33]. Ant and bee colonies have been the subject of many studies involving evolutionary computation as well [9,29,47]. Similarly, as a research setting to study how cooperation can best emerge in multiagent neuroevolution, predator-prey simulation environment was constructed to model hunting behaviors of hyenas. This environment provides immediate motivation and insight from nature; it is also easy to simulate with quantifiable results.

In this environment, a team of predators (hyenas) is evolved using cooperative coevolution to capture fixed-behavior prey (a gazelle or a zebra). The world in this simulation is a discrete toroidal environment with 100×100 grid locations without obstacles, where the prey and predators can move in four directions: east, west, north and south. They move one step at a time, and all the agents take a step simultaneously. To move diagonally, an agent has to take two steps (one in the east-west direction and one in the north-south direction). A predator is said to have caught a prey if it moves into the same location in the world as the prey. The predators are aware of prey positions and the prey are aware of predator positions. Direct communication among predators (in terms of knowledge of other predators' positions) is also introduced in some cases. In all other cases, the predator agents can sense only prey movements and have to use that to coordinate their actions (stigmergic communication). There is no direct communication among the prey. Each predator has as its inputs the x and y offsets of all the prey from that predator. In the case of communicating predators, they also get as input the x and y offsets to the other predators. When fitness rewards from prey capture are shared, all the predators gain fitness even when only one of them actually catches the prey. In cases with individual fitness, only the particular predator that captures the prey gets the reward.

There are two types of prey in the environment - a smaller prey (gazelle) that moves with 0.75 times the speed of the predator and a larger prey (zebra) that has the same speed as the predator. The prey behaviors in these experiments are hard-coded and do not evolve. Each prey simply moves directly away from the current nearest predator. The predators can therefore catch the smaller prey individually, but cannot catch the larger prey by just following the prey around, because their grid world is toroidal. The predators have to surround a zebra from different directions before they can catch it. In cases where both types of prey exist in the field simultaneously, the predators need to decide whether to catch the small prey individually or to coordinate and hunt the larger prey together. The larger prey give more reward than the smaller prey, and the relative reward amounts can be varied.

Thus, three parameters are progressively modified in these experiments: (1) whether only the individual actually catching the prey receives the fitness, or whether it is shared by all individuals, (2) whether the predators can observe one another or not (direct vs. stigmergic communication), and (3) the size of the fitness reward from catching a prey. These experiments are used to contrast the role of each of these parameters in the evolution of cooperation.

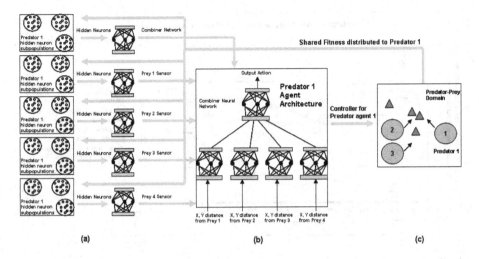

Fig. 1. Multi-Component ESP in the predator-prey domain for predator agent in Experiment 1. A single predator agent (shown in (b)) is composed of five neural networks. Four of these sense one of the prey agents. Their outputs are given to a fifth combiner network that outputs the next move for that predator. Each network is evolved in a separate ESP process, where one subpopulation is evolved for each of the neurons in the network (a). The predator is evaluated in the domain simulation with prey and other predator agents (c). Its fitness is distributed equally among all the networks and among all the neurons that participated in it. In this manner, evolution can discover neurons and networks that cooperate well to form an effective agent.

2.2   The Multi-component ESP Neuroevolution Method

Coevolution is defined as the simultaneous evolution of two or more individuals whose fitness is measured based on their interactions with each other [25]. In cooperative coevolution, the individuals have to evolve to cooperate to perform a task. They share the rewards and punishments of their individual actions equally. It turns out that it is often easier to coevolve components that cooperate to form a solution, rather than evolve the complete solution directly [15,26]. The components will thus evolve different roles in the cooperative task.

For example, in the Enforced SubPopulations (ESP) architecture [15], neurons selected from different subpopulations are required to form a neural network whose fitness is then shared equally among them. Such an approach breaks a complex task into easier sub-tasks, avoids competing conventions among the component neurons and makes the search space smaller. These effects make neuroevolution faster and more efficient.

Similarly, Multi-Component ESP extends this approach to evolve a team of agents (Figure 1). Each agent comprises multiple ESP-type neural networks to sense different objects in the environment. The team's reward from fitness evaluations is shared equally by the component networks of all the agents [30]. The cooperative coevolution approach has been shown to be effective when

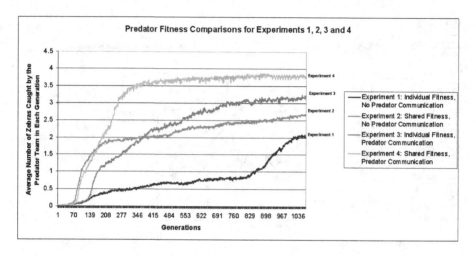

Fig. 2. Average number of prey (zebras) caught (out of four possible) in Experiments 1, 2, 3 and 4. The total number of prey caught by the three predators was averaged over 6000 trials for each generation. Cooperation is slow to evolve with individual rewards and without communication, and is less efficient (Experiment 1). Introduction of reward sharing results in faster and more effective evolution of cooperation (Experiment 2). Knowledge of positions of other predators makes it easier to evolve coordinated hunting strategies (Experiment 3). Evolution of cooperation is strongest when reward sharing and communication are combined (Experiment 4).

coevolving teams of agents. First, Yong and Miikkulainen [51] showed that a team of predators that share fitness can evolve to cooperate to catch prey with or without communication. In their experiments, without communication, the roles the predators evolve are more rigid but more effective; with communication, their roles are less efficient but more flexible. Second, Rawal et al. [30] showed that the Multi-Component ESP architecture can coevolve a team of predators with a team of prey. The individuals cooperate within the team, but the predator team competes with the prey team. Therefore, the Multi-Component ESP architecture will be used to evolve the predators in this paper as well.

In prior work, the outputs of the neural networks within a predator or prey agent were summed to get the final output action. However, preliminary experiments showed that including a combiner network to combine the outputs of these networks was more powerful and resulted in the emergence of more complex behaviors. Hence, this technique was used in this paper (Figure 1). The combiner network weights were evolved using the same technique as the other networks.

2.3   Experimental Setting Results

In the control experiment (Experiment 1), the predators neither communicate nor share fitness. Cooperation does not evolve initially and as a result, they

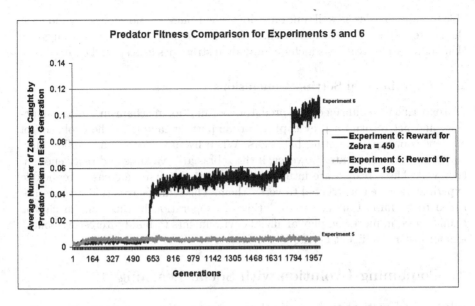

Fig. 3. Average number of zebras caught in Experiments 5 and 6. The total number of prey (out of one possible) caught by the three predators was averaged over 6000 trials for each generation. When the payoff on capturing a zebra is low with respect to the difficulty of catching it (Experiment 5), the predators prefer to hunt the easy-to-catch gazelles individually. When the net return for capturing the zebra is high enough (Experiment 6), the predators evolve to discover cooperative strategies to hunt it. Once it is caught, they continue by hunting gazelles.

rarely catch any zebras. On the other hand, adding reward sharing (Experiment 2) increases the number of prey caught as the predators efficiently evolve to cooperate over the early generations. The average number of zebras caught in each generation in Experiments 1 and 2 are contrasted in Figure 2.

Similarly, adding communication to predators with individual fitness in Experiment 3 results in the predators easily evolving to cooperate, leading to more prey captures (Figure 2). This effect is even stronger with both communication and fitness sharing enabled (Experiment 4; Figure 2), suggesting that these two factors affect different aspects of the evolution process, i.e. how easy it is to establish cooperation, and how worthwhile it is.

Experiments 5 and 6 were designed to answer the question: If there are both gazelles, which can be caught easily but give a lower fitness, and zebras, which need all the predators to cooperate to catch them but give higher fitness, which one is preferred? In Experiment 5, the predators prefer to hunt gazelles instead of evolving to cooperate to capture the zebra. The reward for catching the zebra is not large enough for cooperative behaviors to be selected during evolution. In contrast, in Experiment 6, it is large enough, and the predators slowly evolve to team up to capture this more difficult prey, thus verifying the hypothesis that net return is important in the evolution of cooperation (Figure 3). Interestingly, they are still able to hunt gazelles as well, but only do it when there are no zebras around even though zebras are still hard to catch. This result is important because it suggests that cooperative strategies include individual strategies as a special case.

2.4 Experimental Setting Conclusions

The experiments confirmed that predator coordination mechanism, reward structure, and net return on prey capture are important factors in the evolution of efficient cooperative hunting behaviors. When hyenas survive on gazelles, they do not need to cooperate. However, if the zebras are available and tasty enough, they will. These results are intuitive, but this is the first time easily replicable experiments were constructed to verify them. The same factors that were established to be important in the evolution of cooperation in this domain can be manipulated in more complex artificial environments to build interesting behaviors for other intelligent agents in the future.

3   Combining Evolution with Social Learning

After a brief motivation for social learning in multiagent neuroevolution, the robot foraging domain and the NEAT neuroevolution method are briefly described, followed by results answering the questions posed in section 1.

3.1 Motivation for Social Learning

Evolutionary algorithms (EAs) [14] evaluate agents either in isolation or in direct competition with a subset of the other members of the population. Social and cultural learning algorithms [32] extend EAs by enabling agents to leverage observations of other members of the population to improve their own performance during their lifetime. By learning from others without having to directly experience or acquire knowledge, social learning algorithms have been able to improve the learning rate of EAs in many challenging domains [8,17,46,1,7,48].

Traditionally in social learning algorithms, each agent is either a student or a teacher [28,2]. All actions of the teacher agents are considered to be good examples from which to learn, as they are derived from a high-fitness strategy (i.e. the teacher's policy). However, an agent with high overall fitness may not always choose good actions and agents with low overall fitness may actually perform well in some limited scenarios. Filtering potential observations based on their own merit may therefore be more appropriate and lead to both improved learning rate and stronger final strategies.

This paper presents Egalitarian Social Learning (ESL) as an alternative to the student-teacher paradigm. Agents in ESL are divided into subcultures at the start of each generation and can learn from any other agent in their subcultural group. Learning examples are determined by a user-defined acceptability function that filters out examples leading to low rewards. When an action is accepted, agents mimic it in order to learn a policy similar to that of the observed agent. ESL differs from other social learning algorithms in that the quality of a training example is measured by the reward received rather than the fitness of the agent generating the example.

3.2   The Foraging Domain

The domain used to evaluate ESL is a foraging world in which agents move freely on a continuous toroidal surface. The world is populated with various plants, some of which are nutritious and bear positive reward, while others are poisonous and bear negative reward. These plants are randomly distributed over the surface of the world. The foraging domain is non-competitive and non-cooperative; each agent acts independently of all other agents, with the exception of the teaching signals that pass between them. At the start of each generation, all individuals begin at the center of the world, oriented in the same direction, and confronted with the same plant layout and configuration. Every agent then has a fixed number time steps to move about the surface of the world eating plants— which happens automatically when an agent draws sufficiently close to one— before the evaluation is over.

Agents "see" plants within a 180° horizon via a collection discretized sensors. Each agent has eight sensors for each type of plant, with each sensor covering a different 12.5° sector of the 180° ahead of the agent. Agents cannot see other individuals or plants they have already eaten— all they can see is edible food. The strength of the signal generated by each plant is proportional to its proximity to the agent. Agents also have a sensor by which they can detect their current velocity. As agents can only turn up to 30° in a given timestep, knowledge of velocity is necessary for agents to accurately plan optimal trajectories (e.g. agents may need to slow down in order to avoid overshooting a plant). Each agent is controlled by an artificial neural network that maps from the agent's sensor readings to the desired change in orientation and velocity.

Two separate configurations of the robot foraging world are used in the experiments. The first two experiments use a "simple" world where the toroidal surface is 2000 by 2000 units, with a single plant type of value 100 and 50 randomly distributed instances of the plant. In this world, the agents have a straightforward task of learning to navigate efficiently and gather as many plants as possible. The third set of experiments uses both the simple world and a second, more complex world to evaluate performance. The "complex" world has a surface of 500 by 500 units, with five different plant types of value -100, -50, 0, 50, and 100. For each plant type, 20 instances are created and randomly distributed across the surface. This world presents the agents with a more difficult task as they must efficiently gather nutritious food while simultaneously avoiding the poisonous food.

In all four experiments, 100 different agents are created in each generation. All networks are initialized with fully-connected weights with no hidden neurons and a learning rate of 0.1 is used when performing backpropagation. Agents automatically eat any plant within five units. Each evaluation lasts 1000 timesteps and the results for each experiment are the average of 30 independent runs. The acceptability function for all experiments is to learn from any action yielding a positive reward.

3.3   The NEAT Neuroevolution Method

NeuroEvolution of Augmenting Topologies (NEAT)[39] is an evolutionary algorithm that generates recurrent neural networks. Through a process of adding and removing nodes and changing weights, NEAT evolves genomes that unfold into networks. In every generation, those networks with the highest fitness reproduce, while those with the lowest fitness are unlikely to do so. NEAT maintains genetic diversity through speciation and encourages innovation through explicit fitness sharing.

In the foraging domain, NEAT is used to generate a population of individual neural networks that control agents in the world. The input to each network is the agent's sensors, and the outputs control the agent's velocity and orientation. The fitness of each network is determined by the success of the agent it controls— over the course of a generation, networks that control agents who eat a good deal of rewarding food and very little poison will have high fitness and those that control agents with less wise dietary habits will have low fitness.

In standard NEAT, the networks that are created do not change within one generation. To facilitate social learning, we must perform backpropagation [34] on the networks that NEAT creates in order to train agents on accepted examples. Since NEAT networks are recurrent, ESL enhances NEAT with backpropagation capabilities using the backpropagation through time algorithm [49].

The final fitness of each phenome, then, reflects the performance of the individual that used that phenome and elaborated on it over the course of a generation. This elaboration drives evolution in two alternate ways. In Darwinian evolution, the changes that were made to the phenome only affect selection and are not saved; in Lamarckian, the genome itself is modified.

3.4   Social Learning Results

Three experiments were performed: ESL was first applied to the entire population (without subcultures), and the best way to make use of learning (Darwinian vs. Lamarckian) determined. The effect of maintaining diversity through explicit subcultures was then evaluated. In the third experiment, ESL was compared to the traditional student-teacher model of social learning.

Figure 4 shows the results of applying a monocultural egalitarian social learning algorithm to the foraging domain in both the Lamarckian and Darwinian paradigms. The performance of both algorithms quickly converges, with Lamarckian reaching a higher-fitness solution than Darwinian evolution. In the context of *on-line* evolutionary learning algorithms, previous work [50] showed that Darwinian evolution is likely to be preferable to Lamarckian evolution in dynamic environments where adaptation is essential and the Baldwin effect [38] may be advantageous. However, as adaptation is not necessary for foraging agents

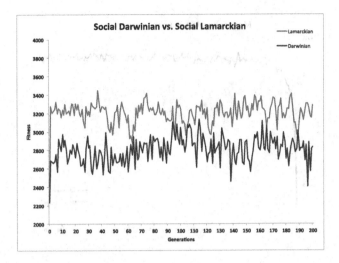

Fig. 4. The effects of Darwinian and Lamarckian evolution when using a monocultural variant of ESL. While both evolutionary paradigms converge rapidly Lamarckian evolution is more effective than Darwinian in the foraging domain. Consequently, Lamarckian evolution is the paradigm used in all remaining experiments.

(i.e. the rewards of each plant type are the same in every generation), in this experiment Lamarckian evolution outperforms Darwinian evolution. Nevertheless, in both cases performance converges to a lower score than that of simple neuroevolution.

On the other hand, monocultural Lamarckian social learning is likely to provide redundant information that may result in getting stuck in local optima. In order to address this problem, subcultural version of egalitarian social learning was designed to promote and protect diversity. At the start of each generation, the population is divided into 10 subcultures of 10 agents each, with each agent's subculture decided at random. During the evaluation, agents only teach and learn from other agents in their own subculture.

Figure 5 shows results comparing monocultural and subcultural learning. Subcultural learning not only reaches a higher peak than the monocultural method, but also arrives at this level of fitness more rapidly than the simple neuroevolution approach. When every mutated organism has the opportunity to train every other, as is the case in monocultural learning, the entire population may be negatively impacted by any one individual. By preventing agents that lead the population towards local optima from impacting the remainder of the population, subcultural learning provides safety and protection from premature convergence.

In the third set of experiments, subcultural ESL is compared to an on-line student-teacher learning algorithm inspired by the NEW TIES system [17].

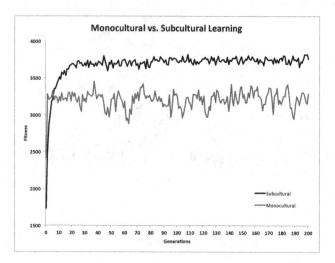

Fig. 5. Monocultural agents learning from the entire population and subcultural agents learning only from their subcultures. Subcultural agents outperform monocultural agents, converging to a much higher ultimate fitness.

The system utilizes a steady-state evolution in which at every timestep each agent probabilistically teaches the lowest-fitness member of the population within some radius, effectively forming geographical subcultures.

Figures 6 and 7 show the results of the subcultural ESL algorithm compared to the student-teacher variant of NEW TIES and simple neuroevolution. Subcultural ESL converges to a near-optimal solution faster than the student-teacher variant in both the simple and the complex world. While in the simple world (Figure 6) this speed-up is slight, in the complex world (Figure 7) the egalitarian approach is more than an order of magnitude faster, reaching a higher fitness by generation 50 than either the student-teacher or simple neuroevolution methods achieve by generation 500.

3.5   Social Learning Conclusions

Unlike traditional social learning algorithms that follow a student-teacher model, ESL teaches agents based on acceptable actions taken by any agent in its subculture. By constraining teaching samples to those from the same subcultural group, ESL promotes diversity in the overall population and prevents premature convergence. Experiments in a complex robot foraging domain demonstrated that this approach is highly effective at quickly learning a near-optimal policy with Lamarckian evolution. The results thus suggest that egalitarian social learning is a strong technique for taking advantage of team behaviors that exist in the evolving population.

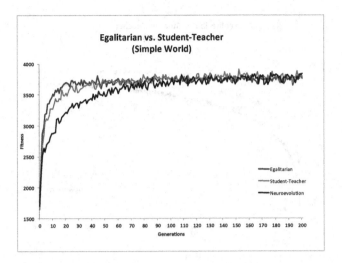

Fig. 6. ESL compared to simple neuroevolution, and student-teacher learning in the simple world. All strategies converge to solutions of similar quality, with egalitarian learning converging in the fewest evaluations.

4 Evaluating Multiagent Performance

4.1 Motivation

The NERO video game [40] was originally developed to demonstrate that neuroevolution could be a powerful tool for constructing solutions to open-ended design problems. A human player provides increasingly challenging goals, and a team of NPCs evolves to meet those goals, eventually excelling in the game. Complex behavior was demonstrated in a number of different challenge situations, such as running a maze, approaching enemy while avoiding fire, and coordinating behavior of small sub-teams. However, the final behavior of entire teams was never evaluated, so it is not clear how complex the behaviors could become in this process and what successful behavior in the game might actually look like. Also, it is not clear whether there is one simple winning strategy that just needs to be refined to do well in the game, or whether there are multiple good approaches; similarly, it is unclear whether winning requires combining individuals with different skills into a single team, or perhaps requires on-line adaptation of team composition or behaviors.

In any case, such evaluations are difficult for two reasons: (1) designing teams takes significant human effort, and covering much of the design space requires that many different designers participate; (2) evaluation of the resulting behaviors takes significant computational effort, and it is not clear how it can be best spent. This paper solves the first problem by crowd-sourcing, i.e. running a NERO tournament online. Students in the 2011 Stanford online AI course[1]

---

[1] www.ai-class.com

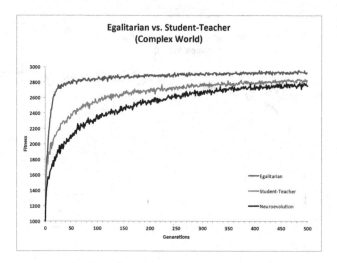

Fig. 7. ESL compared to simple neuroevolution, and student-teacher learning in the complex world. ESL is more than an order of magnitude faster, reaching a higher fitness by generation 50 than either comparison method achieves by generation 500.

were invited to participate. About 85 of them did, many spending considerable effort to produce good teams, thereby resulting in a wide selection of approaches and solutions. The second problem was solved by first testing out different tournament structures, and eventually running a comprehensive round robin tournament of 24,180 games in parallel in a Condor cluster. The results from the tournament were then used to identify complex and interesting behaviors that perform well on the task.

4.2   The NERO Domain

NERO [41] was originally developed as an experimental platform for training teams of agents to accomplish complex tasks based on the rtNEAT [39] method for evolving artificial neural networks. The rtNEAT method is a version of NEAT described in the previous section, with the difference that individuals are evaluated, reproduced, and replaced continuously instead of in generations. This approach allows running evolution in the background continuously in real-time without distracting the human player. The original NERO game was later extended into an open-source version called OpenNERO,[2] which is a general-purpose platform for AI research and education [20]. OpenNERO includes several different environments and AI methods in addition to the NERO game environment itself, but only the NERO environment in OpenNERO was used in this research.

Each NERO agent on a team has a fixed array of 15 sensors that detect agents on the same and opposite teams, placement of nearby walls, distance to

---

[2] opennero.googlecode.com

Fig. 8. A screenshot of a single NERO match. Two teams of agents are shown as bipedal robots in a playing arena with obstacles and boundaries. The teams start opposite each other on the two sides of the obstacle wall in the middle and have to get around this obstacle to damage opponents and earn points.

a flag (if present), current motion, damage to opponents, and damage to the agent itself. Agents control their movement on the field using a two-dimensional control signal $u = < \ddot{r}, \ddot{\theta} >$, where $\ddot{r}$ is the linear acceleration of the agent in the direction of the agent's current orientation θ, and $\ddot{\theta}$ is the agent's angular acceleration.

Training teams in OpenNERO is similar to NERO. The user can dynamically change the virtual environment by adding, scaling, rotating or removing walls, moving a flag, and adding or removing immobile enemy agents. The user can also change the way the fitness function is computed by adjusting a (positive or negative) weight on each of the different available fitness dimensions. The available fitness dimensions are *stand ground* (i.e. minimize $\dot{r}$), *stick together* (minimize distance to the team's center of mass), *approach flag* (minimize distance to a flag on the field, if present), *approach enemy* (minimize distance to the closest enemy agent), *hit target* (successfully fire at an enemy), and *avoid fire* (minimize accrued damage).

For the battle task, two teams—each consisting of 50 NERO agents—occupy a continuous, two-dimensional, virtual playing field of fixed size (see Figure 8).

The playing field contains one central obstacle (a wall), four peripheral obstacles (trees), and four walls around the perimeter to contain all agents in the same general area. Each NERO agent starts a battle with 20 hit-points. At each time slice of the simulation, each agent has the opportunity to fire a virtual laser at the closest target on the opponent's team that is within two degrees of the agent's current orientation. If an agent fires and hits an opponent, the opponent loses one hit-point. The score for a team is equal to the number of hit-points that the opponent team loses in the course of the battle.

A team of NERO agents can be serialized to a flat text file. The text file describes each of the 50 agents on a team. Agents that use rtNEAT serialize to a description of the genotype for each agent, and agents that use Q-learning serialize their (hashed) Q-tables directly to the file. Anyone was allowed to participate in the tournament by submitting online a serialized team of virtual agent controllers for the NERO battle task. The only difference between the teams was in the training of the controllers contributed by the competitors.

The OpenNERO code was extended for this tournament to allow teams to consist of mixtures of rtNEAT (neural network) and reinforcement learning (Q-learning) agents; this distinction is primarily interesting in the sense that rtNEAT agents search for control policies directly, while Q-learning searches in value-function space and then uses value estimates for each state to determine appropriate actions. For rtNEAT–based training, individuals within the population are ranked based on the weighted sum of the Z-scores over the fitness components. For Q-learning–based training, each fitness dimension is scaled to $[0,1]$, and then a linear weighted sum is used to assign a total reward to each individual.

Both types of controllers could be submitted to the online tournament: artificial neural network controllers of arbitrary weight and topology, and hash tables approximating the value function of game states. The competitors could extend and/or modify the available OpenNERO training methods as well as create their own training environments and regimens. It was this training that determined the fitness of each team when pitted against other teams submitted to the tournament.

4.3   Evaluation Results

An online NERO tournament was run in December 2011. About 85 participants submitted 156 teams to the tournament. Of these, 150 teams contained neural network-controlled agents and 11 contained value table-controlled agents. Mixed teams were also allowed; four of the submitted teams contained mixed agent types. Because of the large number of teams, each game was played off-screen and limited to 5 minutes of game time. (In practice, good teams were able to eliminate all opponents in less than 5 minutes.) The team with the highest number of remaining hit points was declared the winner at the end of the match. Ties were extremely rare and were broken by a pseudo-random coin toss. The match-making script allowed matches to be run in parallel on a single machine or to be distributed to a Condor compute cluster [43].

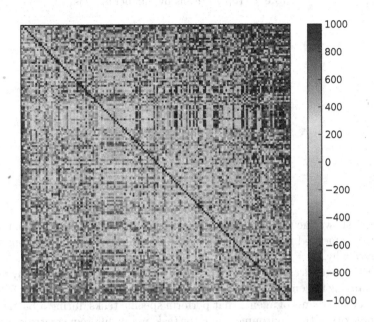

Fig. 9. Results from the round-robin NERO tournament. Teams are sorted by average score differential over all matches. Rows and columns in the matrix represent teams in the tournament, and colors represent score differentials for the respective individual matches between two teams. Red indicates victory by the row team, and blue indicates victory by the column team.

First, several double-elimination tournaments were run with the submitted teams. Repeated runs of the double-elimination tournament revealed that while the set of teams in the top 10 was generally consistent, their ranking was not stable. The teams were then placed in a round-robin tournament to evaluate the overall winner more accurately. In the round-robin tournament, each of the 156 submitted teams was matched against the 155 other teams. Each pair of teams was matched up twice (i.e. $k = 2$), allowing each team to play once as the blue team and once as the red team. This resulted in 24180 separate games, which were processed in parallel on approximately 100 computing nodes, which allowed the entire round-robin tournament to complete in less than 24 hours.

Figure 9 shows the complete results of the round-robin tournament. Black squares along the diagonal represent matches that were not played, blue squares indicate a win by the column team, and red squares indicate a win by the row team. One group of near-duplicate teams was submitted to the tournament; this shows up as the band of similar-colored games about one-third of the way through the matrix. The teams in the Figure are enumerated on both axes in order of increasing average match score differential.

Table 1. Top 10 teams by number of wins

Rank Team	Total wins
1 synth.pop	137
2 synth_flag.pop	130
3 lolwutamidoing	126
3 me - Rambo	126
5 PollusPirata	125
6 Cyber-trout	124
7 CirclingBullies	123
8 SneakySnipers	121
8 Tut	121
10 coward1	120

Table 1 shows the top ten teams. Despite the large number of teams, no single competitor emerged that significantly outperformed all others. It is interesting to analyze why.

In NERO, agents can be "shaped" towards more complex behaviors by progressively changing the environment and the fitness function during training. This process can create teams of agents that perform specific tasks during a battle. Given the complexity of the environment and the task, many different strategies can arise in this process, and they can interact with each other in complex ways. Considering this potential complexity, evolved strategies in the tournament turned out to be surprisingly easy to analyze. Because fitness is evaluated similarly for each team member during training, teams generally consist of agents that perform similar actions in a given world state. In principle, multiple teams can be trained using different shaping strategies, and single agents from those teams then combined into one team by copying the appropriate parts of the serialized team files (as was suggested in the online tournament instructions). However, most teams submitted to the tournament did not (yet) take advantage of this possibility; instead, agents on a single team usually performed similar actions in response to each game state. It was therefore possible to characterize the most common strategies used in the tournament, as outlined below. The example teams and videos of games between then are available at the tournament website[3].

Pack: The most prominent strategy among winning teams was to train agents to move as a group toward the central wall, then follow the wall tightly to go around it, and then proceed towards the opponents on the other side. This strategy shows up in several of the top ten teams, but most notably in synth.pop and me-Rambo. These teams actively pursued their opponents by forming agents into a "pack" that had a lot of firepower and was therefore able to eliminate opponents effectively.

Backpedaling: A second successful strategy was almost exactly the opposite from the "pack" strategy: From the start, most agents backpedaled away from

---

[3] code.google.com/p/opennero/wiki/TournamentResults2011

the opponents and the central wall, and then took shots against the opponents from afar. Backpedaling preserves the ability of the agents to fire at opponents, while increasing the distance to the opposing team and maximizing view of the field. Backing up was often effective against the "pack" strategy because a team could eliminate the opponents one-by-one as they emerged around the edge of the wall. Examples of this strategy included EvilCowards and SneakySnipers.

Encircling: Some teams followed a third strategy, where all agents on the team would run to the far right or left of the wall in a wide arc, and that way try to encircle the enemy agents. Interestingly, although at the outset this strategy seems logical, and was indeed devastating to some opponents, it often lost to the first two strategies. Against the teams that adopted the "pack" strategy, agents following the "encircle" strategy were often not pointed toward their opponents, and thus could be fired upon without recourse. Similarly, teams following the "encircle" strategy tended to fail against the "backpedal" teams because the "encircle" agents again tended to be pointed away from enemy agents too often to fire successfully. Examples of encircling teams include Caipirinha_01_FNM and Artificial Ignorance.

Brownian Motion: Teams that used reinforcement learning agents tended to cluster in the middle of the playing field and move back and forth in Brownian motion. This behavior likely originated from a difficulty in acquiring sufficient data during training, and from a difficulty in approximating the value function for the task, resulting in agents that effectively chose random actions at each time step. However, sometimes this behavior was seen in rtNEAT teams as well, and it was at times surprisingly effective. When other teams approached the cluster, they were not always lined up to shoot—on the other hand, because of the Brownian motion, the shots were not always on target either. So the Brownian motion teams formed a firing squad that was difficult to approach, performing sometimes better than teams that employed a strategy for going around the wall. Examples of Brownian motion teams include Peaceful Barbarians 1 and The729Gang.

Perhaps most interestingly, the strategies do not form a strict dominance hierarchy, but instead are highly cyclic. For instance, the third-place me-Rambo (a "pack" team) reliably defeats the first-place synth.pop (also a "pack" team), apparently due to subtle differences in timing. On the other hand, synth.pop wins over the 24th-place EvilCowards (a "backpedal" team), because the synth.pop pack splits into two and breaches the wall from both edges simultaneously. However, EvilCowards handily defeats me-Rambo, because agents in the me-Rambo train are eliminated one-at-a-time as they come around the wall!

There are many other similar cycles in the tournament graph as well, i.e. there is not a team in the tournament that is objectively better than all other teams. It actually seems that there is not even a single strategy that is better than the others: as e.g. the "pack" strategy becomes highly optimized, it also becomes more vulnerable to the "backpedal" counter-strategy. Such relationships may indeed be inherent and even desirable for complex games.

Based on these observations, a compelling next step might be to construct composite teams of individuals from several different teams. The idea is that such teams could perform well against a variety of strategies. Such an approach was already possible in the online tournament, but not extensively used. With more multi-objective evolutionary methods [35,36], it might also be possible to develop multi-modal behaviors that identify what strategy the opponent is using, and select a counter-strategy accordingly. It might also be possible in principle to adapt to opponents online, while the battle is taking place. Such extensions should result in more versatile multi-agent behavior; they will also make it even more difficult to analyze such behavior in the future.

4.4 Evaluation Conclusions

The results of the online NERO tournament demonstrate that multi-agent behavior can be evaluated quantitatively using tournaments. To fully characterize the behaviors, it is necessary to run round-robin tournaments: There may not be a single best strategy, but behaviors may instead be highly diverse, and perform differently against different opponents. This phenomenon may indeed be an inherent property of multiagent behavior in complex domains, and further computational tools may need to be developed to analyze it fully.

5 Discussion and Future Work

The experiments on cooperation raise an interesting issues about the nature of cooperation. For instance, the predators in Experiment 3 (individual rewards with communication) evolve cooperative hunting strategies efficiently, but they do not have any fitness incentive for cooperation. Instead, they use one another to improve individual fitness. Is this real cooperation? In biological literature, a cooperator is defined as an individual who pays a cost for another individual to receive a benefit [27]. This is a useful working definition in artificial settings as well. Thus in Experiment 3, though not all the predators gain by coordinating their behaviors, it is still considered cooperation.

Social learning is strongly motivated by biological analogy as well. The social intelligence hypothesis [5,19] and the cultural intelligence hypothesis [44] suggest that the need to handle complex social behaviors was the primary selection pressure driving the increase in brain size in primates and humans. These hypotheses are indeed supported by strong empirical evidence in recent years [18]. Further, egalitarianist philosophy advocates treating all individuals in a population as equals, regardless of such factors as background and status [3]. In hunter-gatherer societies, egalitarianism is a common paradigm for managing daily activities and organizing social structures [4]. It is likely that this lack of hierarchy and strict maintenance of equality has been pivotal in the development of human society and in separating humans from other primates [10]. It is interesting to see that the same conclusion follows from computational experiments on social learning: Egalitarianism promotes diversity, which in turn allows the population as a whole to achieve better performance. Given how difficult it may be to verify social and computational intelligence hypotheses directly, computational simulations may prove instrumental in testing and refining it further in the future.

Given that the NERO evaluation with 156 teams took significant supercomputing resources, it is useful to evaluate how this approach might be scaled up. Larger tournaments could be organized by using a hybrid structure; round-robin pools could be run in parallel to identify the proper seeds for top-ranking teams, and then a double-elimination tournament could be used to identify the overall winner. Thanks to the independence of individual matches in round-robin tournaments and within each level of a knockout tournament, it should be possible to scale up to even larger tournaments by running games on more compute nodes or carefully designing a tournament structure to optimize use of computing resources.

On the other hand, the tournament also showed that machine-learning games, where neuroevolution of multiagent behavior play a central role, may indeed be a viable game genre in the future. Several approaches to the game were identified in the tournament, none of them dominating all others. This is precisely what makes such games interesting: There is room for innovation and creativity, and the outcomes often turn out to be surprising. Using such games as a platform, it may also be possible to make significant research progress in multi-agent systems and intelligent agents in general.

6   Conclusion

Multiagent systems can be seen as the next frontier in constructing intelligent behavior through neuroevolution. This paper reviewed three challenges and opportunities in such systems: manipulating the rewards, coordination, and return; combining social learning with evolution; and evaluating performance through tournaments. Significant interactions and complexity were observed in each case, leading to the conclusion that the research is still in the beginning stages, but also that the technology is a good match with the opportunities.

Acknowledgments. Special thanks to Marc Wiseman and Kay Holekamp for their help with hyena data, all participants of the OpenNERO online tournament, Sebastian Thrun and Peter Norvig for allowing us to run the tournament in their online AI class, and Ken Stanley and Adam Dziuk for help with OpenNERO. This research was supported in by Google, Inc. and NSF under grants DBI-0939454, IIS-0915038, and IIS-0757479.

References

1. Acerbi, A., Nolfi, S.: Social learning and cultural evolution in embodied and situated agents. In: IEEE Symposium on Artificial Life, ALIFE 2007, pp. 333–340. IEEE (2007)

2. Acerbi, A., Parisi, D.: Cultural transmission between and within generations. Journal of Artificial Societies and Social Simulation 9(1) (2006)
3. Arneson, R.: Egalitarianism. In: Zalta, E.N. (ed.) The Stanford Encyclopedia of Philosophy. Stanford University (2009)
4. Boehm, C.: Hierarchy in the forest: The evolution of egalitarian behavior. Harvard Univ. Pr. (2001)
5. Byrne, R., Whiten, A.: Machiavellian Intelligence: Social Expertise and the Evolution of Intellect in Monkeys, Apes, and Humans. Oxford University Press, USA (1989)
6. Czirok, A., Vicsek, T.: Collective behavior of interacting self-propelled particles. Physica A 281, 17–29 (2000)
7. de Oca, M., Stutzle, T., Van den Enden, K., Dorigo, M.: Incremental social learning in particle swarms. IEEE Transactions on Systems, Man, and Cybernetics, Part B: Cybernetics 41(2), 368–384 (2011)
8. Denaro, D., Parisi, D.: Cultural evolution in a population of neural networks. In: Marinaro, M., Tagliaferri, R. (eds.) Neural Nets Wirn 1996, pp. 100–111. Springer, Newyork (1996)
9. Dorigo, M., Maniezzo, V., Colorni, A.: Ant system: optimization by a colony of cooperating agents. IEEE Transactions on Systems, Man, and Cybernetics–Part B 26(1), 29–41 (1996)
10. Erda, L., Whiten, A., Mellars, P., Gibson, K.: Egalitarianism and Machiavellian intelligence in human evolution. McDonald Institute for Archaeological Research (1996)
11. Floreano, D., Dürr, P., Mattiussi, C.: Neuroevolution: From architectures to learning. Evolutionary Intelligence 1, 47–62 (2008)
12. Floreano, D., Urzelai, J.: Evolutionary robots with on-line self-organization and behavioral fitness. Neural Networks 13, 431–4434 (2000)
13. Fogel, D.B.: Blondie24: Playing at the Edge of AI. Morgan Kaufmann, San Francisco (2001)
14. Fogel, L., Owens, A., Walsh, M.: Artificial intelligence through simulated evolution. John Wiley (1966)
15. Gomez, F., Miikkulainen, R.: Incremental evolution of complex general behavior. Adaptive Behavior, 317–342 (1997)
16. Gomez, F., Miikkulainen, R.: Active guidance for a finless rocket using neuroevolution. In: Proceedings of the Genetic and Evolutionary Computation Conference, pp. 2084–2095. Morgan Kaufmann, San Francisco (2003)
17. Haasdijk, E., Vogt, P., Eiben, A.: Social learning in population-based adaptive systems. In: IEEE Congress on Evolutionary Computation, CEC 2008 (IEEE World Congress on Computational Intelligence), pp. 1386–1392. IEEE (2008)
18. Herrmann, E., Call, J., Hernández-Lloreda, M., Hare, B., Tomasello, M.: Humans have evolved specialized skills of social cognition: The cultural intelligence hypothesis. Science 317(5843), 1360 (2007)
19. Humphrey, N.: The social function of intellect. Growing Points in Ethology, 303–317 (1976)
20. Karpov, I.V., Sheblak, J., Miikkulainen, R.: OpenNERO: A game platform for AI research and education. In: Proceedings of the Fourth Artificial Intelligence and Interactive Digital Entertainment Conference (2008)
21. Kohl, N., Stanley, K.O., Miikkulainen, R., Samples, M., Sherony, R.: Evolving a real-world vehicle warning system. In: Proceedings of the Genetic and Evolutionary Computation Conference (2006)

22. Kubota, N., Shimojima, K., Fukuda, T.: Virus-evolutionary genetic algorithm - coevolution of planar grid model. In: Proceedings of the Fifth IEEE International Conference on Fuzzy Systems (FUZZ IEEE 1996), pp. 8–11 (1996)

23. Lipson, H., Pollack, J.B.: Automatic design and manufacture of robotic lifeforms. Nature 406, 974–978 (2000)

24. Miikkulainen, R.: Neuroevolution. In: Encyclopedia of Machine Learning. Springer, Berlin (2010)

25. Mitchell, M., Thomure, M.D., Williams, N.L.: The role of space in the success of coevolutionary learning. In: Artificial Life X: Proceedings of the Tenth International Conference on the Simulation and Synthesis of Living Systems, pp. 118–124 (2006)

26. Moriarty, D.E., Miikkulainen, R.: Forming neural networks through efficient and adaptive coevolution. Evolutionary Computation 5(4), 373–399 (1997)

27. Nowak, M.A.: Five rules for the evolution of cooperation. Science 314, 1560–1563 (2006)

28. Parisi, D.: Cultural evolution in neural networks. IEEE Expert 12(4), 9–14 (1997)

29. Pérez-Uribe, A., Floreano, D., Keller, L.: Effects of Group Composition and Level of Selection in the Evolution of Cooperation in Artificial Ants. In: Banzhaf, W., Ziegler, J., Christaller, T., Dittrich, P., Kim, J.T. (eds.) ECAL 2003. LNCS (LNAI), vol. 2801, pp. 128–137. Springer, Heidelberg (2003)

30. Rawal, A., Rajagopalan, P., Miikkulainen, R.: Constructing competitive and cooperative agent behavior using coevolution. In: IEEE Conference on Computational Intelligence and Games (CIG 2010) (August 2010)

31. Reynolds, C.W.: Flocks, herds, and schools: A distributed behavioral model. In: Computer Graphics (SIGGRAPH 1987 Conference Proceedings), vol. 21(4), pp. 25–34 (1987)

32. Reynolds, R.: An introduction to cultural algorithms. In: Proceedings of the Third Annual Conference on Evolutionary Programming, pp. 131–139. World Scientific (1994)

33. Roeva, O., Pencheva, T., Tzonkov, S., Arndt, M., Hitzmann, B., Kleist, S., Miksch, G., Friehs, K., Flaschel, E.: Multiple model approach to modelling of escherichia coli fed-batch cultivation extracellular production of bacterial phytase. Electronic Journal of Biotechnology 10(4), 592–603 (2007)

34. Rumelhart, D., Hinton, G., Williams, R.: Learning representations by back-propagating errors. Nature 323(6088), 533–536 (1986)

35. Schrum, J., Miikkulainen, R.: Evolving agent behavior in multiobjective domains using fitness-based shaping. In: Proceedings of the Genetic and Evolutionary Computation Conference (2010)

36. Schrum, J., Miikkulainen, R.: Evolving multimodal networks for multitask games. In: Proceedings of the IEEE Conference on Computational Intelligence and Games (CIG 2011), pp. 102–109. IEEE, Seoul (2011)

37. Seno, H.: A density-dependent diffusion model of shoaling of nesting fish. Ecol. Modell. 51, 217–226 (1990)

38. Simpson, G.: The baldwin effect. Evolution 7(2), 110–117 (1953)

39. Stanley, K., Miikkulainen, R.: Evolving neural networks through augmenting topologies. Evolutionary Computation 10(2), 99–127 (2002)

40. Stanley, K.O., Bryant, B.D., Miikkulainen, R.: Real-time neuroevolution in the NERO video game. IEEE Transactions on Evolutionary Computation 9(6), 653–668 (2005)

41. Stanley, K.O., Bryant, B.D., Miikkulainen, R.: Real-time neuroevolution in the NERO video game. IEEE Transactions on Evolutionary Computation 9(6), 653–668 (2005)
42. Sutton, R.S., Barto, A.G.: Reinforcement Learning: An Introduction. MIT Press, Cambridge (1998)
43. Thain, D., Tannenbaum, T., Livny, M.: Distributed computing in practice: the condor experience. Concurrency - Practice and Experience 17(2-4), 323–356 (2005)
44. Tomasello, M.: The cultural origins of human cognition. Harvard Univ. Pr. (1999)
45. Valsalam, V., Miikkulainen, R.: Evolving symmetry for modular system design. IEEE Transactions on Evolutionary Computation 15, 368–386 (2011)
46. Vogt, P., Haasdijk, E.: Modeling social learning of language and skills. Artificial Life 16(4), 289–309 (2010)
47. Waibel, M., Floreano, D., Magnenat, S., Keller, L.: Division of labour and colony efficiency in social insects: effects of interactions between genetic architecture, colony kin structure and rate of perturbations. In: Proceedings of the Royal Society B, vol. 273, pp. 1815–1823 (2006)
48. Watkins, C., Dayan, P.: Q-learning. Machine Learning 8(3), 279–292 (1992)
49. Werbos, P.: Backpropagation through time: what it does and how to do it. Proceedings of the IEEE 78(10), 1550–1560 (1990)
50. Whiteson, S., Stone, P.: Evolutionary function approximation for reinforcement learning. The Journal of Machine Learning Research 7, 877–917 (2006)
51. Yong, C., Miikkulainen, R.: Coevolution of role-based cooperation in multi-agent systems. IEEE Transactions on Autonomous Mental Development (2010)

Reverse-Engineering the Human Auditory Pathway

Lloyd Watts

Audience, Inc.
440 Clyde Avenue
Mountain View, CA, 94043
lwatts@audience.com

Abstract. The goal of reverse-engineering the human brain, starting with the auditory pathway, requires three essential ingredients: Neuroscience knowledge, a sufficiently capable computing platform, and a long-term funding source. By 2003, the neuroscience community had a good understanding of the characterization of sound which is carried out in the cochlea and auditory brainstem, and 1.4 GHz single-core computers with XGA displays were fast enough that it was possible to build computer models capable of running and visualizing these processes in isolation at near biological resolution in real-time, and it was possible to raise venture capital funding to begin the project. By 2008, these advances had permitted the development of products in the area of two-microphone noise reduction for mobile phones, leading to viable business by 2010, thus establishing a self-sustaining funding source to continue the work into the next decade 2010-2020. By 2011, advances in fMRI, multi-electrode, and behavioral studies have illuminated the cortical brain regions responsible for separating sounds in mixtures, understanding speech in quiet and in noisy environments, producing speech, recognizing speakers, and understanding and responding emotionally to music. 2GHz computers with 8 virtual cores and HD displays now permit models of these advanced auditory brain processes to be simulated and displayed simultaneously in real-time, giving a rich perspective on the concurrent and interacting representations of sound and meaning which are developed and maintained in the brain, and exposing a deeper generality to brain architecture than was evident a decade earlier. While there is much still to be discovered and implemented in the next decade, we can show demonstrable progress on the scientifically ambitious and commercially important goal of reverse-engineering the human auditory pathway.

As outlined in 2003 [1], the goal of reverse-engineering the human brain, starting with the auditory pathway, requires three essential ingredients: Neuroscience knowledge, a sufficiently capable computing platform, and a long-term funding source. In this paper, we will describe the first successful decade of this multi-decade project, and show progress and new directions leading into a promising second decade.

By 2003, the neuroscience community had a good understanding of the characterization of sound which is carried out in the cochlea and auditory brainstem, including the detection of inter-aural time and level differences (ITD and ILD)

J. Liu et al. (Eds.): WCCI 2012 Plenary/Invited Lectures, LNCS 7311, pp. 47–59, 2012.
© Springer-Verlag Berlin Heidelberg 2012

computed in the superior olivary complex (SOC, MSO, LSO) used for determining the azimuthal location of sound sources, and the essential brainstem foundations for extracting polyphonic pitch (delay lines needed for autocorrelation in the nucleus of the lateral lemniscus (NLL), and combination-sensitive cells in the inferior colliculus (IC)). While there was still significant uncertainty about the full role of the inferior colliculus, medial geniculate body (MGB) of the thalamus, and auditory cortical regions, there was sufficient clarity and consensus of the lower brainstem representations to begin a serious modeling effort [1].

In 2003, on a single-core 1.4 GHz processor, it was possible to build computer models capable of running these processes in isolation at near biological resolution in real-time, e.g., a 600-tap cochlea model spanning a frequency range 20Hz - 20kHz at 60 taps/octave with realistic critical bandwidths, efficient event-driven ITD and normalized ILD computations, and a plausible model of polyphonic pitch [1]. By 2008, these advances had permitted the development of products in the area of two-microphone noise reduction for mobile phones [2][3][4], leading to viable business by 2010, thus establishing a commercial foundation to continue the work into the next decade 2010-2020.

1   Neuroscience Advances in 2003-2010 Illuminate Cortical Architecture

During 2003-2010, new fMRI, multi-electrode, and behavioral studies have illuminated the cortical brain regions responsible for separating sounds in mixtures [5][6], understanding speech in quiet and in noisy environments [7], producing speech [7], recognizing speakers [8], and understanding music [9][10]. Similarly, there is greater clarity in the function of the hippocampus [11] and amygdala [12][13] in the limbic system, relating to long-term memory storage and retrieval, and emotional responses to auditory stimuli [12]. While there is still much to be discovered about the underlying representation of signals in the cortex, it is now possible to see an architectural organization begin to emerge within the auditory pathway, as shown in Figures 1 and 2.

These figures were created by starting with the auditory pathway diagram first published in [1], and then updating the cortical regions to show the speech recognition and production pathways from [7], speaker identification pathway from [8], music pathways inferred from the functional description in [9], and limbic system pathways from [10][11][12], with additional guidance from [14].

Based on Figures 1 and 2, we may make some observations about the human auditory pathway:

- The auditory pathway contains many different representations of sounds, at many different levels. The most fundamental representation is the cochlea representation carried on the auditory nerve, from which all other representations are derived. Any realistic computational model of the human hearing system will have to generate all of the representations and allow them to interact realistically, thus extracting and utilizing all of the information in the auditory signals.

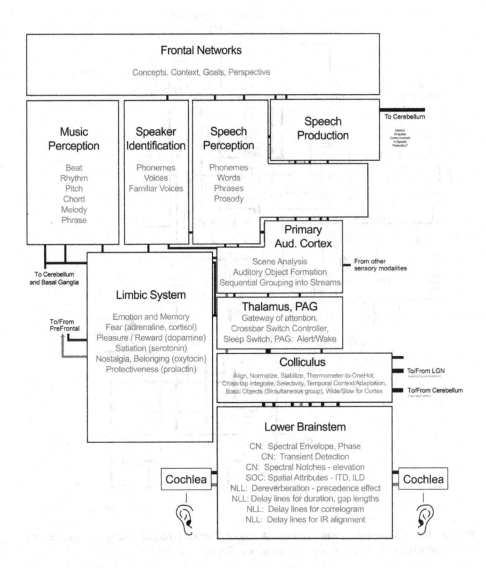

Fig. 1. Block diagram of the Human Auditory Pathway (high-level). Sounds enter the system through the two ears at the bottom of the diagram. The left and right cochleas create the spectro-temporal representation of sounds, which projects onto the cochlear nerve into the lower brainstem, beginning with the cochlear nucleus (CN), then projects to the superior olivary complex (SOC) and nucleus of the lateral lemniscus (NLL). From there, signals project to the inferior and superior Colliculus and Thalamus. The thalamus projects to the Limbic system (emotion and memory) and to primary auditory cortex, which then projects to the specialized pathways for speech recognition, production, speaker identification, and music perception.

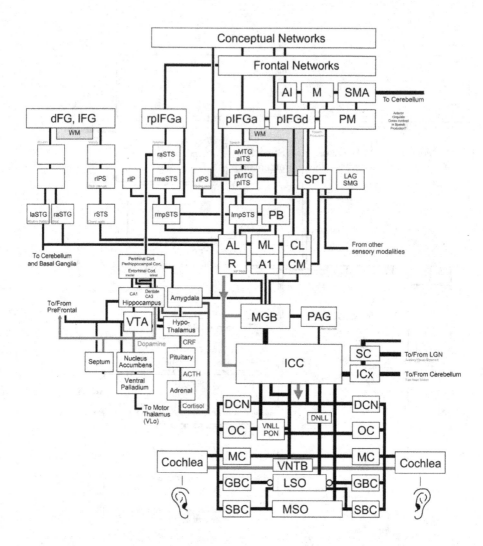

Fig. 2. Block diagram of the Human Auditory Pathway (detail). Original sources in [1][7][8][9][10][11][12]. Please see the Glossary of Terms for full list of abbreviations.

- It would appear that the functions of auditory scene analysis (breaking a mixture of sound up into its constituent sources [15]) must occur within the regions of inferior colliculus, thalamus (MGB), and primary auditory cortex, for the following reasons:

  o Below inferior colliculus, the lower brainstem is only extracting basic attributes of sound; it is too early in the system to have completed auditory scene analysis.

o   Above primary auditory cortex, we see regions specialized for deep analysis of isolated sounds (i.e. speech recognition, speaker identification, music perception). Thus, above auditory cortex, it appears that the auditory scene analysis has been largely completed.

- Thalamus (medial geniculate body (MGB)) functions largely as a wide, controllable cross-bar switch, to allow signals to be routed to cortex (selective attention) or cut off (not paying attention, or during sleep) [16]. However, some signals are capable of waking us up from sleep (i.e. baby cry), suggesting that some rudimentary signal classification is being done below the Thalamus, apparently in the inferior colliculus and peri-acqueductal gray (PAG) [17].

- The cortical speech recognition part of the human auditory pathway includes a phonological network (lmpSTS), lexical network (pMTG/pITS), and combinatorial network (aMTG/aITS) [7]. These elements are roughly analogous to the phoneme classifier, word recognizer, and language model of a conventional speech recognizer. However, as emphasized in [1], conventional modern speech recognizers do not include an auditory scene analysis engine to separate sounds in a mixture into their constituent sources prior to recognition. Instead, a conventional speech recognizer performs the front-end (Fast Fourier Transform and cepstrum) and projects them immediately to the back-end, which can only work well when the input signal is already isolated speech. The lack of an auditory scene analysis engine is the primary reason that modern speech recognizers exhibit poor noise robustness relative to human listeners, especially when the background noise consists of competing speech.

- There is a notably parallel structure between the speech recognition pathway [7] and the speaker identification pathway [8] – note that each has three major stages between primary auditory cortex and inferior frontal gyrus (IFG).

- Finally, the new block diagrams in Figures 1 and 2 indicate some important interactions between the auditory pathway and other parts of the brain. On the right side of both diagrams, there are additional connections:

  o   To/From Cerebellum (at bottom right, from ICx): This connection is to trigger reflexive head movement in response to directional sound.

  o   To/From LGN (at lower right, from SC): This is a bidirectional connection to allow a calibration and spatial alignment between the visual system and auditory system [18].

  o   From other sensory modalities (middle right, to SPT (sylvian parietal-temporal junction)): This is the pathway by which lip-reading can assist the speech recognition pathway in the correct perception of spoken phoneme-level sounds [7], especially in noise where the auditory input may be corrupted [14].

  o   To Cerebellum (upper right, from SMA): This is the motor output for speech production.

These four external interfaces indicate that the auditory pathway does not act in isolation – it interacts with the visual and motor pathways to create a whole-brain system that can hear, see, move, and talk.

2   Compute Capacity in 2012 Is Capable of Comprehensive Simulation and Visualization of the Multi-representation System

In early 2012, high-end gaming notebook computers have 2.0 GHz microprocessors with 8 virtual cores, about 11.4 times the compute capability of the 1.4 GHz single-core machines of 2003. In 2003, it took the entire machine to compute any one of the basic brainstem representations of sound, by itself. In 2012, it is possible to compute all of the representations simultaneously, including new ones which had not been developed in 2003. In 2003, the highest resolution display on a notebook computer was XGA (1024x768 pixels), which was only enough to display a single representation at once. In 2012, with a 1080p HD display (1920x1080 pixels), it is possible to compute and display all of the existing representations simultaneously, as shown in Figure 4.

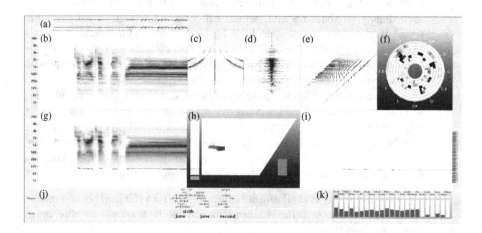

Fig. 3. Output of real-time, high-resolution functioning model of major auditory pathway elements. (a) Waveforms, at the level of the eardrums at the two ears. (b) Cochlea energy, as seen by the Multipolar Cells in the Cochlear Nucleus. (c) Inter-aural time difference (ITD), as computed by the medial superior olive (MSO). (d) Inter-aural level difference (ILD), as computed by the lateral superior olive (LSO) and normalized in the inferior colliculus (IC). (e) Correlogram, as computed in the nucleus of the lateral lemniscus (NLL) and inferior colliculus (IC). (f) Pitch Chroma Spiral (cortical pitch representation). (g) Pitch-adaptive spectral smoothing, with formant tracking (cortical speech representation). (h) Vocal Articulator mapping, in the sylvian parietal-temporal junction (SPT). (i) Polyphonic pitch. (j) Speech recognition. (i) Speaker identification.

There is still much to be done – in particular, the highest-level recognition functions (speech recognition, speaker ID) currently implemented are introductory placeholders based on fairly basic technologies. And currently, the representations are running simultaneously, but they are not yet interacting with each other. The true promise of integrating all of the representations together so that they can help each other is still to be done. But it is clear that we have sufficient neuroscience knowledge of a powerfully multi-representation system, and a sufficiently capable computing platform to be able to build the next level of the integrated system and visualize its output.

3   Next Steps in Neuroscience Research for the Next Decade 2010-2020

Neuroscientists are now beginning to explore the interactions between the scene analysis, speaker tracking, and speech recognition functions. One excellent example of this is the recent work by Dr. Eddie Chang at the University of California at San Francisco, in which subjects are asked to listen to a mixture of commands spoken by two different speakers (one male, one female), pick out a keyword spoken by one of them, and report the following command by the correct speaker [6], as shown in Figure 5.

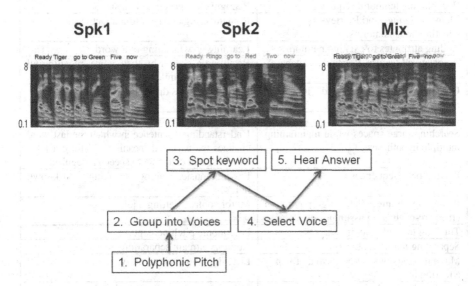

Fig. 4. Dr. Eddie Chang's task can be understood in the context of the whole auditory pathway. For the subjects to get the correct answer, they must separate the voices, presumably on the basis of polyphonic pitch, since the subjects are unable to reliably perform the task if there is not a clear pitch difference. Then they must spot the keyword, then track the voice that spoke the keyword, and then listen for the command in the chosen voice while ignoring the other voice, all while under time pressure.

Dr. Chang's task exercises the major elements of the auditory pathway – polyphonic pitch detection, grouping and separation into voices, word spotting, selective attention to the correct voice, and listening for the correct answer. And he is able to make direct multi-electrode recordings from the relevant brain regions of awake functioning human beings – his neurosurgery patients who have volunteered to participate in his study. This is a major recent advancement in auditory neuroscience, already shedding light on the detailed mechanisms of auditory attention, stream separation, and speech recognition, with much promise over the next decade 2010-2020.

While the architectural advances from the last decade's fMRI studies are very important and encouraging, a notable foundational weakness still remains: what is the general computational and learning strategy of the cortical substrate? In 2012, it is safe to say that there is no clear consensus, although there are many sophisticated models with persuasive proponents [19][20][21][22], including Hierarchical Bayesian Models [23], Hierarchical Temporal Memories [24], and Deep Belief Networks [25]. From my own work on modeling the human auditory pathway, it is apparent that the cortex must be capable of at least the following set of functions:

Table 1. Functions performed in cortex

Cortical Capability	Example
Finding patterns in sensory input	Recognizing sounds of speech
Recognizing temporal sequences	Recognizing speech and music
Memory Storage and Retrieval Creating new memories	Remembering and recalling a fact
Adding attributes to existing memories	Learning new meaning of a word
Associative Memory, Relational Database	Recalling a person by their voice, recalling all people with a similar voice
Organizing short-term and long-term memory (with hippocampus)	Memory updates during sleep
Learning	Learning a language or a song
Searching large spaces while maintaining multiple hypotheses	Understanding a sentence in which the last word changes the expected meaning. Getting a joke. Viterbi search in a modern speech recognizer.
Playing back sequences	Playing music, singing, speaking well-known phrases.
Predicting future, detecting prediction errors, re-evaluating assumptions	Motor control, getting a joke.
Tracking multiple moving targets	Polyphonic pitch perception
Separating multiple objects	Auditory Stream separation
Making decisions about what to pay attention to	Listening in a cocktail party
Local cross-correlations	Stereo Disparity in the visual system

Note that there are computer programs that can do each of the above things, *in isolation*, at some level of ability. For example, music sequencers can play back long and complicated sequences of musical notes. The Acoustic Model part of a modern speech recognizer has been trained to estimate the likelihood of phonemes, given speech input. Back-propagation and Deep Belief Networks are examples of programs that learn. Google's web crawl and hash table updates are examples of organizing associative memories for fast recall. Creating new memories and adding new attributes to existing memories are routine operations on linked lists. Stereo disparity algorithms have been around since the early 1990's [26].

In principle, I see nothing in the brain that could not be implemented on a sufficiently fast computer with enough memory, although matching the low power consumption of the brain will favor a parallel/slow architecture over the conventional fast/serial architecture. It is common to regard the cortical columns as basic units of computation [19][21][22][24], and in principle, I see no reason why these columns (or groups of columns) could not be reasonably modeled by a sufficiently capable microprocessor running a suitable program, provided the microprocessors can communicate adequately with each other. But the key question is:

In such a model, should each cortical processor be running the same program?

I believe the answer is *No*. The highly differentiated functions performed in the different cortical regions shown in Figure 2 and listed in Table 3, suggest that, while the cortical structure (hardware) may be quite uniform across the cortex, the functions performed in the mature brain in each region (software) must be quite specialized for each region. For example, the functions of stream separation performed in auditory cortex are extremely different than the functions of phoneme recognition performed in the left medial posterior Superior Temporal Sulcus (lmpSTS), which in turn are extremely different from the functions of working memory for extracting the meaning of sentences in and near the posterior Inferior Frontal Gyrus (pIFG). And all of these are fundamentally different from the functions of controlling movement in motor cortex or computing the cross-correlations for determining stereo disparity in visual cortex.

It is not clear whether the functional specialization in the mature cortex is the result of a uniform cortical structure in which different regions learn their specialized function solely because of their unique inputs (i.e., wiring determines function), or if there is some other additional way that the specialized functions in each region are determined during development – perhaps genetic [27][28][29]. For example, recent evidence from 2001-2009 points to mutations in the FOXP2 gene as causing severe speech and language disorders [30][31][32][33][34], including defects in processing words according to grammatical rules, understanding of more complex sentence structure such as sentences with embedded relative clauses, and inability to form intelligible speech [35].

I am emphasizing this point because the observation that the cellular structure of cortex appears uniform has led to a widely accepted hypothesis that there must be a single learning or computational strategy that will describe the development and operation of all of cortex. For this hypothesis to be true, the learning or computational strategy would have to be capable of developing, from a generic substrate, a wide variety of very different functional specialties, including functions well-modeled as correlators, hierarchical temporal memories, deep belief networks, associative memories, relational databases, pitch-adaptive formant trackers, object trackers, stream separators, phoneme detectors, Viterbi search engines, playback sequencers, etc.

In any case, so far, to even come close to matching the functions that are observed by my neuroscience collaborators working in mature brains, I have found it necessary to write very specialized programs to model each functional area of the mature brain.

4   Non-technical Issues: Collaboration and Funding for 2010-2020

In 2003 [1] and 2007 [2], I outlined the importance of collaboration with leading neuroscientists, and of finding a funding model that would sustain the multi-decade project of reverse-engineering the brain, beginning with the auditory pathway. The basic science work and early prototypes were done in 1998-2000 at Interval Research, and in 2000-2003 in the early days of Audience. From 2004-2010, the focus was on building a viable business to commercialize the practical applications of research into the auditory pathway. In 2010-2011, we revisited the neuroscience community and found substantial progress had been made in the cortical architecture of the auditory pathway, and Moore's Law has ensured that compute capacity has grown ten-fold as expected. It remains to be seen what new insights and products will emerge from the next phase of scientific exploration over the next few years, but we can at least say that after the first decade, the neuroscience, compute capacity and funding aspects of the project have all advanced in sync with each other, as hoped in [1], and the multi-decade project is still on track.

5   Conclusions

The goal of reverse-engineering the human brain, starting with the auditory pathway, requires three essential ingredients: Neuroscience knowledge, a sufficiently capable computing platform, and a long-term funding source to sustain a multi-decade project. All of these were available on a small scale at the time of founding Audience in 2000, enough to begin the project in earnest. By 2010, neuroscience knowledge had advanced dramatically, giving major insights into cortical architecture and function, compute capacity had grown ten-fold, and a commercial foundation had been established to allow the project to continue into the next decade 2010-2020. While there is still much work to do, and many risks remain, the multi-decade project still appears to be on track.

References

1. Watts, L.: Visualizing Complexity in the Brain. In: Fogel, D., Robinson, C. (eds.) Computational Intelligence: The Experts Speak, pp. 45–56. IEEE Press/Wiley (2003)
2. Watts, L.: Commercializing Audi-tory Neuroscience. In: Frontiers of Engineering: Reports on Leading-Edge Engineering from the 2006 Symposium, pp. 5–14. National Academy of Engineering (2007)
3. Watts, L.: Advanced Noise Reduction for Mobile Telephony. IEEE Computer 41(8), 90–92 (2008)
4. Watts, L., Massie, D., Sansano, A., Huey, J.: Voice Processors Based on the Human Hearing System. IEEE Micro, 54–61 (March/April 2009)
5. Mesgarani, N., David, S.V., Fritz, J.B., Shamma, S.A.: Influence of context and behavior on stimulus reconstruction from neural activity in primary auditory cortex. J. Neurophysiol. 102(6), 3329–3339 (2009)

6. Mesgarani, N., Chang, E.: Robust cortical representation of attended speaker in multitalker speech perception. Submitted to Nature (2011)

7. Hickok, G., Poeppel, D.: The cortical organization of speech processing. Nature Reviews Neuroscience 8(5), 393–402 (2007)

8. von Kriegstein, K., Giraud, A.L.: Distinct functional substrates along the right superior temporal sulcus for the processing of voices. NeuroImage 22, 948–955 (2004)

9. Peretz, I., Zatorre, R.: Brain organization for music processing. Annual Review of Psychology 56, 89–114 (2005)

10. Levitin, D.: This is Your Brain on Music. Dutton Adult (2006)

11. Andersen, P., Morris, R., Amaral, D., Bliss, T., O'Keefe, J.: The Hippocampus Book. Oxford University Press (2007)

12. LeDoux, J.: The Emotional Brain. Simon & Schuster (1998)

13. Whalen, P., Phelps, E.: The Human Amygdala. The Guilford Press (2009)

14. Hervais-Adelman, A.: Personal communication (2011)

15. Bregman, A.: Auditory Scene Analysis. MIT Press (1994)

16. http://en.wikipedia.org/wiki/Thalamus

17. Parsons, C., Young, K., Joensson, M., Brattico, E., Hyam, J., Stein, A., Green, A., Aziz, T., Kringelbach, M.: Ready for action: A role for the brainstem in responding to infant vocalisations. Society For Neurosciences, Poster WW23 299.03 (2011)

18. Hyde, P., Knudsen, E.: Topographic projection from the optic tectum to the auditory space map in the inferior colliculus of the barn owl. J. Comp. Neurol. 421(2), 146–160 (2000)

19. Calvin, W.: The Cerebral Code. MIT Press (1998)

20. Douglas, R., Martin, K.: In: Shepherd, G. (ed.) The Synaptic Organization of the Brain, 4th edn., pp. 459–510. Oxford University Press (1998)

21. Mountcastle, V.B.: Introduction to the special issue on computation in cortical columns. Cerebral Cortex 13(1), 2–4 (2003)

22. Dean, T.: A computational model of the cerebral cortex. In: The Proceedings of Twentieth National Conference on Artificial Intelligence (AAAI 2005), pp. 938–943. MIT Press, Cambridge (2005)

23. George, D., Hawkins, J.: A Hierarchical Bayesian Model of Invariant Pattern Recognition in the Visual Cortex. In: Proceedings of the International Joint Conference on Neural Networks (2005)

24. Hawkins, J., Blakeslee, S.: On Intelligence. St. Martin's Griffin (2005)

25. Hinton, G.E., Osindero, S., Teh, Y.: A fast learning algorithm for deep belief nets. Neural Computation 18, 1527–1554 (2006)

26. Okutomi, M., Kanade, T.: A Multiple-Baseline Stereo. IEEE Transactions on Pattern Analysis and Machine Intelligence 15(4), 353–363 (1993)

27. Dawson, G., Webb, S., Wijsman, E., Schellenberg, G., Estes, A., Munson, J., Faja, S.: Neurocognitive and electrophysiological evidence of altered face processing in parents of children with autism: implications for a model of abnormal development of social brain circuitry in autism. Dev. Psychopathol. 17(3), 679–697 (2005), http://www.ncbi.nlm.nih.gov/pubmed?term=%22Dawson%20G%22%5BAuthor%5D

28. Dubois, J., Benders, M., Cachia, A., Lazeyras, F., Leuchter, R., Sizonenko, S., Borradori-Tolsa, C., Mangin, J., Hu, P.S.: Mapping the Early Cortical Folding Process in the Preterm Newborn Brain. Cerebral Cortex 18, 1444–1454 (2008)

29. Kanwisher, N.: Functional specificity in the human brain: A window into the functional architecture of the mind. Proc. Natl. Acad. Sci. USA (2010)

30. Lai, C., Fisher, S., Hurst, J., Vargha-Khadem, F., Monaco, A.: A forkhead-domain gene is mutated in a severe speech and language disorder. Nature 413(6855), 519–523 (2001)

31. MacDermot, K., Bonora, E., Sykes, N., Coupe, A., Lai, C., Vernes, S., Vargha-Khadem, F., McKenzie, F., Smith, R., Monaco, A., Fisher, S.: Identification of FOXP2 truncation as a novel cause of developmental speech and language deficits. Am. J. Hum. Genet. 76(6), 1074–1080 (2005)
32. http://www.nytimes.com/2009/11/12/science/12gene.html
33. Konopka, G., Bomar, J., Winden, K., Coppola, G., Jonsson, Z., Gao, F., Peng, S., Preuss, T., Wohlschlegel, J., Geschwind, D.: Human-specific transcriptional regulation of CNS devel-opment genes by FOXP2. Nature 462, 213–217 (2009)
34. http://www.evolutionpages.com/FOXP2_language.htm
35. Vargha-Khadem, et al.: Praxic and nonverbal cognitive deficits in a large family with a genetically transmitted speech and language disorder. Proc. Nat. Acad. Sci. USA 92, 930–933 (1995)

Glossary of Terms

Abbreviation	Full Name	Function
SBC	Spherical Bushy Cell	Sharpen timing, phase locking for ITD comparison
GBC	Globular Bushy Cell	Condition for ILD amplitude comparison
MC	Multipolar Cell	Detect amplitude independent of phase
OC	Octopus Cell	Broadband transient detection
DCN	Dorsal Cochlear Nucleus	Elevation processing
MSO	Medial Superior Olive	ITD comparison
LSO	Lateral Superior Olive	ILD comparison
VNTB	Ventral Nucleus of the Trapezoid Body	Control efferent signals to cochlea OHCs (top-down gain control loop)
MNTB	Medial Nucleus of the Trapezoid Body	Inverter between GBC and LSO to allow amplitude subtraction operation
VNLL	Ventral Nucleus of the Lateral Lemniscus	Prepare for broad system-wide reset in ICC (triggered temporal integration?)
PON	Peri-Olivary Nuclei	
DNLL	Dorsal Nucleus of the Lateral Lemniscus	Precedence effect processing of spatial information, compensate for reverberation
ICC	Inferior Colliculus (Central)	Scaling, normalizing (L-R)/(L+R), align data structure, selectivity
ICx	Inferior Colliculus (Exterior)	Audio visual alignment
SC	Superior Colliculus	Audio visual alignment
MGB	Medial Geniculate Body (Thalamus)	Attentional relay, sleep switch
PAG	Peri-acqueductal Gray	Wake from sleep from sounds like baby cry
LS	Limbic System (includes Amygdala, Hippocampus, hypothalamus, Pituitary gland, adrenal gland)	Fast-acting fear pathway, memory controller, hash table generator
A1	Primary Auditory Cortex	Primary area of Auditory cortex
R	Rostral part of Auditory Cortex	
CM	Caudal Medial part of AC	
AL	Anterior Lateral part of AC	Extraction of spectral shape – pitch-adaptive spectral smoothing, or preparations for it
ML	Medial Lateral part of AC	
CL	Caudal Lateral part of AC	
STS	Superior Temporal Sulcus	Phonological network (phonemes, speech components). Possible site of pitch-adaptive spectral smoothing and formant detection

PB	ParaBelt region	Pitch, noise
pMTG	Posterior Medial Temporal Gyrus	Lexical network (words, vocabulary, HMM)
pITS	Posterior Inferior Temporal Sulcus	Lexical network (words, vocabulary, HMM)
aMTG, aITS	Anterior Medial Temporal Gyrus, Anterior Inferior Temporal Sulcus	Combinatoric network (sentences, grammar, HMM)
SPT	Sylvian Parietal-Temporal junction	Sensori-motor interface
LAG, SMG	Left Angular Gyrus Super Modular Gyrus	Activated in degraded/challenging speech conditions
rmpSTS	Right medial posterior Superior Temporal Sulcus	Voice recognition
rmaSTS	Right medial anterior Superior Temporal Sulcus	Non-familiar voices
raSTS	Right anterior Superior Temporal Sulcus	Familiar voices
IP	Inferior Parietal	
pIFGa	Posterior Inferior Frontal Gyrus (anterior part)	Syntax and Semantics in speech comprehension, working memory for speech
pIFGd	Posterior Inferior Frontal Gyrus (dorsal part)	Phonemes in speech production
PM	Pre-Motor Cortex	
AI	Anterior Insula	Modulation of speech production (disgust)
M	Motor Cortex	
SMA	Supplemental Motor Area	Interface between Motor Cortex and Cerebellum, subvocalization, rhythm perception and production
rSTS	Right Superior Temporal Sulcus	Chord and scale in music
rIPS	Right Inferior Parietal Sulcus	Pitch intervals in music
lIPS	Left Inferior Parietal Sulcus	Gliding pitch in speech
raSTG	Right anterior Superior Temporal Gyrus	Beat in music
laSTG	Left anterior Superior Temporal Gyrus	Rhythm pattern in music
dFG, IFG	Dorsal Frontal Gyrus, Inferior Frontal Gyrus	Working memory for pitch, tones harmonic expectations/violations
	Cerebellum, basal ganglia	Auditory intervals (lateral cerebellum, basal ganglia), Motor timing (medial cerebellum, basal ganglia)

Unpacking and Understanding Evolutionary Algorithms

Xin Yao

CERCIA, School of Computer Science
University of Birmingham
Edgbaston, Birmingham B15 2TT, UK
x.yao@cs.bham.ac.uk
http://www.cs.bham.ac.uk/~xin

Abstract. Theoretical analysis of evolutionary algorithms (EAs) has made significant progresses in the last few years. There is an increased understanding of the computational time complexity of EAs on certain combinatorial optimisation problems. Complementary to the traditional time complexity analysis that focuses exclusively on the problem, e.g., the notion of NP-hardness, computational time complexity analysis of EAs emphasizes the relationship between algorithmic features and problem characteristics. The notion of EA-hardness tries to capture the essence of when and why a problem instance class is hard for what kind of EAs. Such an emphasis is motivated by the practical needs of insight and guidance for choosing different EAs for different problems. This chapter first introduces some basic concepts in analysing EAs. Then the impact of different components of an EA will be studied in depth, including selection, mutation, crossover, parameter setting, and interactions among them. Such theoretical analyses have revealed some interesting results, which might be counter-intuitive at the first sight. Finally, some future research directions of evolutionary computation will be discussed.

1 Introduction

Evolutionary computation refers to the study of computational systems that are inspired by natural evolution. It includes four major research areas, i.e., evolutionary optimisation, evolutionary learning, evolutionary design and theoretical foundations of evolutionary computation.

1.1 Evolutionary Optimisation

Evolutionary optimisation includes a wide range of topics related optimisation, such as global (numerical) optimisation, combinatorial optimisation, constraint handling, multi-objective optimisation, dynamic optimisation, etc. Many evolutionary algorithms (EAs) have been used with success in a variety of application domains that rely on optimisation. For example, in the area of global optimisation, fast evolutionary programming (FEP) and improved FEP [1] were used

J. Liu et al. (Eds.): WCCI 2012 Plenary/Invited Lectures, LNCS 7311, pp. 60–76, 2012.

successfully in modelling and designing new aluminium alloys [2]. Self-adaptive differential evolution with neighbourhood search (SaNSDE) [3] was used to calibrate building thermal models [4].

In the area of combinatorial optimisation, memetic algorithms were designed for tackling capacitated arc routing problems [5], which were inspired by the real world problem of route optimisation for gritting trucks in winter road maintenance [6]. Multi-objective EAs (MOEAs) were used very effectively for module clustering in software engineering [7] and optimal testing resource allocation in modular software systems [8]. Advantages of using the MOEAs were clearly demonstrated in these two cases.

1.2 Evolutionary Learning

Evolutionary learning appears in many different forms, from the more classical learning classifier systems to various hybrid learning systems, such as neural-based learning classifier systems [9], evolutionary artificial neural networks [10], evolutionary fuzzy systems [11], co-evolutionary learning systems [12], etc. While most of the learning problems considered in evolutionary learning are also investigated in the broader domain of machine learning, co-evolutionary learning has stood out as a learning paradigm that is rather unique to evolutionary computation.

1.3 Evolutionary Design

Evolutionary design is closely related to optimisation, especially in engineering domains, such as digital filter design [13] and shape design [14]. However, there is one different consideration in evolutionary design, which is evolutionary discovery. There has been a strong interest in using EAs as a discovery engine, in discovering novel designs, rather than just treating EAs as optimisers. A lot of work has appeared in using interactive evolutionary computation for creative design, e.g., traditional Batik design [15] and others.

1.4 Theoretical Foundations of Evolutionary Computation

In spite of numerous successes in real-world applications of EAs, theories of evolutionary computation have not progressed as fast as its applications. However, there have been significant advances in the theoretical foundation of evolutionary computation in the last decade or so. There have been a number of theoretical analyses of different fitness landscapes in terms of problem characterisation, as well as rigorous analysis of computation time used by an EA to solve a problem.

In global (numerical) optimisation, the analysis of EA's convergence has routinely been done. It was an active research topic in 1990s. Later on, convergence rates (to a local optimum) were also studied in depth. In recent years, there has been significant growth in the computational time complexity analysis of EAs for combinatorial optimisation problems, which really bridges the gap between the analysis of EAs in the evolutionary computation field and the analysis of heuristics in theoretical computer science in general. After all, EAs are algorithms and can/should be analysed just like we analyse any other algorithms.

This book chapter will focus on the computational time complexity analysis of EAs for combinatorial optimisation problems. It will not cover other theories of evolutionary computation, which are equally important for the field. Even for the computational time complexity analysis of EAs, we will not be able to cover everything within a limited numner of pages. The choice of the topics covered in this chapter is highly biased by the author's own experience. One of the objectives of this chapter is to illustrate that there are some interesting theoretical results in evolutionary computation, which may help us in understanding why and when EAs work/fail. Such theoretical results are of interest in their own right. They also help to guide the design of better algorithms in the future.

The rest of this chapter is organised as follows. Section 2 introduces drift analysis as an easy-to-understand approach to analyse computational time complexity of EAs. General conditions under which an EA solves (or fails to solve) a problem within a polynomial time will be given. Section 3 presents a problem classification, which tries to identify hard and easy problem instances for a given EA. Such problem classification helps us to get a glimpse at potential characteristics that make a problem hard/easy for a given EA. Section 4 analysed the role of population in EAs. It is interesting to discover that a common belief — the large the population, the more powerful an EA would be — is not necessarily true. There are proven cases where a large population could be harmful. Section 5 investigates the impact of crossover on EA's computation time on the unique input output problem, a problem that occurs in finite state machine testing. This was the first time that crossover was analysed in depth on a non-artificial problem. All previous analyses on crossover were done on artificial problems. Section 6 examines the interactions of different components of an EA. Rather than analysing search operators (such as crossover and mutation) or selection mechanisms individually, this section is focused on the interactions between mutation and selection. It is shown that even parameter settings can have a significant on EA's computation, even when exactly the same EA was used. Section 7 discusses some recent results on analysing estimation of distribution algorithms (EDAs), which have rarely been studied in terms of computational time complexity analysis. A more general problem classification is also given. Finally, Section 8 concludes this chapter with some remarks and future research directions.

2   Evolutionary Algorithms and Drift Analysis

This section reviews some basic techniques used in analysing EAs as given previously [16]. The combinatorial optimization problem considered in this chapter can be described as follows: Given a finite state space S and a function $f(\mathbf{x}), \mathbf{x} \in S$, find

$$\mathbf{x}^* = \arg\max\{f(\mathbf{x}); \mathbf{x} \in S\},$$

where $\mathbf{x}^*$ is a state with the maximum function value, i.e., $f_{\max} = f(\mathbf{x}^*)$.

The EA for solving the combinatorial optimization problem can be described as follows:

1. *Initialization*: generate, either randomly or heuristically, an initial population of $2N$ individuals, $\xi_0 = (\mathbf{x}_1, \cdots, \mathbf{x}_{2N})$, and let $k \leftarrow 0$, where $N > 0$ is an integer. For any population ξ_k, define $f(\xi_k) = \max\{f(\mathbf{x}_i); \mathbf{x}_i \in \xi_k\}$.
2. *Generation*: generate a new intermediate population by crossover and mutation (or any other operators for generating offspring), and denote it as $\xi_{k+1/2}$.
3. *Selection*: select and reproduce $2N$ individuals from the combined population of $\xi_{k+1/2}$ and ξ_k, and obtain another new intermediate population ξ_{k+s}.
4. If $f(\xi_{k+s}) = f_{\max}$, then terminate the algorithm; otherwise let $\xi_{k+1} = \xi_{k+s}$ and $k \leftarrow k + 1$, and go to step 2.

The EA framework given above is very general because it allows for any initialization methods, any search operators and any selection mechanisms, to be used. The only difference from some EAs is that selection is applied after, not before, the search operators. However, the main results introduced in this chapter, i.e., Theorems 1 and 2, are independent of any such implementation details. In fact, the results in this section hold for virtually any stochastic search algorithms. They serve as the basis for many more specific results using specific EAs on specific problems.

2.1   Modelling EAs Using Stochastic Processes

Assume $\mathbf{x}^*$ is an optimal solution, and let $d(\mathbf{x}, \mathbf{x}^*)$ be the distance between a solution $\mathbf{x}$ and $\mathbf{x}^*$, where $\mathbf{x} \in S$. If there are more than one optimal solution (that is, a set S^*), we use $d(\mathbf{x}, S^*) = \min\{d(\mathbf{x}, \mathbf{x}^*) : \mathbf{x}^* \in S^*\}$ as the distance between individual $\mathbf{x}$ and the optimal set S^*. For convenience, we can denote the distance as $d(\mathbf{x})$, which satisfies $d(\mathbf{x}^*) = 0$ and $d(\mathbf{x}) > 0$ for any $\mathbf{x} \notin S^*$.

Given a population $X = \{\mathbf{x}_1, \cdots, \mathbf{x}_{2N}\}$, let

$$d(X) = \min\{d(\mathbf{x}) : \mathbf{x} \in X\},$$

which is used to measure the distance of this population to the optimum. The drift of the random sequence $\{d(\xi_k), k = 0, 1, \cdots\}$ at time k is defined by

$$\Delta(d(\xi_k)) = d(\xi_{k+1}) - d(\xi_k).$$

Define the stopping time of an EA as $\tau = \min\{k; d(\xi_k) = 0\}$, which is the **first hitting time** on an optimal solution. Our interest now is to investigate the relationship between the expected first hitting time and the problem size n, i.e., the computational time complexity of EAs in our context. In this chapter, we will establish the conditions under which an EA is guaranteed to find an optimal solution in polynomial time on average and conditions under which an EA takes at least exponential time on average to find an optimal solution. Such theoretical results help us to gain a better understanding of when and why an EA is expected to work well/poorly.

2.2   Conditions for Polynomial Average Computation Times

Theorem 1 ([16]). *If* $\{d(\xi_k); k = 0, 1, 2, \cdots\}$ *satisfies the following two conditions,*

1. *there exists a polynomial of problem size* n, $h_0(n) > 0$, *such that*

$$d(\xi_k) \leq h_0(n)$$

   for any population ξ_k, *and*
2. *for any* $k \geq 0$, *if* $d(\xi_k) > 0$, *then there exists a polynomial of problem size* n, $h_1(n) > 0$, *such that*

$$E[d(\xi_k) - d(\xi_{k+1}) \mid d(\xi_k) > 0] \geq \frac{1}{h_1(n)},$$

then starting from any initial population ξ_0 *with* $d(\xi_0) > 0$,

$$E[\tau \mid d(\xi_0) > 0] \leq h(n),$$

where $h(n)$ *is a polynomial of problem size* n.

The first condition in the theorem implies that all populations occurred during the evolutionary search process are reasonably close to the optimum, i.e., their distances to the optimum is upper bounded by a polynomial in problem size. The second condition implies that, on average, the EA always drifts towards the optimum with at least some reasonable distance, i.e., the drifts are lower bounded by $\frac{1}{h_1(n)}$, where $h_1(n) > 0$ is a polynomial. The theorem basically says that the stochastic process defined by the EA can reach the optimum efficiently (in polynomial time) if the search is never too far away from the optimum and the drift towards the optimum is not too small.

Using the same intuition and analytical methods, as first proposed by Hajek [17], we can establish conditions under which an EA will take at least exponential time to reach an optimum.

2.3   Conditions for Exponential Average Computation Time

Theorem 2 ([16]). *Assume the following two conditions hold:*

1. *For any population* ξ_k *with* $d_b < d(\xi_k) < d_a$, *where* $d_b \geq 0$ *and* $d_a > 0$,

$$E[e^{-(d(\xi_{k+1}) - d(\xi_k))} \mid d_b < d(\xi_k) < d_a] \leq \rho < 1,$$

   where $\rho > 0$ *is a constant.*
2. *For any population* ξ_k *with* $d(\xi_k) \geq d_a$, $d_a > 0$,

$$E[e^{-(d(\xi_{k+1}) - d_a)} \mid d(\xi_k) \geq d_a] \leq D,$$

   where $D \geq 1$ *is a constant.*

If $d(\xi_0) \geq d_a$, $D \geq 1$ and $\rho < 1$, then there exist some $\delta_1 > 0$ and $\delta_2 > 0$ such that

$$E[\tau \mid d(\xi_0) \geq d_a] \geq \delta_1 e^{\delta_2(d_a - d_b)}$$

The first condition in the above theorem indicates that (d_b, d_a) is a very difficult interval to search. When this condition is satisfied, $d(\xi_{k+1}) > d(\xi_k)$. In other words, the offspring population is on average drifting away from the optimum, rather than getting closer to it. The second condition indicates that a population in the interval $[d_a, +\infty)$ will not, on average, drift towards the optimum too much because it is always quite far away from the optimum, i.e., $(d(\xi_{k+1})) \geq d_a - \ln D$.

Although the above two general theorems were first proved more than a decade ago [16], they still serve as foundations of many later results for EAs on specific problems, e.g., the subset sum problem [16], maximum matching [18, 19], vertex cover [20], unique input-output sequence [21], etc. The analytical techniques, i.e., drift analysis, advocated here is very intuitive and offer a general approach to analysing different EAs on different problems, which avoids the need to develop different and complicated analytical techniques for different EAs and problems.

3   Problem Classification: EA-hard vs EA-easy

Traditional complexity theories, such as NP-hardness, characterise the inherent complexity of a problem, regardless of any algorithms that might be used to solve them. However, we might not always encounter the worst case in practical cases. For a hard problem, we are interested in understanding what instance classes are hard and what instances are actually easy. When we analyse an algorithm, we want to know which problem instance classes are more amenable to this algorithm and which are not. Different instance classes of a problem pose different challenges to different algorithms. In evolutionary computation, we are particularly interested in problem characteristics that make the problem hard or easy for a given algorithm. A problem instance class may be very hard for one algorithm, but easy for another. Analysing the relationship between problem characteristics and algorithmic features will shed light on the essential question of when to use which algorithm in solving a difficult problem instance class. In order to emphasise such an algorithm-specific complexity concept, we introduce EA-hard and EA-easy problem instance classes in this section. For simplicity, we will just use the term problems to mean problem instance classes here.

Given an EA, we can divide all optimisation problems into two classes based on the mean number of generations (i.e., the mean first hitting time) needed to solve the problems [22].

Easy Class: For the given EA, starting from *any* initial population, the mean number of generations needed by the EA to solve the problem, i.e., $E[\tau|\xi_0]$, is *at most* polynomial in the problem size.

Hard Class: For the given EA, starting from *some* initial population, the mean number of generations needed by the EA to solve the problem, i.e., $E[\tau|\xi_0]$, is *at least* exponential in the problem size.

Theorem 3 ([22]). *Given an EA, a problem belongs to the* **EA-easy Class** *if and only if there exists a distance function $d(\xi_k)$, where ξ_k is the population at generation k, such that for any population ξ_k with $d(\xi_k) > 0$,*

1. *$d(\xi_k) \leq g_1(n)$, where $g_1(n)$ is polynomial in the problem size n, and*
2. *$E[d(\xi_k) - d(\xi_{k+1})|\xi_k] \geq c_{low}$, where $c_{low} > 0$ is a constant.*

Although the above theorem is closely related to Theorem 1, it shows stronger necessary and sufficient conditions. Similarly, the following theorem, related to Theorem 2, establishes the necessary and sufficient conditions for a problem to be hard for a given EA.

Theorem 4 ([22]). *Given an EA, a problem belongs to the* **EA-hard Class** *if and only if there exists a distance function $d(\xi_k)$, where ξ_k is the population at generation k, such that*

1. *for some population ξ_{k_1}, $d(\xi_{k_1}) \geq g_2(n)$, where $g_2(n)$ is exponential in the problem size n, and*
2. *for any population ξ_k with $d(\xi_k) > 0$, $E[d(\xi_k) - d(\xi_{k+1})|\xi_k] \leq c_{up}$, where $c_{up} > 0$ is a constant.*

The above two theorems can be used to verify whether a problem is hard/easy for a given EA. The key steps are to prove whether the two conditions hold. These conditions give us some important insight into problem characteristics that make a problem hard/easy for a given EA.

4   Is a Large Population Always Helpful?

Population has always been regarded as a crucial element of EAs. There have been numerous empirical studies that showed the benefit of a large population size. Whenever a problem becomes more challenging, one tends to increase the population size in an attempt to make the EA more 'powerful'. However, such an intuition might not be correct in all cases. He and Yao [23] first compared (1+1) EAs and (N+N) EAs theoretically. They showed cases where (N+N) EAs are indeed more efficient than (1+1) EAs, i.e., populations do help. They also showed somewhat surprising cases where (N+N) EAs might actually perform worse than (1+1) EAs, i.e., having a population actually makes an EA less efficient.

More recently, Chen *et al.* [24] investigated the population issue further and used the solvable rate as an improved performance measure of EAs. The solvable rate is a more precise performance measure than the mean first hitting time, because it considers a probability distribution, rather than just a mean.

Let $\tau = \min\{t|\mathbf{x}^* \in \xi_t\}$ be the first hitting time, where $\mathbf{x}^*$ is the global optimum and ξ_t is the population at the tth generation. The solvable rate κ is defined by

$$\kappa = P\left(\tau \prec Poly(n)\right),$$

where the event $\tau \prec Poly(n)$ means that there exists some polynomial function (of the problem size n) $h(n)$ such that $\tau < h(n)$ for any $n > n_0 > 0$.

Consider the following multi-modal TRAPZEROS problem with its global optimum at $\mathbf{x}^* = (1, ..., 1)$.

$$\text{TRAPZEROS}(\mathbf{x}) \triangleq \begin{cases} 2n + \sum_{i=1}^{n} \prod_{j=1}^{i}(1 - x_j), & \text{if } (x_1 = 0) \wedge (x_2 = 0); \\ 3n + \sum_{i=1}^{n} \prod_{j=1}^{i} x_j, & \text{if } (x_1 = 1) \wedge (x_2 = 1) \wedge \left(\prod_{i=3}^{\ln^2 n + 2} x_i = 1\right); \\ n + \sum_{i=1}^{n} \prod_{j=1}^{i} x_j, & \text{if } (x_1 = 1) \wedge (x_2 = 1) \wedge \left(\prod_{i=3}^{\ln^2 n + 2} x_i = 0\right); \\ 0, & \text{if } (x_1 = 0) \wedge (x_2 = 1); \\ 1, & \text{if } (x_1 = 1) \wedge (x_2 = 0); \\ 0, & \text{Otherwise.} \end{cases}$$

Consider the following $(N + N)$ EA used to solve the above problem.

Initialization: The N initial individuals are generated uniformly at random. Generation counter $k := 0$.

Mutation: For each individual in population ξ_k, one offspring is generated by flipping each bit independently with a uniform probability $1/n$, where n is the problem size (chromosome length). The offspring population is $\xi_k^{(m)}$.

Selection: Select the best N individuals from $\xi_k \cup \xi_k^{(m)}$ to form the next generation ξ_{k+1}. $k := k + 1$ and go to the mutation step.

This algorithm is very generic except for the lack of crossover, which we will discuss in the next section. The following results compare the EA's performance theoretically when $N = 1$ and $N > 1$.

Theorem 5 ([24]). *The first hitting time of the $(1 + 1)$ EA on TRAPZEROS is $O(n^2)$ with the probability of $\frac{1}{4} - O\left(\frac{\ln^2 n}{n}\right)$. In other words, the solvable rate of the $(1 + 1)$ EA on TRAPZEROS is at least $\frac{1}{4} - O\left(\frac{\ln^2 n}{n}\right)$.*

This theorem shows that $(1 + 1)$ EA can solve the problem in polynomial time with an almost constant probability.

Theorem 6 ([24]). *The first hitting time of the $(N + N)$ EA, where $N = O(\ln n)$ and $N = \omega(1)$, on TRAPZEROS is $O\left(\frac{n^2}{N}\right)$ with a probability of $1/Poly(n)$, where $1/Poly(n)$ refers to some positive function (of the problem size n), whose reciprocal is bounded from above by a polynomial function of the problem size n. In other words, the solvable rate of the $(N + N)$ EA on TRAPZEROS is at least $1/Poly(n)$.*

When the population size increases from 1 to greater than 1, but not too much greater (i.e., $N = O(\ln n)$), there is no significant gain in terms of EA's computation time, although the upper bound is decreased marginally from $O(n^2)$ to $O\left(\frac{n^2}{N}\right)$. Note that we do not have a near constant solvable rate anymore when the population size is greater than 1.

Theorem 7 ([24]). *The first hitting time of the* $(N + N)$ *EA, where* $N = \Omega(n/\ln n)$, *on* TRAPZEROS *is super-polynomial with an overwhelming probability. In other words, the solvable rate of the* $(N + N)$ *EA on* TRAPZEROS *is super-polynomially close to 0.*

Surprisingly, when the population size is very large, i.e., $N = \Omega(n/\ln n)$, the $(N + N)$ EA is no longer able to solve the TRAPZEROS in polynomial time. A large population size is actually harmful in this case!

Although the above study [24] was carried out on a specific problem using a specific type of EAs, it has actually revealed some interesting problem characteristics under which the $(N + N)$ EA may perform poorly: when the basin of attraction for a local optimum has relatively high fitness in comparison with most areas in the entire search space, a large population may be harmful, since it may lead to a large probability of finding individuals at the local basin. The search process towards and staying at the local basin can quickly eliminate other promising individuals that could lead to the global optimum later. When such congregation at the local basin happens, only large search step sizes can help to find promising individuals again, resulting in a long computation time towards the global optimum.

The weakness of the $(N + N)$ EA without crossover on the above problem characteristic can partially be tackled by employing larger search step sizes. Either an appropriately designed crossover operator or some adaptive/self-adaptive mutation schemes could work well with a large population in this case, as long as they can provide large search step sizes in exploring the correct attraction basin even if the whole population has been trapped in a local basin.

5   Impact of Crossover

The previous section used an artificial problem to gain some insight into the role of population in EAs. Crossover was not considered. This section introduces a real-world problem and analyses when crossover can be beneficial in improving EA's computation time.

Unique input-output sequences (UIO) have important applications in conformance testing of finite state machines (FSMs) [25]. In spite of much experimental work, few theoretical results exist [21, 26]. One significant result that does exist is the rigorous analysis of crossover's impact on EA's performance on one type of UIO problems [27].

Following [27], a finite state machine (FSM) is defined as a quintuple, $M = (I, O, S, \delta, \lambda)$, where $I(O)$ is the set of input (output) symbols, S is the set of states, $\delta : S \times I \to S$ is the state transition function, and $\lambda : S \times I \to O$ is the output function. A unique input-output sequence (UIO) for a state s in M is a string x over I such that $\lambda(s, x) \neq \lambda(t, x), \forall t, t \neq s$. In other words, x identifies state s uniquely. Although the shortest UIO in the general case can be exponentially long with respect to the number of states [25], our objective here is to search for an UIO of length n for state s in an FSM, where the fitness of an input sequence is defined as a function of the state partition tree induced by the input sequence [26]. In other words, given an FSM M with m states, the associated fitness function $UIO_{M,s} : I^n \to \mathcal{N}$ is defined as

$$UIO_{M,s}(x) := m - \gamma_M(s, x),$$

where

$$\gamma_M(s, x) := |\{t \in S | \lambda(s, x) = \lambda(t, x)\}|.$$

For theoretical analysis, a special FSM instance class, i.e., the TWOPATHS problem, is introduced here [27].

For instance size n and constant ϵ, $0 < \epsilon < 1$, a TWOPATHS FSM has input and output symbols $I := \{0,1\}$ and $O := \{a,b,c\}$, respectively, and $2(n+1)$ states $S = R \cup Q$, where $R := \{s_1, s_2, \ldots, s_{n+1}\}$ and $Q := \{q_1, q_2, \ldots, q_{n+1}\}$. The output function λ is

$$\lambda(q_i, x) := \begin{cases} c, & \text{if } i = n+1 \text{ and } x = 0 \\ a, & \text{otherwise} \end{cases}$$

$$\lambda(s_i, x) := \begin{cases} b, & \text{if } i = n+1 \text{ and } x = 1 \\ a, & \text{otherwise} \end{cases}$$

The state transition function δ is

$$\delta(s_i, 0) := \begin{cases} q_{(1-\epsilon)n+3}, & \text{if } i = (1-\epsilon)n+1, \\ s_1, & \text{otherwise} \end{cases}$$

$$\delta(s_i, 1) := \begin{cases} q_1, & \text{if } i = n+1 \\ s_{i+1}, & \text{otherwise} \end{cases}$$

$$\delta(q_i, 1) := q_1$$

$$\delta(q_i, 0) := \begin{cases} s_1, & \text{if } i = n+1 \\ q_{i+1}, & \text{otherwise} \end{cases}$$

We can use the following $(N+1)$ steady state EA (SSEA) [27] to solve the above problem.

Initialisation: Initialise N individuals uniformly at random from $\{0,1\}^n$ to form the initial population P_0. $i = 0$.

Reproduction: Perform one of the following two choices

1-point Crossover: With probability $p_c(n)$, select $\mathbf{x}$ and $\mathbf{y}$ uniformly at random from population P_i. Select k from $\{1, \ldots, n\}$ uniformly at random. Perform 1-point crossover between $\mathbf{x}$ and $\mathbf{y}$ and obtain

$$\mathbf{x}' := x_1 x_2 \cdots x_{k-1} y_k y_{k+1} \cdots y_n,$$

$$\mathbf{y}' := y_1 y_2 \cdots y_{k-1} x_k x_{k+1} \cdots x_n.$$

If $\max\{f(\mathbf{x}'), f(\mathbf{y}')\} \geq \max\{f(\mathbf{x}), f(\mathbf{y})\}$, then $\mathbf{x} := \mathbf{x}', \mathbf{y} := \mathbf{y}'$.

Mutation Only: With probability $1 - p_c(n)$, select $\mathbf{x}$ from P_i uniformly at random. Flip each bit of $\mathbf{x}$ independently with probability $1/n$. If the result is no worse than $\mathbf{x}$, the mutant replaces $\mathbf{x}$.

$i := i + 1$: and go to the Reproduction step.

Given the UIO problem and the SSEA as described above, the following results show clearly the significant impact crossover has on SSEA's performance.

Theorem 8 ([27]). *For a sufficiently large constant $c > 0$, if the $(N+1)$ SSEA with a constant crossover probability $p_c > 0$ and population size N, $2 \leq N = Poly(n)$, is restarted every cN^2n^2 generations on* TwoPaths, *then the expected first hitting time is $O(N^2n^2)$.*

In other words, the optimum can be found within polynomial time as long as crossover is used. The following theorem shows that it is no longer possible to find the optimum in polynomial time if crossover is not used.

Theorem 9 ([27]). *If the crossover probability $p_c = 0$, the probability that the $(N + 1)$ SSEA with population size $N = Poly(n)$ finds the optimum of* TwoPaths *within 2^{cn} generations, where c is a constant, is upper-bounded by $e^{-\Omega(n)}$.*

It is important to note that these two theorems only state the benefits of this crossover operator for the TwoPaths problem. The conclusions should not be generalised to other problems without new proofs, because different search operators are effective on different problems. There are problems on which crossover will not be beneficial.

6   Interaction between Operators/Parameters

The performance of an EA is determined by its operators, parameters and interactions among them. While there have been studies on individual operators, such as crossover described in the previous section, and parameters, such as population size as discussed in Section 4, only one study [28] exists, which analyses the interaction of two operators, i.e., mutation and selection. It was shown in this work that neither mutation nor selection alone could determine the performance of an EA [28]. It was their combined effect that determined EA's performance. While this might sound intuitive, it was the first time that a rigorous analysis was given.

Let's investigate a non-elitist population-based EA with the linear ranking scheme [28], which captures many features of the EAs used in practice.

Initialisation: Generate N individuals at random for the initial population P_0. Each individual $P_0(i)$ is generated by sampling $\{0,1\}^n$ uniformly at arandom, $i \in \{1, 2, \ldots, N\}$. $t := 0$.

Evolutionary Cycle: Repeat the following until certain stopping criterion is met.

1. Sort P_t according to fitness f such that

$$f(P_t(1)) \geq f(P_t(2)) \geq \cdots \geq f(P_t(N)).$$

2. For $i \in \{1, 2, \ldots, N\}$,

    (a) Sample r from $\{1, \ldots, N\}$ using the linear ranking scheme.

    (b) $P_{t+1} := P_t(r)$.

    (c) Flip each bit in $P_{t+1}(i)$ with probability χ/n.

  3. $t := t + 1$.

In the above algorithm, N is the population size and χ determines the mutation probability. Both are fixed during evolution. To illustrate the importance of selection-mutation balance in this algorithm, the following problem is considered.

For any constants $\sigma, \delta, 0 < \delta < \sigma < 1 - 3\delta$, and integer $k \geq 1$, the fitness function considered here is [28]

$$\text{SELPRES}_{\sigma,\delta,k}(x) := \begin{cases} 2n, & \text{if } x \in X_\sigma^*, \\ \sum_{i=1}^n \prod_{j=1}^i x_j, & \text{otherwise} \end{cases}$$

where the set of optimal solutions X_σ^* contain all bitstrings $\mathbf{x} \in \{0,1\}^n$ satisfying

$$\|x[1, k+3]\| = 0,$$

$$\|x[k+4, (\sigma-\delta)n - 1]\| = 1,$$

$$\|x[(\sigma+\delta)n, (\sigma+2\delta)n - 1]\| \leq 2/3.$$

Theorem 10 ([28]). *For any constant integer $k \geq 1$, let T be the runtime of the non-elitist population-based EA with linear ranking selection. Its population size N satisfies $n \leq N \leq n^k$. It has a constant selection pressure of η, where $1 < \eta \leq 2$. The bit-wise mutation rate is χ/n. On function $\text{SELPRES}_{\sigma,\delta,k}$, for any constant $\epsilon > 0$,*

1. If $\eta < \exp(\chi(\sigma - \delta)) - \epsilon$, then for some constant $c > 0$,

$$Pr(T \geq e^{cn}) = 1 - e^{-\Omega(n)}.$$

2. If $\eta = \exp(\chi\sigma)$, then

$$Pr(T \leq n^{k+4}) = 1 - e^{-\Omega(n)}.$$

3. If $\eta > \frac{2\exp(\chi(\sigma+3\delta))-1}{1-\delta}$, then

$$E(T) = e^{\Omega(n)}.$$

A couple of observations can be made from the above theorem. First, the theorem shows an interesting relationship between selection pressure η and mutation rate χ. Neither determines the EA's computation time by itself. If selection pressure is high, it can be compensated by a high mutation rate to achieve the balance between the two, i.e., $\eta = \exp(\chi\sigma)$. If selection pressure is too low, we can lower the mutation rate accordingly to maintain an efficient EA. This theorem also suggests that trying to increase the mutation rate in order to increase evolvability and the ability of escaping from local optima may not work well for some problems, unless selection pressure is also increased appropriately.

Second, the EA's computation time is very sensitive to the ratio between η and χ. The ratio needs to be in a very narrow range around $\eta = \exp(\chi\sigma)$ to achieve EA's efficiency, i.e., polynomial computation time. Given a mutation rate, either a slightly small selection pressure or a moderately larger one will lead to exponential computation time. This is a rather unique example that unpacks the relationship between the EA and the problem, and sheds light into how the parameter interactions affect EA's performance on this problem.

7   Estimation of Distribution Algorithms (EDAs)

Although estimation of distribution algorithms (EDAs) were proposed and studied in the field of evolutionary computation, they are actually very different from other EAs. Instead of using any search operators, EDAs rely on model-building and sampling.

Initialisation: Generate N individuals using the initial probability distribution. $t := 0$.
Iterations: Repeat the following until the stopping criterion is met.
    1. M individuals are selected from the population of N individuals;
    2. A probability distribution is estimated from these M individuals;
    3. N individuals are sampled from this estimated probability distribution;
    4. $t := t + 1$.

Similar to Section 3, given an EDA, we can classify all problem instance classes into hard and easy cases [29].

EDA-easy Class. For a given EDA, a problem is EDA-easy *if and only if*, with the probability of $1 - 1/SuperPoly(n)$, the first hitting time needed to reach the global optimum is polynomial in the problem size n.
EDA-hard Class. For a given EDA, a problem is ED-hard *if and only if*, with the probability of $1/Poly(n)$, the first hitting time needed to reach the global optimum is superpolynomial in the problem size n.

Note the hardness definition here is EDA-dependent, because we are interested in the relationship between algorithms and problems. The above definition is similar to but different from that in Section 3 because the probabilities are used here, not mean first hitting times as in Section 3.

    We define formally an optimisation problem as $I = (\Omega, f)$, where Ω is the search space and f the fitness function. $\mathcal{P} = (\Omega, f, \mathcal{A})$ indicates an algorithm $\mathcal{A}$ on a fitness function f in the search space Ω. P_t^* indicates the probability of generating the global optimum in one sampling at the t-th generation.

    Given a function $f(n)$, where $f(n) > 1$ always holds and when $n \to \infty$, $f(n) \to \infty$, denote

1. $f(n) \prec Poly(n)$ and $g(n) = \frac{1}{f(n)} \succ \frac{1}{Poly(n)}$ if and only if $\exists a, b > 0, n_0 > 0$: $\forall n > n_0, f(n) \leq an^b$.

2. $f(n) \succ SuperPoly(n)$ and $g(n) = \frac{1}{f(n)} \prec \frac{1}{SuperPoly(n)}$ if and only if $\forall a, b > 0 : \exists n_0 : \forall n > n_0, f(n) > an^b$.

Theorem 11 ([29]). *For a given $\mathcal{P} = (\Omega, f, \mathcal{A})$, if the population size N of the EDA $\mathcal{A}$ is polynomial in the problem size n, then*

1. *if problem I is* **EDA-easy** *for $\mathcal{A}$, then $\exists t' \leq \lceil E[\tau(\mathcal{P})|\tau(\mathcal{P}) \prec Poly(n)] \rceil + 1$ such that*

$$P_{t'}^* \succ \frac{1}{Poly(n)};$$

2. *if $\forall t \prec Poly(n), P_t^* \prec \frac{1}{SuperPoly(n)}$, then problem I is* **EDA-hard** *for $\mathcal{A}$.*

Because the hardness definition used here is algorithm dependent. A problem that is easy for one EDA can be hard for another EDA or another EA. Chen *et al.* [29] described one problem that is EA-easy but EDA-hard. An example of EA-hard bu EDA-easy problems is yet to be found. Such theoretical comparison of problem hardness under different algorithms can often lead to a better understanding of the algorithms and shed light on the issue of what algorithmic features are most effective in tackling certain problem characteristics.

8 Concluding Remarks

Although most research in evolutionary computation relies on computational studies, there have been an increasing number of theoretical results in recent years. Significant progresses in analysing the computational time complexity of EAs have been made. Not only have there been a large number of papers on evolutionary computation theories in journals in evolutionary computation, artificial intelligence and theoretical computer science, there are also published books [30, 31]. One of the three sections, Section C, of the well-established *Theoretical Computer Science (TCS)* journal is entirely devoted to Natural Computing. According to Elsevier (http://www.journals.elsevier.com/theoretical-compu ter -science/most-cited-articles/), two of the top three most cited TCS papers published since 2007 are on evolutionary computation theories. This chapter only reviewed a tiny part of the results in evolutionary computation theory.

In spite of significant progresses, there is still much work to be done in developing better theories for evolutionary computation. There are several future research directions that seem to be particularly attractive and important.

First, the analysis of EDAs has been very few. The work by Chen *et al.* [29] investigated UMDAs only and on two artificial problems. More work is needed to analyse other EDAs on non-artificial problems. In particular, it will be very interesting to study when an EDA is likely to outperform an EA and why [32]. It is also interesting to analyse the impact of different probabilistic models on EDA's performance. Is it possible to improve EDA's performance by using a more powerful probabilistic model? When will a more powerful probabilistic model help?

Second, many real world problems are dynamic. Yet the analysis of EAs on dynamic problems is lagging behind applications. The existing work in this topic area is still in its infancy [33, 34]. There is a need for more theoretical work to complement computational studies in this area.

Third, all the work reviewed here focused on the time of finding the global optimal solution. In practice, good approximate solutions are often sufficient. Theoretical analysis of evolutionary approximation algorithms has shown some promising results [18–20, 35]. It has been shown that EAs from a random initial population can perform just as well as tailored heuristics for certain combinatorial optimisation problems. Can we find an example that an EA finds an approximate solution to a problem more efficiently than a human-designed heuristic?

Fourth, there has been some interest in algorithm portfolios in evolutionary computation [36, 37]. Computational studies have shown very promising results. However, it is unclear whether or not such type of algorithms offers any theoretical advantages over conventional ones. This is certainly an interesting challenge for theoretical research.

Fifth, the work reviewed in this chapter is all related to combinatorial optimisation. Yet EAs are equally often used in global (numerical) optimisation. There has been excellent work on the convergence and convergence rates of various EAs. However, theoretical analysis of EA's scalability has been few, in spite of recent surge in the interest of large scale optimisation [38–40]. It is still unclear what the relationship is between the optimisation time and the problem size (in terms of dimensionality) for different EAs on different problems. In fact, it is not entirely clear what a good measure for the optimisation time should be. The convergence time may not be very interesting from a practical point of view as we may not find the exact global optimum in finite time. It is more interesting to analyse the computation time towards a near optimum. Maybe we should explore the potential links to Blum *et al.*'s seminal work [41].

Acknowledgement. This work was partially supported by an EPSRC grant (No. EP/I010297/1).

References

1. Yao, X., Liu, Y., Lin, G.: Evolutionary programming made faster. IEEE Transactions on Evolutionary Computation 3, 82–102 (1999)
2. Li, B., Lin, J., Yao, X.: A novel evolutionary algorithm for determining unified creep damage constitutive equations. International Journal of Mechanical Sciences 44(5), 987–1002 (2002)
3. Yang, Z., Tang, K., Yao, X.: Self-adaptive differential evolution with neighborhood search. In: Proceedings of the 2008 IEEE Congress on Evolutionary Computation (CEC 2008), pp. 1110–1116. IEEE Press, Piscataway (2008)
4. Yang, Z., Li, X., Bowers, C., Schnier, T., Tang, K., Yao, X.: An efficient evolutionary approach to parameter identification in a building thermal model. IEEE Transactions on Systems, Man, and Cybernetics — Part C (2012), doi:10.1109/TSMCC.2011.2174983

5. Tang, K., Mei, Y., Yao, X.: Memetic algorithm with extended neighborhood search for capacitated arc routing problems. IEEE Transactions on Evolutionary Computation 13, 1151–1166 (2009)
6. Handa, H., Chapman, L., Yao, X.: Robust route optimisation for gritting/salting trucks: A CERCIA experience. IEEE Computational Intelligence Magazine 1, 6–9 (2006)
7. Praditwong, K., Harman, M., Yao, X.: Software module clustering as a multi-objective search problem. IEEE Transactions on Software Engineering 37, 264–282 (2011)
8. Wang, Z., Tang, K., Yao, X.: Multi-objective approaches to optimal testing resource allocation in modular software systems. IEEE Transactions on Reliability 59, 563–575 (2010)
9. Dam, H.H., Abbass, H.A., Lokan, C., Yao, X.: Neural-based learning classifier systems. IEEE Transactions on Knowledge and Data Engineering 20, 26–39 (2008)
10. Yao, X., Islam, M.M.: Evolving artificial neural network ensembles. IEEE Computational Intelligence Magazine 3, 31–42 (2008)
11. Cordón, O., Gomide, F., Herrera, F., Hoffmann, F., Magdalena, L.: Ten years of genetic fuzzy systems: current framework and new trends. Fuzzy Sets and Systems 141(1), 5–31 (2004)
12. Chong, S.Y., Tino, P., Yao, X.: Measuring generalization performance in co-evolutionary learning. IEEE Transactions on Evolutionary Computation 12, 479–505 (2008)
13. Salcedo-Sanz, S., Cruz-Roldán, F., Heneghan, C., Yao, X.: Evolutionary design of digital filters with application to sub-band coding and data transmission. IEEE Transactions on Signal Processing 55, 1193–1203 (2007)
14. Zhang, P., Yao, X., Jia, L., Sendhoff, B., Schnier, T.: Target shape design optimization by evolving splines. In: Proc. of the 2007 IEEE Congress on Evolutionary Computation (CEC 2007), pp. 2009–2016. IEEE Press, Piscataway (2007)
15. Li, Y., Hu, C., Yao, X.: Innovative batik design with an interactive evolutionary art system. J. of Computer Sci. and Tech. 24(6), 1035–1047 (2009)
16. He, J., Yao, X.: Drift analysis and average time complexity of evolutionary algorithms. Artificial Intelligence 127, 57–85 (2001)
17. Hajek, B.: Hitting time and occupation time bounds implied by drift analysis with applications. Adv. Appl. Probab. 14, 502–525 (1982)
18. He, J., Yao, X.: Maximum cardinality matching by evolutionary algorithms. In: Proceedings of the 2002 UK Workshop on Computational Intelligence (UKCI 2002), Birmingham, UK, pp. 53–60 (September 2002)
19. He, J., Yao, X.: Time complexity analysis of an evolutionary algorithm for finding nearly maximum cardinality matching. J. of Computer Sci. and Tech. 19, 450–458 (2004)
20. Oliveto, P., He, J., Yao, X.: Analysis of the (1+1)-ea for finding approximate solutions to vertex cover problems. IEEE Transactions on Evolutionary Computation 13, 1006–1029 (2009)
21. Lehre, P.K., Yao, X.: Runtime analysis of the (1+1) ea on computing unique input output sequences. Information Sciences (2010), doi:10.1016/j.ins.2010.01.031
22. He, J., Yao, X.: A study of drift analysis for estimating computation time of evolutionary algorithms. Natural Computing 3, 21–35 (2004)
23. He, J., Yao, X.: From an individual to a population: An analysis of the first hitting time of population-based evolutionary algorithms. IEEE Transactions on Evolutionary Computation 6, 495–511 (2002)

24. Chen, T., Tang, K., Chen, G., Yao, X.: A large population size can be unhelpful in evolutionary algorithms. Theoretical Computer Science (2011), doi:10.1016/j.tcs.2011.02.016

25. Lee, D., Yannakakis, M.: Principles and methods of testing finite state machines — a survey. Proceedings of the IEEE 84(8), 1090–1123 (1996)

26. Lehre, P.K., Yao, X.: Runtime analysis of (1+1) ea on computing unique input output sequences. In: Proc. of the 2007 IEEE Congress on Evolutionary Computation (CEC 2007), pp. 1882–1889. IEEE Press, Piscataway (2007)

27. Lehre, P.K., Yao, X.: Crossover can be constructive when computing unique input-output sequences. Soft Computing 15, 1675–1687 (2011)

28. Lehre, P.K., Yao, X.: On the impact of mutation-selection balance on the runtime of evolutionary algorithms. IEEE Transactions on Evolutionary Computation (2011), doi:10.1109/TEVC.2011.2112665

29. Chen, T., Tang, K., Chen, G., Yao, X.: Analysis of computational time of simple estimation of distribution algorithms. IEEE Transactions on Evolutionary Computation 14, 1–22 (2010)

30. Neumann, F., Witt, C.: Bioinspired Computation in Combinatorial Optimization: Algorithms and Their Computational Complexity. Springer, Berlin (2010)

31. Auger, A., Doerr, B. (eds.): Theory of Randomized Search Heuristics: Foundations and Recent Developments. World Scientific, Singapore (2011)

32. Chen, T., Lehre, P.K., Tang, K., Yao, X.: When is an estimation of distribution algorithm better than an evolutionary algorithm? In: Proceedings of the 2009 IEEE Congress on Evolutionary Computation, pp. 1470–1477. IEEE Press, Piscataway (2009)

33. Droste, S.: Analysis of the (1+1) ea for a dynamically changing onemax-variant. In: Proceedings of the 2002 IEEE Congress on Evolutionary Computation, pp. 55–60. IEEE Press, Piscataway (2002)

34. Rohlfshagen, P., Lehre, P.K., Yao, X.: Dynamic evolutionary optimisation: An analysis of frequency and magnitude of change. In: Proceedings of the 2009 Genetic and Evolutionary Computation Conference, pp. 1713–1720. ACM Press, New York (2009)

35. Yu, Y., Yao, X., Zhou, Z.-H.: On the approximation ability of evolutionary optimization with application to minimum set cover. Artificial Intelligence (2012), doi:10.1016/j.artint.2012.01.001

36. Fukunaga, A.S.: Genetic algorithm portfolios. In: Proceedings of the 2000 IEEE Congress on Evolutionary Computation, pp. 16–19. IEEE Press, Piscataway (2000)

37. Peng, F., Tang, K., Chen, G., Yao, X.: Population-based algorithm portfolios for numerical optimization. IEEE Transactions on Evolutionary Computation 14, 782–800 (2010)

38. Yang, Z., Tang, K., Yao, X.: Large scale evolutionary optimization using cooperative coevolution. Information Sciences 178, 2985–2999 (2008)

39. Yang, Z., Tang, K., Yao, X.: Scalability of generalized adaptive differential evolution for large-scale continuous optimization. Soft Computing 15, 2141–2155 (2011)

40. Li, X., Yao, X.: Cooperatively coevolving particle swarms for large scale optimization. IEEE Transactions on Evolutionary Computation (2011), doi:10.1109/TEVC.2011.2112662

41. Blum, L., Shub, M., Smale, S.: On a theory of computation and complexity over the real numbers: NP-completeness, recursive functions and universal machines. Bulletin of the American Mathematical Society 21, 1–46 (1989)

Representation in Evolutionary Computation

Daniel Ashlock[1], Cameron McGuinness[1], and Wendy Ashlock[2]

[1] University of Guelph, Guelph, Ontario, Canada, N1G 2W1
{dashlock,cmcguinn}@uoguelph.ca
[2] York University, Toronto, Ontario, Canada, M3J 1P3
washlock@cse.yorku.ca

The *representation* of a problem for evolutionary computation is the choice of the data structure used for solutions and the variation operators that act upon that data structure. For a difficult problem, choosing a good representation can have an enormous impact on the performance of the evolutionary computation system. To understand why this is so, one must consider the *search space* and the *fitness landscape* induced by the representation. If someone speaks of the fitness landscape of a problem, they have committed a logical error: problems do not have *a* fitness landscape. The data structure used to represent solutions for a problem in an evolutionary algorithm establishes the set of points in the search space. The topology or connectivity that joins those points is induced by the variation operators, usually crossover and mutation. Points are connected if they differ by one application of the variation operators. Assigning fitness values to each point makes this a fitness landscape. The question of the type of fitness landscape created when a representation is chosen is a very difficult one, and we will explore it in this chapter.

The primary goal of this chapter is to argue for more research into representation in evolutionary computation. The impact of representation is substantial and is not studied enough. The genetic programming community has been using *parameter sweeps* [18] which compare different choices of operations and terminals within a genetic programming environment. This is a big step in the right direction, but even this work ignores the issue of whether genetic programming is appropriate for a given problem. One of the implications of the No Free Lunch Theorem of Wolpert and Macready is that the quality of a given optimizer is problem specific. This includes the choice of representation.

There are reasons that representation has not been explored. While there can be huge rewards from exploring different representations, there is also a substantial cost. One must implement alternate representations; one must run well-designed experiments with them which probably include parameter tuning for each representation; and then one must find a way to compare the solutions. This last task seems simple – could not one simply examine final fitness numbers? While the first answer to this question is clearly yes, it may be that a problem requires diverse solutions or robust solutions. The recent explosion of research in multicriteria optimization with evolutionary algorithms means that issues like the diversity of solutions produced are important.

J. Liu et al. (Eds.): WCCI 2012 Plenary/Invited Lectures, LNCS 7311, pp. 77–97, 2012.

We will examine the question of representation through a series of examples involving: a simple toy optimization problem, the problem of evolving game playing agents, real optimization problems, and, finally, a problem drawn from automatic content generation for games.

1   Representation in Self-avoiding Walks

The *self-avoiding walk* (SAW) problem involves traversing a grid, given instructions for each move, in such a way that every square is visited. Fitness is evaluated by starting in the lower left corner of the grid and then making the moves specified by the chromosome. The sequence of moves made is referred to as the *walk*. If a move is made that would cause the walk to leave the grid, then that move is ignored. The walk can also revisit cells of the grid. Fitness is equal to the number of squares visited at least once when the walk is completed. The problem is called the *self-avoiding* walk problem because optimal solutions for a number of moves equal to the number of squares minus one do not revisit squares; they are self-avoiding walks. Figure 1 shows the 52 global optima for the 4×4 SAW problem. In addition to a diverse set of optimal solutions, the SAW problem has many local optima when the grid is large enough.

The SAW problems has a number of nice qualities as an evolutionary computation test problem:

- The problem has a large number of cases, one for each possible size of grid. While problem difficulty does increase with grid size, it is also different for grids of the same size with different dimensions such as 4×4 and 2×8.
- Even for quite long genes, the solutions have a simple two-dimensional representation. This makes visualizing final populations easy. Visualizations of the final walk also make it easy to compare between different representations.
- The problem, when the grid is large enough, has a large number of both global and non-global optima. This starts, roughly, when both dimensions are larger than 3. Table 1 gives the number of global optima.
- The global optima are not symmetrically distributed. Some have many other optima nearby, while others are far from other optima. This means that, even though they have the same fitness, they differ in how easy they are to locate. The notion of *nearby* used here is Hamming distance.

Having made a case that the SAW has desirable properties for a test problem, the next step is to construct multiple representations for it. We will examine three representations, one of them the obvious choice, and all implemented as strings over some alphabet. Other than changing the representation, all experiments will be performed using a population of 100 strings using two-point crossover and a mutation operator that changes two of the characters in the string. The problem case used is the 4×4 SAW.

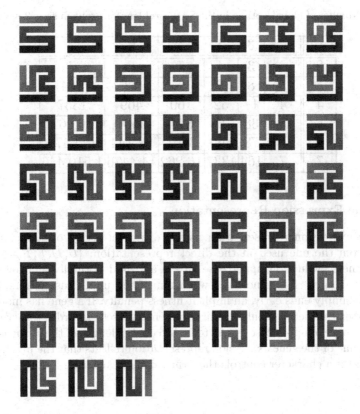

Fig. 1. The optimal solutions to the 4 × 4 SAW problem

The Direct Representation

The direct representation uses a character string over the alphabet $\{U, D, L, R\}$, which stand for *Up, Down, Left,* and *Right*. The string is of length fifteen and the fitness function simply executes the moves in order, recording the number of squares visited. Since evaluation starts in the lower left square with that square already visited the minimum fitness is one and the maximum is 16. The string length is equal to the minimum number of moves required to visit all the squares.

The Relative Representation

The relative representation uses a character string of length 15 over the alphabet $\{F, R, L\}$ which stand for *forward, turn right and then move forward,* and *turn left and then move forward*. Like the direct representation, the relative representation keeps track of the square it currently occupies. It adds to that information the direction it is currently facing. Fitness evaluation starts with the drawing agent facing upward. Fitness evaluation is otherwise like the direct representation.

Table 1. Number of global optima in the SAW problem for problem sizes $2 \leq n, m \leq 7$

n/m	2	3	4	5	6	7
2	2	3	4	5	6	7
3	3	8	17	38	78	164
4	4	17	52	160	469	1337
5	5	38	160	824	3501	16,262
6	6	78	469	3501	22,144	144,476
7	7	164	1337	16,262	14,4476	1,510,446

The Gene Expression Representation

The gene expression representation uses a character string over an alphabet derived from the one used in the direct representation: $\{U, D, L, R, u, d, l, r\}$. During fitness evaluation upper case letters are used normally and lower case letters are ignored. If a gene has fewer than fifteen upper case letters, fitness evaluation simply ends early, an implicit fitness penalty. If a gene has more than fifteen upper case letters, only the first fifteen are used. In order to permit the average number of upper case letters to be fifteen, the length of the string is set to 30. The name of the gene expression representation reflects that the upper/lower case status of a character controls the *expression* of each gene loci.

Results

A simple assessment of the impact of changing the representation is given in Figure 2. The time to solution for sets of 1000 replicates done for all three representations was sorted and then graphed. The performance of the representations is strongly stratified with the direct representation exhibiting the worse performance (longer times to solution), the gene expression representation coming in second, and the relative representation coming in first. For the replicates with the longest time to solution (right end of the sorting order), the gene expression representation takes over for first place.

The goal of demonstrating that the choice of representation makes a difference has been met for the SAW problem. Let us now consider what caused the change in performance. The size of the search space for the relative representation is 3^n, while for the direct representation the size is 4^n, meaning that evolution has a smaller job to do. The relative representation encodes far fewer walks than the direct one. In particular, the relative representation is incapable of moving back to the square it just came from, a move that always results in a suboptimal solution. This gives the relative representation a substantial advantage: it retains all the optimal solutions in the direct representation while excluding many sub-optimal ones. This is an example of building domain information into the representation.

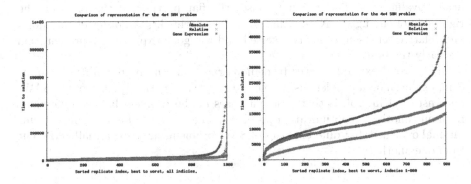

Fig. 2. This figure shows the impact of changing representation on the time to solution for the 4x4 SAW problem. The graphs display the sorted times to solution for 1000 independent evolutionary replicates. The left panel displays all 1000 replicates while the right one displays only the first 900.

The best average performer is the gene expression representation, though this is due to a small number of bad results for the relative representation. The size of the search space here is 8^{30}, enormously larger than direct or relative representation. This demonstrates that the size of the search space, while potentially relevant, cannot possibly tell the whole story. Both the direct and relative representation uniquely specify a sequence of moves. The gene expression representation has billions of different strings that yield the same sequence of moves. It also specifies some sequences of moves the other two representations cannot, but these all contain fewer than fifteen moves and have intrinsically bad fitness.

To understand the good performance of the gene expression representation, it is necessary to consider the fitness landscape. A mutation of a gene in the direct or relative representation changes one move in the walk represented by that gene. Some of the mutations in the gene expression representation have the same effect, but those that change a capital letter into lower case or vice versa have the effect of inserting or deleting characters from the walk specified by the gene. This means that the gene expression representation has *edit* mutations that can insert, delete, or change the identity of a character in the walk the gene codes for. The other two representations can only change the identity of characters.

If we consider the space of encoded walks, rather than genes, the gene expression representation has more connectivity. If we think of optima as hills in the fitness landscape, then using the gene expression representation has the effect of merging some of the hills. Since the number of optimal results remains constant, this means the only effect is to eliminate local optima.

One weakness of this demonstration of the impact of representation on the SAW problem is that only one case of the problem was examined. Figure 3 shows the result of performing the same experiments for the 5×5 case of the SAW problem. The algorithm was set to halt if it did not find a solution in 1,000,000 fitness evaluations. This is what causes the flat portions of the plots on the right side of the figure. Notice that, in this experiment, the order of the direct and relative representations is reversed and the gene expression representation is clearly the best.

This second example serves to demonstrate that the representation issue is complex, even on a problem as simple as the SAW. Another quality of the SAW, demonstrated in [15], is that different cases of the problem behave differently from one another as optimization problems. The results in Figure 3 provide additional evidence that different sizes of SAW problems are substantially different from one another.

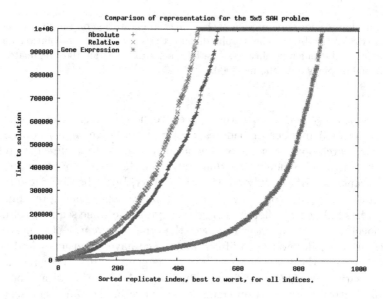

Fig. 3. This figure shows the impact of changing representation on the time to solution for the 5x5 SAW problem. The graphs display the sorted times to solution for 1000 independent evolutionary replicates.

2   Representation in Game-Playing Agents

The game used to demonstrate the impact of representation on the evolution of game playing agents is the iterated prisoner's dilemma. The *prisoner's dilemma* [13] is a widely known abstraction of the tension between cooperation and conflict. In the prisoner's dilemma two agents each decide simultaneously, without communication, whether to cooperate (C) or defect (D). If both players cooperate, they receive a payoff of C; if both defect, they receive a payoff of D. If one cooperates and the other defects, then the defector receives the *temptation* payoff T, while the cooperator receives the *sucker* payoff S. In order for a simultaneous two-player game to be prisoner's dilemma two conditions must hold:

$$\mathcal{S} \leq \mathcal{C} \leq \mathcal{D} \leq \mathcal{T} \tag{1}$$

and

$$(\mathcal{S} + \mathcal{T}) \leq 2\mathcal{C} \tag{2}$$

The first of these simply places the payoffs in their intuitive order while the second requires that the average score for both player's in a unilateral defection be no better than mutual cooperation.

A situation modeled by the prisoner's dilemma is that of a drug dealer and an addict exchanging money for drugs in simultaneous blind drops to avoid being seen together by the police. Cooperation consists of actually leaving the money or drugs; defection consists of leaving something worthless like an envelope full of newspaper clippings in place of money or an inert white powder in place of the drugs. If the exchange is made only once, then neither party has an incentive to do anything but defect. If the drop is to be made weekly, into the indefinite future, then the desire to get drugs or money next week strongly encourages cooperation today. This latter situation is an example of the *iterated* prisoner's dilemma. When play continues, the potential for future retribution opens the door to current cooperation. The payoff values used in the experiments described here are $S = 0$, $D = 1$, $C = 3$, and $T = 5$, largely because these values have been used in many other studies in which prisoner's dilemma agents were evolved [2,10,4,9].

Earlier research [5,1] compared ten different representations for the iterated prisoner's dilemma. These experiments all used populations of 36 agents whose fitness was computed as the average score in a round-robin tournament of 150 rounds of iterated prisoner's dilemma between each pair of players. Each agent had access to their opponent's last three plays, and perhaps more in the case of state conditioned representations. Evolution was run for 250 generations with the crossover and mutation operators kept as similar as possible given the differing representations.

Representations

The representations studied for the iterated prisoner's dilemma are as follows:

Finite State Machines. Two types of finite state machines are used: directly encoded finite state machines with 16 states (**AUT**) and finite state machines represented with a developmental encoding [19]. The number of states in the machine is variable but not more than twenty. These finite state machines are referred to by the tag **CAT**.

Function Stacks. The tags **F40**, **F20**, and **F10** are for *function stacks*, a linear genetic programming representation based on a directed acyclic graph. The data type is boolean and the operations available are logical And, Or, Nand, and Nor. The constants true and false are available as are the opponent's last three actions.

We use the encoding true=defect, false=cooperate. The numbers 10, 20, and 40 refer to the number of nodes in the directed acyclic graph.

Tree-Based Genetic Programming. We use standard tree-based genetic programming [20] with the same encoding as the function stacks and the same Boolean functions with access to the opponent's last three actions. These are referred to with the tag **TRE**. The tag **DEL** corresponds to Boolean parse trees, identical to TRE,save that a one-time-step delay operator is incorporated into the operation set.

Markov Chains. The tag **MKV** is used for Markov chains implemented as look-up tables indexed by the opponent's last three actions that gives the probability of cooperation. Once this probability has been found a random number is used to determine the agent's action. The tag **LKT** is used for look-up tables indexed by the opponent's last three actions. The lookup tables are like the Markov chains if the only probabilities permitted are 0 and 1.

ISAc Lists. A different linear genetic programming representation denoted by **ISC** are If-Skip-Action lists [12]. An ISAc list executes a circular list of Boolean tests on data items consisting of the opponent's last three actions and the constants "cooperate" and "defect" until a test is true. Each Boolean test has an action associated with it, the action for the true test is the agent's next action. On the next round of the game execution starts with the next test. The lists of tests used here have a length of 30.

Neural Nets. The tag **CNN** is used for feed-forward neural nets with a per-neuron bias in favor of the output signifying cooperation; they access the opponent's last three actions and have a single hidden layer containing three neurons. The tag **NNN** are feed-forward neural nets identical to CNN save that they have no bias in favor of cooperation or defection.

2.1 Results

The metric used to compare representations is the probability the final population, at generation 250, is essentially cooperative. We measure this as having an average payoff of 2.8 or more. This is a somewhat arbitrary measure, carefully justified only for the **Aut** representation. For finite state automata, a series of initial plays between two players must be followed by a repeating sequence of plays caused by having reached a closed loop in the (finite) space of states. When fitness evaluation consists of 150 rounds of iterated prisoner's dilemma and the automata have no more than sixteen states, an average score of 2.8 or more corresponds to having no defections in the looped portion of play.

Figure 4 shows the probability that different representations will be cooperative. This result is, in a sense, appalling. The outcome of the basic experiment to demonstrate that cooperation arises [22] has an outcome that can be dialed

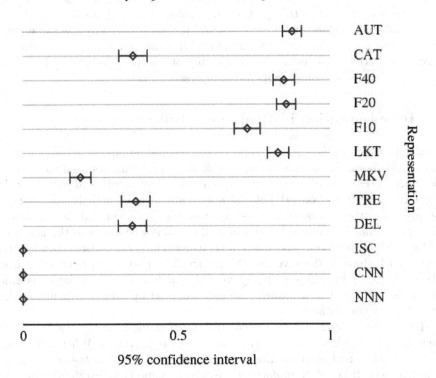

Fig. 4. Shown are 95% confidence intervals on the probability that the final generation of an evolutionary algorithm training prisoner's dilemma playing agents, with different representations, will be cooperative

from 95% cooperative to no cooperation by changing the representation. This shows not only that representation has an impact but that it can be the dominant factor. In other words, an experiment using competing agents that does not control for the effects of representation may have results entirely dictated by the choice of representation.

There are a number of features of these experiments that make the situation worse. In order to check the importance of changing the parameters of a single representation, the function stack (Boolean directed acyclic graph genetic programming) representation was run with 10, 20, and 40 nodes. Notice that the 10-node version of this representation is significantly less cooperative than the others.

In a similar vein, the **AUT** and **CAT** are alternate encodings of the same data structure. In spite of this they have huge differences in their degree of cooperativeness. There is a mathematical proof in [1] that the function stack representation encodes the same space of strategies as finite state machines.

A slightly different version of this fact applies to the two neural net (**CNN** and **NNN**) and the lookup table (**LKT**) representations. All of these are a map from the opponent's last three actions to a deterministic choice of an action. It is not difficult to show all three representation encode *exactly* the same space of 2^8 strategies in different ways. In spite of this the neural net representations have an experimentally estimated probability of zero of cooperating; the lookup tables are among the most cooperative representations.

3   Representation in Real Optimization

Real parameter optimization is one of the earliest applications of evolutionary computation. Evolution strategies [14] were originally designed to optimize parameters that described an airfoil and also has had substantial success at designing nozzles that convert hot water into steam efficiently. Real parameter optimization also substantially pre-dates evolutionary computation; it is one of the original applications of the differential calculus with roots in the geometry of the third century B.C. and modern treatments credited to Isaac Newton and Gottfried Leibniz in the seventeenth century. The natural representation from the calculus, as functions mapping m-tuples of numbers to a single parameter to be optimized, is a natural one only adopted by some techniques within evolutionary computation.

The largest difference between real optimization and representations built on character strings is the set of available mutation operators. When we change the value of a real number, that change is a probability distribution on the real numbers. It could be uniform, Gaussian, or some more exotic distribution. Evolutionary programming [16] pioneered the use of mutation operators that use covariance across parameters to permit evolution to modify mutation operators to respect the local search gradient. One of the representations we will examine completely avoids the issue of selecting the correct distribution for a mutation operator, while the other two retain their critical dependence on that choice. Correct choice of the type of mutation operator in real parameter optimization is very important, but it is not the subject we are concerned with in this article.

3.1   Representations

We will examine three possible representations for real-parameter optimization. There are many others. Some of the earliest work in real optimization [17] represented sets of real parameters as strings of bits with blocks of bits first mapped onto an integer value and then the integer value used to pick out a value from an equally spaced set for a given parameter value. This representation required techniques, like Grey-coding, to ensure that some bits were not far more important than others. This is a nice, early example of finding a more effective representation.

The Standard Direct Representation. Current evolutionary real optimization often operate on vectors of real numbers holding the parameters to be optimized in some order. We will call treating such a vector of real numbers as a string of values, using crossover operators analogous to the string like ones, the *standard direct representation*. This is the first of our three representations.

The Gene Expression Representation for Real Parameters. The gene expression representation, used on the SAW problem in Section 1, can easily be adapted to real parameter optimization. We first lengthen the vector of real parameters by double and then add an *expression layer* in the form of a binary gene with one loci for each parameter in the vector of reals. Before the vector of reals is sent to the fitness function, an *expression step* is performed. Suppose that n real parameters are required. Only those real values with a one in the corresponding position in the expression layer are used. If fewer than n real parameters are expressed in this fashion, then the individual receives fitness that is the worst possible. If n or more parameters are expressed, then the first n, in the order they appear in the data structure, are used. In this case, the usual fitness for those n parameters is the fitness of the entire data structure.

The Sierpinski Representation. The *Sierpinski representation* first appears in [11] and was used in [3] to located parameters for interesting subsets of the Mandelbrot set. The Sierpinski representation is inspired by the *chaos game*, an iterative averaging algorithm for generating the Sierpinski triangle or gasket, shown in Figure 5. The game starts at any vertex of the triangle. The game then iteratively moves half way toward a vertex of the triangle selected uniformly at random and plots a point. The points have been colored by averaging a color associated with each vertex into a color register each time a particular corner was selected. This visualizes the importance of each vertex to each plotted point.

If, instead of the three vertices of a triangle, we use the 2^n points that are the vertices of an n-dimensional box, then a series of averaging moves toward these points specify a collection of points that densely covers the interior of the box [11]. Strings of averaging moves form the representation for evolutionary search. Formally:

Definition 1. The Sierpinski representation. *Let $G = \{g_0, g_1, \ldots, g_{k-1}\}$ be a set of points in $\mathbb{R}^n$ called the* generator *points for the Sierpinski representation. Typically these are the vertices of a box in some number of dimensions. Associate each of the points, in order, with the alphabet $\mathcal{A} = \{0, 1, \ldots, k - 1\}$. Let the positive integer r be the* depth *of representation. Then for $s \in \mathcal{A}^r$ the point represented by s, p_s, is given by Algorithm 31.*

Definition 2. *The* **normalized Sierpinski representation** *(NSR) is achieved by insisting that the last character of the string always be the first generator.*

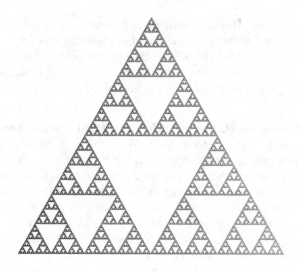

Fig. 5. The Sierpinski triangle or gasket

Algorithm 31. Sierpinski Unpacking Algorithm

Input: A string $s \in \mathcal{A}^r$;The set G of k generator points;
An averaging weight α
Output: A point in $\mathbb{R}^n$
Details:
Set $x \leftarrow g_{s[r-1]}$
for($i \leftarrow r - 2; i \geq 0; i \leftarrow i - 1$)
    $x \leftarrow \alpha \cdot g_{s[i]} + (1 - \alpha) \cdot x$
end for
return(x)

The following lemma is offered without proof (but is elementary).

Lemma 1. *Let a string s of length r be a name for a point $x = p_s$. Suppose that $\alpha = 0.5$ and that the last character (initial generator) of s is always the same. Then s is the sole name of x of length r.*

The Sierpinski representation reduces the problem of real optimization to that of evolving a string. Lemma 1 tells us that each string in the normalized version of the representation corresponds to a unique point. An important feature of the Sierpinski representation is that it searches only inside the convex hull of the generators. This has good and bad points; the generators can be used to direct search, but the search *cannot* use a mutation operator to locate an optima outside of the initial boundaries in which the population was initialized – something both the other representations can do.

3.2   Comparing the Direct and Sierpinski Representations

We will compare the direct and Sierpinski representations on the problem of optimizing the function:

$$h(x_0, x_1, \ldots, x_{n-1}) = sin(\sqrt{x_0^2 + x_1^2 + \cdots + x_{n-1}^2}) \cdot \prod sin(x_i) \qquad (3)$$

in $n = 5$ dimensions. This function possesses an infinite number of optima of varying heights and is thus good for testing the ability of an algorithm to locate a diversity of optima.

Notice that the Sierpinski representation stores points as strings of characters. This means that we can store and retrieve points in a dictionary – with logarithmic time for access – and can compare points for "nearness" by simply checking their maximum common prefix. In particular, if we are searching a space with multiple optima, it becomes very easy to database optima that the algorithm has already located. The MOSS, given as Algorithm 32, was first specified in [11].

Algorithm 32. Multiple Optima Sierpinski Searcher (MOSS)

Input: A set of generator points G
An averaging parameter α
A depth of representation r
A depth of exclusion d
A multi-modal function f to optimize
Output: A collection of optima
Details:
Initialize a population of Sierpinski representation strings
Run a string-EA until an optimum x is found
Initialize a dictionary D with the string specifying x
Repeat
    Re-run the EA, awarding minimal fitness to any string
       with the same d-prefix as any string in the dictionary
    Record the new optimum's string in D
Until(Enough optima are found)

The MOSS algorithm creates zones of low fitness around the optima it has already located. The size of the zones is determined by the exclusion depth and have a shape identical to the convex hull of the generators. Each increase in the exclusion depth decreases the size of the holes around known optima by one-half.

Table 2 compares 100 runs of the standard algorithm with 100 runs of the MOSS algorithm supported by the Sierpinski representation. The goal, in this case, is to locate as many optima as possible. The table gives a tabulation of optima located stratified by the number of times they were located. The results are striking: the MOSS algorithm located far more optima (969) than the standard

Table 2. Relative rate of location among the 1000 populations optimizing Equation 3 for the original and MOSS algorithms

Times located	Number of Optima Original	MOSS	Times located	Number of Optima Original	MOSS
1	122	938	8	6	0
2	77	31	9	1	0
3	56	0	10	3	0
4	40	0	11	1	0
5	19	0	12	1	0
6	15	0	13	1	0
7	14	0			

algorithm (356) and never located a given optima three times. The standard algorithm located six optima more than ten times each. The average quality of the optima located is higher for the standard algorithm, because it locates high quality optima multiple times. The two representations compared are not, in an absolute sense, better or worse. Rather, each has situations in which it is better. The strength of the Sierpinski representation is locating a diversity of optima; it makes databasing optima easy and so enables the MOSS algorithm.

3.3   Comparison of the Direct and Gene Expression Representations

We compare the standard direct and gene expression representations on the function:

$$g_n(x_1, x_2, \ldots, x_n) = \frac{1}{20n} \sum_{k=0}^{n} x_k + \sum_{k=0}^{n} sin(\sqrt{k} \cdot x_k) \qquad (4)$$

in two through seven dimensions. This problems has many local optima and the small linear trend means that the better optima are further afield.

For each dimension and representation, 400 replicates of an evolutionary algorithm were run and a 95% confidence interval on the quality of the optima located were constructed. This confidence interval was constructed at both 100,000 fitness evaluations and 1,000,000 fitness evaluations. The results are given in Table 3.

The advantage of using the gene expression representation is largest in lower dimensions. It ceases to be significant when we compare the results in $d = 7$ dimensions for the shorter evolutionary runs. The significance returns in the longer evolutionary runs. This demonstrates that the gene expression representation made better use of additional time.

The fitness landscape for this problem is easy to understand - it has a lot of hills and the small linear trend means that searching further afield will always locate better optima. This lets us draw the following conclusion: the gene

Table 3. Mean value of best optima located, averaged over replicates, and best optimum located in any replicate for the polymodal function. This table compares the standard direct representation and the gene expression representations for two different lengths of evolution.

	100,000 fitness evaluations				1,000,000 fitness evaluations			
	Gene Expression		Direct		Gene Expression		Direct	
Dimension	Mean	Best	Mean	Best	Mean	Best	Mean	Best
2	2.77 ± 0.02	3.25	2.59 ± 0.02	2.98	3.04 ± 0.03	4.17	2.58 ± 0.02	2.98
3	3.74 ± 0.01	4.11	3.54 ± 0.02	3.97	3.94 ± 0.02	4.45	3.55 ± 0.02	3.97
4	4.71 ± 0.01	5.05	4.54 ± 0.01	4.97	4.88 ± 0.01	5.33	4.52 ± 0.01	4.97
5	5.68 ± 0.02	5.98	5.51 ± 0.01	5.85	5.80 ± 0.02	6.31	5.52 ± 0.01	5.87
6	6.58 ± 0.02	6.95	6.49 ± 0.01	6.82	6.73 ± 0.01	7.10	6.51 ± 0.01	6.90
7	7.54 ± 0.04	7.93	7.50 ± 0.01	7.85	7.68 ± 0.02	8.08	7.50 ± 0.01	7.80

expression representation is better at exploration, while the standard direct representation is better at exploitation. It is easy to test this hypothesis in the opposite direction by optimizing a different function. Recall that mutations to the binary expression layer amount to inserting or deleting values from the sequence of real parameters. The gene expression representation preforms badly when optimizing a unimodal function with its mode selected so that no two of its coordinates are the same (data not shown). It has a far worse mean time to solution than the direct representation in low dimensions and completely fails to locate the optimum in higher dimensions.

As with the Sierpinski representation, the gene expression representation is neither better nor worse than the standard direct one. Each has its own appropriate domain of applicability. This is additional support for the thesis of this chapter, that we should study representation more assiduously. Each new representation is an additional item in our toolbox. Both the comparisons made in this section demonstrate that there is a significant impact to the choice of representation.

4   Representation in Automatic Content Generations

In this section we look at the problem of evolving a maze. Examples of the types of mazes we are evolving are shown in Figures 8 and 9. We will use the same evolutionary algorithm for each of five representations. The representations used in this study are defined in [6,8,21]. Since none of the cited publications used the same fitness function on all five representations we use a new fitness function. The mazes we are evolving are specified on a grid. The mazes have an entrance in the center of each wall and two internal checkpoints. Figure 7 designates long and short distances. These distances are the lengths of the shortest paths between

Fig. 6. Shown are a raw (left) and rendered (right) maze of the sort specified by evolving which squares on a grid are obstructed

the specified points. The fitness function is the sum of the long distances divided by the sum of the short distances, except that any maze where we cannot move from each door to both checkpoints is awarded a fitness of zero. Distances are computed using a simple dynamic programming algorithm [6].

We will demonstrate that the visual character of the mazes changes substantially when the representation is changed. All the evolutionary algorithms use a population of 100 mazes stored as strings of values with a 1% mutation rate and two-point crossover. Evolution proceeds for 500,000 fitness evaluations. The representations are as follows:

First Direct Representation. Open and blocked squares within a rectangular grid are specified directly as a long, binary gene.

Chromatic Representation. A direct representation, in which the squares within a grid are assigned colors from the set { red, orange, yellow, green, blue, violet}. These colors are specified directly as a long gene over the alphabet $\{R, O, Y, G, B, V\}$. An agent can move between adjacent squares if they are (i) the same color or (ii) adjacent in the above ordering.

Height-Based Representation. A direct representation, in which the squares within a grid are assigned heights in the range $0 \leq h \leq 10.0$. An agent can move between adjacent squares if their heights differ by 1.0 or less.

Indirect positive representation. The chromosome specifies walls that are placed on an empty grid to form the maze. The walls can be horizontal, vertical, or diagonal. In this representation walls are explicit, and rooms and corridors are implicit.

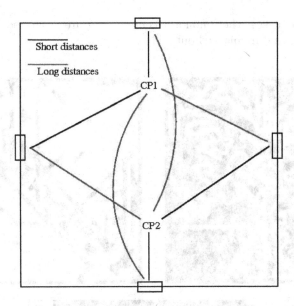

Fig. 7. This figure shows the four doors and internal checkpoints within a maze. The fitness function maximizes the quotient of the sum of the red distances and the sum of the black distances. This permits the evolution of a diverse collection of mazes with similar properties.

Indirect Negative Representation. The chromosome specifies material to remove from a filled grid to form the maze. In this representation, rooms and corridors are explicit, and walls and barriers are implicit.

All of the representations, except the indirect negative representation, use a technique called *sparse initialization* to compensate for the fact that, when the data structure is filled in uniformly at random, it is quite likely to have zero fitness, because there is no path between at least one door and at least one checkpoint. Sparse initialization biases the initial population to have high connectivity. Sparse initialization takes the following forms. For the direct representation, only 5% of the squares are filled in. For the chromatic representation all squares are initialized to green or yellow. For the height-based representation the heights are initialized to a Gaussian random value with mean three and height one. For the indirect positive representation all walls start at length three. Using sparse initialization places the burden of building the maze onto the variation operators. The initialization to a highly connected state biases the trajectory of evolution.

4.1 Results

Figure 8 gives examples of evolved mazes for the direct and indirect negative and positive representations. Figure 9 gives the examples for the chromatic and height representations. Since it is very hard to see paths in these mazes, they are accompanied by a key where non-adjacent squares are separated by walls and inaccessible squares are blacked out.

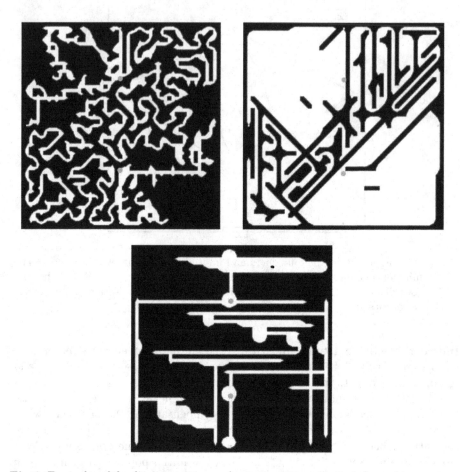

Fig. 8. Examples of the direct, positive, and negative representations for making mazes. The checkpoints are shown as green circles.

The results in this section speak for themselves. Even though they are evolved to satisfy the same distance-based fitness function the overall appearance of the mazes is very different. The appearance is entirely dependent on the choice of representation. The two most similar representations are the chromatic and height based. The keys to these mazes look similar. The actual mazes, though, look quite different. The type of representation that should be chosen, in this example, depends strongly on the goals of the user of the maze. The mazes shown here are simple. In [7] more complex design criteria for mazes are given. In [8], individual evolved maze tiles are used to build scalable maps.

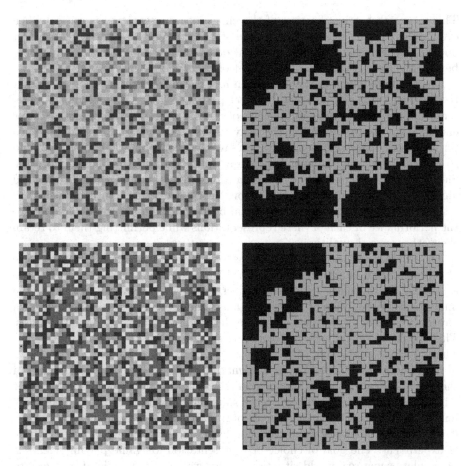

Fig. 9. Examples of height-based and chromatic mazes, together with their keys. The checkpoints are shown as black circles. The minimum height in the height-based mazes is colored red and colors move down the rainbow order of colors as heights increase.

5   Discussion and Conclusions

In all four examples in this chapter it has been demonstrated that representation has a substantial impact on the outcome of an evolutionary algorithm. The example using self avoiding walks showed that changing the representation changed the time to solution, but in different ways for different cases of the problem. This demonstrates that, even within a simple problem domain, the best choice of representation is problem specific. The SAW problem is the simplest system that has, so far, shown this sort of complex response to the change of representation. This makes it a good sandbox for developing tools for exploring the issue of representation.

In the section on evolving agents to play the iterated prisoner's dilemma we saw that the choice of representation can dominate the behavior of a simulation.

This means that a justification of the choice of representation is critical when evolving competing agents. This is the strongest evidence of a need to better understand representation within evolutionary computation. This example goes beyond the issue of performance to that of validity of results.

The examples given in the section on real parameter optimization show that the representation can be chosen to meet particular goals. The Sierpinski representation permits log-time databasing of optima already located and so makes enumeration of optima within the convex hull of its generators a simple matter. The gene expression representation favors exploration over exploitation and so is good for an environment with many optima of differing quality. Both the Sierpinski and gene expression representations are potentially valuable in hybrid algorithms. Each could generate locations near optima that are then finished by a standard hill-climber. Adding a special purpose local optimizer could enhance the performance of each of these representations while permitting them to retain their other special qualities.

The experiments with representations for maps of mazes show that the choice of representation can be used to control the appearance of the output of the algorithm. In the maze evolution project there is no point to comparing the final fitness of the different representations. The needs of a game designer for a particular type of appearance dominate the need to obtain a global best fitness. This speaks to an important point: one should carefully consider one's goals when choosing a representation.

Representation for the mazes is a matter of controlling the appearance of the maze. In real optimization the final goal may be global best fitness, but it might also be obtaining a diverse set of solutions. This latter goal becomes more important if the goal is multi-criteria optimization. With game playing agents, it was demonstrated that representation has a dominant effect on a simple type of experiment. The goal of choice of representation for game playing agents is to simulate some real-world situation. So far we have very little idea of *how* to choose an humaniform or ant-like representation for simulation of conflict and cooperation. This is a wide-open area for future research. The SAW problems are a toy problem that exhibit complex representational effects.

We hope that this chapter has convinced the reader of the importance of considering representation. We conclude by noting that for every example we chose to discuss here we have five others for which there was no room. We invite and appeal to the reader to join in the effort of understanding representation in evolutionary computation.

References

1. Ashlock, D.: Training function stacks to play iterated prisoner's dilemmain. In: Proceedings of the 2006 IEEE Symposium on Computational Intelligence in Games, pp. 111–118 (2006)
2. Ashlock, D., Ashlock, W., Umphry, G.: An exploration of differential utility in iterated prisoner's dilemma. In: Proceedings of the 2005 IEEE Symposium on Computational Intelligence in Bioinformatics and Computational Biology, pp. 271–278 (2006)

3. Ashlock, D., Bryden, K.M., Gent, S.: Multiscale feature location with a fractal representation. In: Intelligent Engineering Systems Through Artificial Neural Networks, vol. 19, pp. 173–180 (2009)
4. Ashlock, D., Kim, E.-Y.: Fingerprinting: Automatic analysis and visualization of prisoner's dilemma strategies. IEEE Transaction on Evolutionary Computation 12, 647–659 (2008)
5. Ashlock, D., Kim, E.Y., Leahy, N.: Understanding representational sensitivity in the iterated prisoner's dilemma with fingerprints. Transactions on Systems, Man, and Cybernetics–Part C: Applications and Reviews 36(4), 464–475 (2006)
6. Ashlock, D., Lee, C., McGuinness, C.: Search-based procedural generation of maze-like levels. IEEE Transactions on Computational Intelligence and AI in Games 3(3), 260–273 (2011)
7. Ashlock, D., Lee, C., McGuinness, C.: Simultaneous dual level creation for games. IEEE Computational Intelligence Magazine 6(2), 26–37 (2011)
8. Ashlock, D., McGuinness, C.: Decomposing the level generation problem with tiles. In: Proceedings of IEEE Congress on Evolutionary Computation, pp. 849–856. IEEE Press, Piscataway (2011)
9. Ashlock, D., Rogers, N.: A model of emotion in the prisoner's dilemma. In: Proceedings of the 2008 IEEE Symposium on Computational Intelligence in Bioinformatics and Computational Biology, pp. 272–279 (2008)
10. Ashlock, D.A., Kim, E.Y.: Fingerprint analysis of the noisy prisoner's dilemma. In: Proceedings of the 2007 Congress on Evolutionary Computation, pp. 4073–4080 (2007)
11. Ashlock, D.A., Schonfeld, J.: A fractal representation for real optimization. In: Proceedings of the 2007 Congress on Evolutionary Computation, pp. 87–94 (2007)
12. Ashlock, D., Joenks, M.: ISAc lists, a different representation for program induction. In: Genetic Programming 1998, Proceedings of the Third Annual Genetic Programming Conference, pp. 3–10. Morgan Kaufmann, San Francisco (1998)
13. Axelrod, R.: The Evolution of Cooperation. Basic Books, New York (1984)
14. Beyer, H.-G.: The Theory of Evolution Strategies. Springer, Berlin (2001)
15. Bryden, K.M., Ashlock, D., Corns, S., Willson, S.: Graph based evolutionary algorithms. IEEE Transactions on Evolutionary Computation 5(10), 550–567 (2005)
16. Fogel, L.J.: Intelligence through Simulated Evolution: Forty Years of Evolutionary Programming. John Wiley, Hoboken (1999)
17. Goldberg, D.E.: Genetic Algorithms in Search, Optimization, and Machine Learning. Addison-Wesley Publishing Company, Inc., Reading (1989)
18. Greene, C.S., Moore, J.H.: Solving complex problems in human genetics using gp: challenges and opportunities. SIGEVOlution 3, 2–8 (2008)
19. Kim, E.Y.: Analysis of Game Playing Agents with Fingerprints. PhD thesis, Iowa State University (2005)
20. Koza, J.R.: Genetic Programming. The MIT Press, Cambridge (1992)
21. McGuinness, C., Ashlock, D.: Incorporating required structure into tiles. In: Proceedings of Conference on Computational Intelligence in Games, pp. 16–23. IEEE Press, Piscataway (2011)
22. Miller, J.H.: The coevolution of automata in the repeated prisoner's dilemma. Journal of Economic Behavior and Organization 29(1), 87–112 (1996)

Quo Vadis, Evolutionary Computation?
On a Growing Gap between Theory and Practice

Zbigniew Michalewicz[*]

School of Computer Science,
University of Adelaide,
Adelaide, SA 5005, Australia
zbyszek@cs.adelaide.edu.au

Abstract. At the Workshop on Evolutionary Algorithms, organized by the Institute for Mathematics and Its Applications, University of Minnesota, Minneapolis, Minnesota, October 21 – 25, 1996, one of the invited speakers, Dave Davis made an interesting claim. As the most recognised practitioner of Evolutionary Algorithms at that time he said that all theoretical results in the area of Evolutionary Algorithms were of no use to him – actually, his claim was a bit stronger. He said that if a theoretical result indicated that, say, the best value of some parameter was such-and-such, he would never use the recommended value in any real-world implementation of an evolutionary algorithm! Clearly, there was – in his opinion – a significant gap between theory and practice of Evolutionary Algorithms.

Fifteen years later, it is worthwhile revisiting this claim and to answer some questions; these include: What are the practical contributions coming from the theory of Evolutionary Algorithms? Did we manage to close the gap between the theory and practice? How do Evolutionary Algorithms compare with Operation Research methods in real-world applications? Why do so few papers on Evolutionary Algorithms describe real-world applications? For what type of problems are Evolutionary Algorithms "the best" method? In this article, I'll attempt to answer these questions – or at least to provide my personal perspective on these issues.

1 Inspiration

Since the publication of my first book on Genetic Algorithms (Michalewicz, 1992) exactly twenty years ago, I have been thinking about theory and practice of these algorithms. From my early experiments it was clear that it would be necessary to extend the binary representation of genetic algorithms by some other data structures (hence the 'data structure' term in the title of the book) and incorporate problem-specific knowledge into the algorithm. These initial thoughts were later supplemented by various research activities as well as real world experiences – working on many

---

[*] Also at the Institute of Computer Science, Polish Academy of Sciences and at the Polish Japanese Institute of Information Technology, Warsaw, Poland, and the Chairman of the Board at SolveIT Software Pty Ltd.

J. Liu et al. (Eds.): WCCI 2012 Plenary/Invited Lectures, LNCS 7311, pp. 98–121, 2012.

projects for very large companies (e.g. Ford, General Motors, BMA Coal, Orlando Wines, Bank of America, BHP Billiton, Viterra. Dentsu, Rio Tinto, Xstrata, Fortescue). This chapter summarizes my dual experience – academia vs. industry – and I hope it would be useful for the current and the future generations of researchers in the Evolutionary Computation (EC) community. After all, it took me 20 years sitting on both sides of the fence to collect my thoughts!

However, the inspiration for this chapter (and my talk at the IEEE CEC'12) came from reading a relatively recent short article by Jeff Ullman on how to advise PhD students (Ullman, 2009). By the way, my PhD (from 1981) was in database systems (at that time I have not even heard about evolutionary/genetic algorithms) and Jeff Ullman was one of my heroes – his articles and books at that time set directions for (usually theoretical) research in many areas of database management – from database design to concurrency control (of course, he has made also substantial contributions in other areas of computer science). At that time I considered him as one of the leading researchers in theoretical computer science – it is why I found his thoughts (written up recently – six years or so after his retirement) a bit surprising!

In his article he addressed the issue of advising PhD students, and many comments he made were applicable to a much wider audience. For example, he discussed a standard way to write and publish a paper: "*Look at the last section [of some paper], where there were always some 'open problems.' Pick one, and work on it, until you are able to make a little progress. Then write a paper of your own about your progress, and don't forget to include an 'open problems' section, where you put in everything you were unable to do.*" Indeed, it is hard to disagree – and this is what we often experience reading papers written by researchers from the Evolutionary Computation community… For example, one researcher proposes a new method for something and the method includes a few parameters – the follow-up researcher tunes these parameters and the next researcher proposes adaptive method for handling them. The number of examples of such papers (whether experimental or theoretical) is extensive.

However, Jeff Ullman does not believe that such approach is a good one: "*Unfortunately this approach, still widely practiced today, encourages mediocrity. It gives the illusion that research is about making small increments to someone else's work. But worse, it almost guarantees that after a while, the work is driven by what **can** be solved, rather than what **needs** to be solved. People write papers, and the papers get accepted because they are reviewed by the people who wrote the papers being improved incrementally, but the influence beyond the world of paper-writing is minimal.*"

This is probably a fair introduction and a summary of this chapter – I will argue that the gap between the theory and practice of Evolutionary Algorithms (EA) is getting wider, partially because of reasons identified by Jeff Ullman, and partially because complexity of business problems increased many times over the last 15 years (mainly due to the globalization and integration processes)[1], whereas the theory still studies the same types of problems (TSP, JSSP, Knapsack Problem, sphere

---

[1] Scott Wooldridge, Vice President Industry Business at Schneider Electric Australia said (private communication): "The world is becoming more dynamic and more interconnected each year. Managers and executives are finding that changes in supply and demand can be sudden and unexpected. The complexity of doing business is increasing at a rapid rate and it is becoming hard to anticipate events, interpret their potential impact and understand the implication of various decisions at both an operational and strategic level."

function, etc.). Indeed, the influence of many research papers in the EC community, beyond the world of paper-writing, is minimal. Of course, the problem is much broader – note that publication quantity is often used as a proxy of academic success and influences funding decisions, however the research questions that need answering don't always lend themselves to rapid incremental publications. Moreover, there is no substantial reward for research that demonstrates linkage with the real world, thus no real incentive for academia to solve the problems that industry is really interested in.

His further thought was the following: "*In the first years of computer science as an academic discipline, many theses were 'theoretical,' in the sense that the contribution was mostly pencil-and-paper: theorems, algorithms and the like, rather than software. While much of this work was vulnerable to the problem just described – paper building on paper – it is quite possible for a theoretical thesis to offer a real contribution.*" Then he discussed as a case from many years ago where a "theoretical" piece of work was not based on what some paper left open, but rather on an expressed need of the real-world – and it was a great success in the real-world environment. Another example given was that of Sergey Brin and Larry Page, who saw the need for a better search engine and the key ways that goal could be reached. Clearly, "*...there needs to be an exposure to problems that are at the frontier, and that are needed by a 'customer.' Sometimes, they can find a customer in industry [...]. Summer internships can be a great opportunity. However, advisors should encourage students to intern at a strong industrial research group, one where the goals are more than minor tweaks to what exists. Whether the thesis is theoretical or an implemented solution, students need to be guided to understand who will consume their contribution if they are successful. And the answer cannot be 'people will read the paper I will write, and they will use the open problems in it to help form their own theses.' Especially when dealing with theoretical theses, the chain of consumption may be long, with idea feeding idea, until the payload is delivered. Yet if we let students ignore the question of whether such a chain and payload plausibly exist, we are doing them a disservice.*"

The EC community should be immune from the problem of identifying possible payloads as evolutionary algorithms are directly applicable to a variety of real-world problems. Indeed, many EA papers indicate a strong connection between the presented approach and real-world applicability. But there is a huge difference between a "strong connection" and an "implementation" (or practice) – and it would be necessary for the EC community to make this transition sooner than later. Otherwise, EAs would be perceived as one of many methods one can just try in some special circumstances – whereas these algorithms have a potential to deliver a significant return on investment for many hard real-world problems (see further sections of this chapter).

To progress our discussion on the gap between theory and practice of evolutionary algorithms, first we have to address two fundamental questions: (1) what is an evolutionary algorithm? and (2) what is practice (i.e. a real-world application)? These two questions are important because without some discussion of them it would be difficult to discuss the current gap[2] between their theory and practice.

---

[2] A continuing gap between theory and practice can be viewed as a positive feature of any dynamic, exciting, growing field, however, it seems that in the case of Evolutionary Computation this gap is too large...

For example (this emerged during my recent private correspondence with Hans-Georg Beyer, the current editor-in-chief of *Evolutionary Computation Journal*) some researchers believe that for operational-type problems (i.e. problems that require optimisation with some regular frequency) "*...you may do better by design a problem-specific optimisation algorithm. And this will be quite often not an EA due to different reasons.*" However, the question is how to distinguish between "a problem-specific optimisation algorithm" and "an EA with problem-specific variation operators and some other non-standard features"?

So these two topics are discussed in the following two sections of this chapter (sections 2 and 3), whereas section 4 talks about the gap between theory and practice. Section 5 concludes the chapter.

2   What Is an Evolutionary Algorithm?

The field of meta-heuristics has a rich history. Many meta-heuristics have emerged during the past 30 years; many of them have been inspired by some aspects of nature, ranging from the cooling of metal to the movements of ants. Meta-heuristics methods include a variety of hill climbing techniques (deterministic and stochastic), ant colonies, artificial immune systems, differential evolution, particle swarms, simulated annealing, tabu search, cultural algorithms, evolutionary and co-evolutionary algorithms. These meta-heuristics can be classified into some categories based on different criteria. For example, some meta-heuristics process single solution (e.g. simulated annealing) whereas some others process a set of solutions (and are called population-based methods, e.g. evolutionary algorithms). Some meta-heuristics can be deterministic (e.g. tabu search), some other are stochastic (e.g. simulated annealing). Some meta-heuristic generate complete solutions by *modifying* complete solutions (e.g. evolutionary algorithms), whereas some other *construct* new solutions at every iteration (e.g. ant systems). Many of these meta-heuristics offer some unique features (e.g. use of memory, use of 'temperature', use of methods for exchange information between individuals in population-based methods). Further, even within a single meta-heuristic, there are many variants which incorporate different representations of solutions and different operators for generating new solutions.

However, there is one common denominator: all meta-heuristics strive to create high quality solutions by making a series of improvements during their iterative process. Whether they start with (randomly generated) low quality solutions or they use smart initialisation methods to take advantage of the problem-specific knowledge, they aim to improve solution quality during the search process. At any iteration a meta-heuristic method must make some 'selection' decisions: which solutions of the current iteration should be kept for further processing and which should be discarded? And this selection step is quite important as it is often responsible for maintaining balance between exploration and exploitation of the search space – e.g. strong selective pressure harms exploratory capabilities of the algorithm and often results in a premature convergence. Of course, different meta-heuristics address this question in their own way. For example, evolutionary algorithms usually split selection process into two independent activities: (1) selection of parents and (2) selection of survivors.

So, what is an evolutionary algorithm and how does it differ from other meta-heuristics? Well, this is actually a very good question! Of course, one can explain evolutionary algorithms as the ones based on concepts of natural evolution. It is also possible to discuss 'mutation' and 'crossover' operators, parents and offspring, not to mention the Darwinian principle of natural selection and survival of the fittest. However, from a high-level perspective things are not so clear. Say, you are solving a new challenging optimization problem and you have designed a new method. So you have selected some appropriate representation for candidate solutions for the problem and agreed on the evaluation function which would measure the quality of solutions. Further, you came with (more or less clever) heuristic method for finding a few initial solutions, and (after analysing the characteristics of the problem) you have also designed a few variation operators which would be responsible for modifying the current solutions (thus creating new set of candidate solutions). You have also incorporated some simple selection method (say, ranking method, where the better individual has better chances to be selected) at any iteration. You have extended the system by some additional non-standard features (e.g. special repair method or decoder to deal with problem-specific constraints). Finally, you experimented with the system and as the result you have tuned a few parameters of the method (e.g. population size, rates of operators).

Have you just created an evolutionary algorithm? Note that if your answer is 'yes', meaning "yes, I have created a variant of an evolutionary algorithm", the consequence might be that any meta-heuristic method which searches for a solution in iterative manner can be labelled as a variant of 'evolutionary algorithm'. For example, simulated annealing can be considered as a variant of $(1 + 1)$ evolutionary algorithm with an adaptive selection method; tabu search can be considered as $(1 + 1)$ evolutionary algorithm with memory-based selection method. In general, there are many iterative stochastic search methods – are all of these 'evolutionary algorithms'? Recall also, that some 'evolutionary algorithms' have been 'extended' by memory structures (e.g. when they operate in dynamic environments or to keep the search history – Chow & Yuen, 2011) or by a parameter called 'temperature' (to control mutation rates). And it is possible to provide many other examples!

Consider, for example, the Newton method – the method for finding successively better approximations to the roots of a real-valued function. For one-dimensional functions, the Newton method generates an initial individual x_0 and generate the next point x_1 (provided the function is reasonably well-behaved):

$$x_1 = x_0 - f(x_0)/f'(x_0)$$

where $f'(x)$ is derivative of $f(x)$. Geometrically, x_1 is the intersection with the x-axis of a line tangential to f at $f(x_0)$. The process is repeated:

$$x_{n+1} = x_n - f(x_n)/f'(x_n)$$

until a sufficiently accurate (i.e. near-optimum) value is reached. So the method works as follows: one starts with an initial guess which is reasonably close to the true root (i.e. intelligent initialisation), then a variation operator (problem-specific mutation) is applied to generate offspring. The selection method always selects offspring for further processing. These steps are repeated for several iterations (we should say: generations), until a termination condition is met (the error drops below a predefined threshold).

Did Newton discover an 'evolutionary algorithm'? Before you answer, recall that $(1 + 1)$ evolutionary strategy does not require population of solutions as it is processing just one individual (which is compared with its only offspring). Recall, that many 'evolutionary algorithms' assume a deterministic selection method; many 'evolutionary algorithms' take advantage of smart initialisation and problem-specific operators. Yes, I realise that stochastic component is missing, but it would be relatively easy to modify the method from deterministic one to stochastic. For example, we can easily modify the formula for generating the 'offspring':

$$x_{n+1} = x_n - f(x_n)/f'(x_n) + N(0,\delta)$$

Note also that the argument that evolutionary algorithms in numerical domains are derivative-free (hence applicable to discontinuous functions) will not hold here as there have been many evolutionary algorithms proposed with problem-specific variation operators, which take advantage from the knowledge of the shape of the landscape. Not to mention that – as for evolutionary algorithms – we can test the modified Newton method on different landscapes, investigate its convergence rates, investigate its scalability for higher dimensions, investigate the issue of deceptiveness, etc.

Newton's method emerged in 17th century, however, his method was probably derived from a similar but less precise method by Vieta. And the essence of Vieta's method can be found in the work of the Persian mathematician, Sharaf al-Din al-Tusi. So we can rewrite the history of evolutionary algorithms moving their roots centuries earlier as there are many examples of search methods based on (smart) initialisation, (problem-specific) mutation, selection, iterations (generations), termination conditions... It seems that the era of evolutionary algorithms[3] started with just a new terminology – all pioneers of EAs have been using 'right' vocabulary – generations (instead of iterations), mutation and/or crossover (instead of just 'variation operator'), the Darwinian selection – survival of the fittest (instead of just selection), etc. So what does it take to call an iterative search algorithm – an 'evolutionary algorithm'? Terminology used? A stochastic component? Or something else?

These remarks are also applicable to the whole field of modern heuristic methods. Over the last few years we saw emergence of a variety of new methods; these include bee colony optimisation, honey-bee mating optimisation, glow-worm swarm

---

[3] I do believe that the *modern* era of evolutionary algorithms started at the 4th International Conference on Genetic Algorithms (San Diego, 1991) when Evolution Strategies and Genetic Programming approaches were presented to the participants of the conference, together with other approaches based on different data structures (like the original Evolutionary Programming approach).

optimisation, intelligent water drops, firefly algorithm, the monkey search, cuckoo search, galaxy-based search algorithms, spiral optimisation... Which of these are evolutionary algorithms? For example, in cuckoo search each egg in a nest represents a solution, and a cuckoo egg represents a new solution. The aim is to use the new and potentially better solutions (cuckoos) to replace a not-so-good solution in the nests (in the simplest form, each nest has one egg only and the algorithm can be extended to more complicated cases in which each nest has multiple eggs representing a set of solutions). Cuckoo search is based on three idealized rules: (1) each cuckoo lays one egg at a time, and dumps its egg in a randomly chosen nest; (2) the best nests with high quality of eggs will carry over to the next generation; and (3) the number of available hosts nests is fixed, and the egg laid by a cuckoo is discovered by the host bird with some probability. Clearly, it would be easy to rewrite all these in 'evolutionary' terminology (e.g. including the island population model)...

This very issue was recently addressed by Fred Glover[4] in his article *Sex and Metaheuristic Metaphors,* where a 'new' meta-heuristic method, the *Courtship Algorithm,* is proposed. The algorithm is loaded with some principles that people have applied to courtship – and these principles translate directly into rules for meta-heuristic methods. These include: (a) *Have we met before?* (i.e. use memory to exploit recurrences), (b) *Your place or mine?* (i.e. consider the benefits of regional exploration), (c) *Variety is spice* (i.e. employ diversification at all levels), (d) *Once is not enough* (i.e. iterate over good options), (e) *No need to be timid* (i.e. well-timed aggressive moves can pay off), or (f) *Don't stop now!* (i.e. take advantage of momentum). Of course, they are many more of such useful principles one can use – reading this article I recalled that in 1999 I published a paper (Hinterding et al. 1999) titled *Your Brains and My Beauty: Parent Matching for Constrained Optimisation* – very much in line with the Courtship Algorithm! So, is Evolutionary Algorithm a special case of Courtship Algorithm or vice versa?

It seems that many researchers are just playing games with terminology. In particular, it is always possible to extend the definition of something to include everything else, and people often do this, although it obviously doesn't establish any real generality to the "extended definition". Moreover, the architecture of so-called "special cases" may be more intricate and advanced than the architecture that has been constructed for the "general case" (as where a general-case design of a building may have nothing of the complexity and intricacy that must go into the design of a skyscraper). A similar sort of thing also occurs in speaking of probabilistic and deterministic methods. We can say that deterministic methods are a special case of probabilistic methods, because deterministic methods result by making all probabilities 1 and 0. But this blurs a great deal of the special essence of each category of method.

It seems that Evolutionary Algorithms, in the broad sense of this term, provide just a general framework on how to approach complex problems. All their components, from the initialisation, through variation operators and selection methods, to constraint-handling methods, might be problem-specific. Some researchers (but not that many, I believe) in the Evolutionary Computation community agree – and view an

---

[4] See http://optimaldecisionanalytics.typepad.com/

EA as a meta-heuristic in the broad sense of the term by providing a framework for creating/instantiating problem specific heuristic methods [De Jong, 2002]. Evolutionary Algorithms, to be successfully applied, must be tailored to a specific domain in order to be useful and provide decent results in specified time; without any doubt, incorporating domain knowledge leads to more effective EA searches. Probably this is why so many authors talk about 'variants', 'modified versions', 'extensions', or just 'knowledge-enhanced' evolutionary algorithms in their application-oriented papers – for which theoretical results usually are not relevant.

3   What Is 'Practice'? What Is a Real-World Application?

Clearly, the 'practice' of evolutionary algorithms is connected with real-world applications – after all, this is what the term 'practice' means (at least in the context of evolutionary algorithms). But as it was the case with the definition of evolutionary algorithms, the meaning of the term 'real-world application' is not that clear at all!

The Evolutionary Computation community over the last 30 years has been making claims that their methods are ideal for hard problems – problems, where other methods usually fail. As most real-world problems are very hard and complex, with nonlinearities and discontinuities, complex constraints and business rules, noise and uncertainty, EAs should provide a great benefit to the real-world community.

Even today, many research papers point to ability of evolutionary algorithms to solve real-world problems. For example, I reviewed the last two issues (Spring and Summer 2011) of the *Evolutionary Computation Journal* and the last two issues (October and December 2011) of the *IEEE Transactions on Evolutionary Computation*. In the introductory paragraphs of some papers included in these issues I found the following sentences:

- *"Evolutionary algorithms are a wide class of solution methods that have been successfully applied to many optimisation problems,"*
- *"Despite the lack of theoretical foundation, simplicity of Evolutionary Algorithms has attracted many researchers and practitioners,"*
- *"EAs are randomized search heuristics that solve problems successfully in many cases,"*
- *"...but in practice they optimise many challenging problems effectively,"*
- *"Randomized search algorithms have been very successful is solving combinatorial optimisation problems,"*
- *"For the past three decades, many population based search techniques have surfaced to become a mainstay of optimisation,"*
- *"However, as many real-world optimisation problems are black-box problems of which a priori problem knowledge is not available, the use of meta-heuristics started to prevail,"*
- *"EAs are used for solving many various problems,"* or
- *"Recently, researchers are increasingly directing their focus on solving multi-objective optimisation problems (MOPs), since the solution of these problems is of great importance in the areas of economics, engineering, biology, medicine, materials, and so on."*

Some papers focus on classic problems (e.g. graph partitioning, maximum clique) – and in such cases the authors make a connection with real world:

- *"Graph partitioning is one of the fundamental combinatorial optimisation problems which is notable for its applicability to a wide range of domains, such as very large scale integration (VLSI) design, data mining, image segmentation, and so on,"* and

- *"The maximum clique (MC) problem in graphs is a paradigmatic combinatorial optimisation problem with many relevant applications, including information retrieval, computer vision, and social network analysis."*

The influence of benchmark problems on the EA research manifests itself in statements like *"When genetic algorithms (GAs) are applied to combinatorial problems, permutation representation is usually adopted"* which is very true for many benchmark problems and very false for real-world problems.

It is my opinion that these sentences speak loud and clear about desires of many researchers to see a stronger connection between their research and the real-world (or their belief that this is the case!). However, these statements are harmful for the community as they give a misleading perspective on the current spread of evolutionary methods in businesses and industries and provide unjustified psychological comfort. Where are all these applications that the authors refer to? Where do you see the prevalence of EAs in business and industry? When did these techniques become the 'mainstay of optimisation'? Where are all these success stories? What does it mean that EAs have been successfully applied to many optimisation problems? In my recent essay (Michalewicz, 2012) I addressed this very issue and concluded that (for many reasons) there are not that many real world implementations of EAs after all…

Further, it seems that many researchers are interpreting the phrase "real-world application" in a quite arbitrary way. I have looked through proceedings of many international conferences on evolutionary algorithms (e.g. GECCO, IEEE CEC, PPSN) which often include special sessions on "real-world applications". I have checked many journals (e.g. *Journal of Evolutionary Algorithms, IEEE Transactions on Evolutionary Computation*) with special attention to their special issues on real-world applications (e.g. scheduling) and a few edited volumes which aimed at real-world applications. The main problem is that all these "real-world applications" look like "Guinea pigs" – not really from Guinea, and not really pigs… It seems that researchers are interpreting the phrase ""real-world applications" in very arbitrary way. However, it is relatively easy to group all these "real-world applications" into four broad categories:

1. Applications used in some business/industry on daily (regular) basis.
2. Applications tested on real data (e.g. taken from a hospital, city council, a particular business unit).
3. Applications tested on some well-known model (e.g. TSP, VRP, JSSP) of a real-world problem.
4. Other applications (e.g. numerical optimisation, constraint-handling, multi-objective optimisation).

Of course, this classification is not that precise – one can imagine a border case between categories 1 and 2, where an application might be used intermittently, e.g. in some relatively infrequent planning or modelling task. However, the question is: Are the applications in all these four categories real-world applications? Clearly, category 1 is beyond any dispute. But the case of category 2 is not that clear – it is true that real-world data were used in experimentation; however, application as such is not real-world application – as no one uses it. It is rather an indication, that such approach – in the real-world setting – may have a merit. Category 3 is also muddy – there is no question that some models, e.g. for vehicle routing, scheduling, distribution, are models of real-world environments, however, these applications do not operate in real-world environment (and most likely, they never will). And most other novel applications (category 4), whether they aim at handling constraints, many objectives, or the issue of noise – while very important and directly applicable for real-world settings, are hardly real-world applications.

Clearly, there are also EA-based "solvers" and a variety of EA-based "tools" available; however, there is a huge difference between these and real-world software applications. Solvers and tools are sometimes useful, but they have very limited applicability and a limited audience. Some individuals in some companies use solvers which are based on modern heuristic methods (and Evolutionary Algorithms in particular) for solving some small-scale optimisation problems; however, due to the limited usefulness of such tools I heard quite often negative statements from various researchers employed in major companies, e.g. *"I tried a Genetic Algorithm on my problem, but I did not get any results...They do not work"*.

Some senior researchers agree. Garry Greenwood, the current editor-in-chief of the *IEEE Transactions on Evolutionary Computation*, wrote recently:[5] *"I believe the MATLAB GA toolbox has been devastating to the EC field and has done far, far more damage than good. I wish it had never been created. It was designed to do searches over an extremely broad class of optimisation problems. No domain knowledge could be incorporated so the search results on non-toy problems were often mediocre. More than one person has told me he tried a GA using that MATLAB toolbox and wasn't impressed with the results. Hence, he was convinced GAs weren't very good. The EC community has taken a bad rap for years because people formed a negative opinion of GAs based on one encounter with that damn MATLAB toolbox."*

There are, of course, papers describing some interesting applications (e.g. applications of Covariance Matrix Adaptation Evolution Strategies – CMA-ES[6]), however, many of these are not real-world applications, they are usually restricted to continuous variables only, and they hardly address complex business problems. There are also many papers which aim at classic Operation Research (OR) problems, e.g. traveling salesman problems, job shop scheduling problems, graph colouring problems, variety of vehicle routing problems, knapsack problems, packing and cutting problems – but all these have very limited significance for today's real-world problems.

---

[5] Private correspondence.

[6] See http://www.lri.fr/~hansen/cmaapplications.pdf

Let's illustrate the point of expectations of the real-world community in the context of supply chain – understood as "…a combination of processes, functions, activities, relationships and pathways along which products, services, information and financial transactions move in and between enterprises, in both directions (Gattorna, 2010). One can visualise a supply chain as a combination of multiple logistic networks which involve many upstream and downstream organisations – and clearly, the potential of significant improvement in performance of such network is much greater than within a single silo[7]. Note, for example, that the issue of scheduling production lines (e.g. maximising the efficiency, minimizing the cost) has direct relationships with inventory and stock-safety levels, replenishments strategies, transportation costs, deliver-in-full-on-time, to name a few. Moreover, optimising one silo of the operation may have negative impact on upstream and/or downstream silos[8]. Thus businesses these days need "global solutions" for their whole operation, not silo solutions. This was recognised over 30 years ago by Operations Research community; in 1979 R. Ackoff wrote: *"Problems require holistic treatment. They cannot be treated effectively by decomposing them analytically into separate problems to which optimal solutions are sought."* Only an integrated view of supply chain operations would increase visibility and transparency across end-to-end supply chains. Managers these days are looking for applications that, for example, will enable fully integrated, financially-driven Sales & Operations Planning with, for example, transportation costs, working capital requirements, and stock outs as priorities. They need to create within such applications consistent what-if scenarios to facilitate the planning, scheduling, and optimisation of multi-site sourcing, production, storage, logistics, and distribution activities – all these in time changing (dynamic) environments.

It seems to me that (from a high level perspective) most real-world problems fall into two categories: (1) design/static problems, and (2) operational/dynamic problems. The first category includes a variety of hard problems, like TSP, VRP, JSSP, graph coloring, knapsack problem, and millions of others. Some of them are really hard – and I believe that 99% of research aims at these type of problems. I do believe, however, that the problems from the second category are (a) much harder, (b) the standard methods usually fail for them, and (c) they represent a huge opportunity for EAs. Let me explain further.

The systems to address problems in the 2nd category are really decision-support systems that require continuous flow of data, predictive components, almost immediate recommendations for recovering from sudden changes, etc. Not to mention that they should handle the 1st category instances as they have to solve static versions of the problem as well. This is altogether a different game, and the differences between problems in the 1st and 2nd categories are huge. Problems in the 2nd category are usually multi-silo problems as opposed to single silo for the 1st category problems. Problems in the 2nd category usually deal with many variables, nonlinear relationships, huge varieties of constraints (e.g. constraints in real-world settings often include

---

[7] We use the term 'silo' for one component of a supply chain.

[8] Of course, there are some real-world scientific and engineering problems where a good solution to a silo problem is beneficial (e.g. a variety of design problems), but they do not represent complexities of today's business problems.

'if-then' conditions), business rules, many (usually conflicting) objectives – and all of these are set in a dynamic and noisy environment. These problems belong to an optimization class, as they require recommendations for "the best" decision at the moment (whichever complex characteristic of what "the best" may mean – often with risk factors included). An additional feature of these systems is that they do not operate on crisp values but rather on probability distributions. For example, one thing is to assume that the travel from one point to another takes 36 hours, another thing is to accept a probability distribution with mean 36 and standard deviation determined from past data. Not to mention that these probability distributions change over time (the system learns) as new data are coming in (e.g. due to some improvements, six months later the mean is 35.5 hours, and standard deviation is much smaller). Further, the response time is expected to be much shorter – applications for the design-type problems are not time-critical, whereas applications for operational-type problems are. Finally, robustness of a solution is as important as its quality... In comparison with such problems, NP-hard problems like TSP seem to be toy problems (which they are, in a sense).

Clearly, these are types of problems which should be addressed by Evolutionary Computation community, as they represent hard real-world problems where other methods usually fail.[9] These are the problems for which meta-heuristic methods in general (and EA-based methods in particular) should be the methods of choice.

4   Theory versus Practice

As indicated in the abstract of this chapter, at the Workshop on Evolutionary Algorithms, organized by the Institute for Mathematics and Its Applications, University of Minnesota, Minneapolis, Minnesota, October 21 – 25, 1996, one of the invited speakers, Dave Davis made a claim that all theoretical results in the area of Evolutionary Algorithms were of no use to a practitioner. At that time there was – in his opinion – a significant gap between theory and practice of Evolutionary Algorithms.

So where are we 15 years later with this issue? Did we manage to close the gap between the theory and practice? It seems that there is still a significant mismatch between the efforts of hundreds of researchers who have been making substantial contributions to the theory of evolutionary computation over the years and the number of real-world applications which are based on concepts of evolutionary algorithms – it seems also, that this gap is still growing.

I believe that there are two main reasons for this phenomenon; these are:

1. The growing complexity of real-world problems
2. The focus of research community on issues which are secondary for real-world applications

---

[9] A recent report (private correspondence) generated by several Operations Research experts on optimisation of a particular supply chain with 37 nodes and 69 arcs, 862,000 variables and 1.6 million constraints, would require 18 hours per objective by mixed integer programming.

The first point was already touched on in the previous section of the chapter. For many reasons today's business problems are of much higher complexity than 30 years ago and the real-world is searching for techniques which would address their problems – problems which are loaded with nonlinearities and/or discontinuities, many (possibly conflicting) objectives, variety of business rules, soft and hard constraints, and noise.

However, it seems to me that many researchers in the EC community are quite disconnected with today's business, and they still believe that[10]: *"In business applications, the goal function is often quite simple, a linear one. Most of the modelling is put into the inequality constraints which very often can be linearized and integer conditions can be relaxed. As a result, one often ends up with an LP[11]. And if not, then the modelling is changed in such a manner that the resulting problem is again an LP. This is the main reason why CPLEX (and some others) is so prevalent."*

I think that a perception that in business applications the goal function is often quite simple, was right 30 years ago, but today it is just wrong and there is nothing further from the truth. The real complexity exists mainly in real-world problems (and not in artificial benchmarks). The business goal functions usually are extremely complex, not to mention constraints and business rules, which are often conditional, calendarised, dynamic, etc. and they interact with many objectives in complex ways. And the main reasons of the popularity of LP methods are the ease of their use and the knowledge, how to use them – which is not the case with EAs. The main reason of their popularity is *not* the quality of results... Many business units dream about replacing their software which incorporates LP as their optimisation engine, but they do not see alternatives. There is also an amazing amount of inertia in the business that has nothing to do with what technology really works – and it is much easier to *plug-in* CPLEX than to *develop* an EA for a problem at hand...

Real-world optimisation problems involve large number of variables, numerous objectives, constraints, and business rules, all contributing in various ways to the quality of solutions. The complexity of such problems makes it virtually impossible for human domain experts to find an optimal solution. Further, manual adjustments of scenarios (what-if scenarios and trade-off analysis), which are needed for strategic planning, become an expensive or unaffordable exercise. The main reason behind this complexity is that large-scale business problems consist of several interconnected components, which makes many standard approaches ineffective. Even if we know exact and efficient algorithms for solving particular components or aspects of an overall problem, these algorithms only yield solutions to sub-problems, and it remains an open question how to integrate these partial solutions to obtain a global optimum for the whole problem. Moreover, optimising one silo of the operation may have negative impact on upstream and/or downstream silos. For example, at PPSN'10 I gave a keynote talk on a powerful EA-based enterprise[12] software application that was

---

[10] This is the exact quote from my private correspondence with one of senior members of the EC community.

[11] Linear Programming.

[12] Enterprise software addresses the needs of organization processes and data flow, often in a large distributed environment (e.g. supply-chain management software).

developed recently[13] to address many of wine production challenges present in different parts of the wine supply chain. This enterprise software application for wine industries consists of a suite of software modules (these include predictive modelling for grape maturity, using weather forecasts and readings on Baum, PH, and TA, vintage planning, crush scheduling, tank farm optimisation, bottling-line sequencing, and demand planning) that can optimise the end-to-end wine supply chain.. When deployed together, these applications can optimise all planning & scheduling activities across a winery's entire supply chain.

Each of these individual software application deals with a silo problem; for example, the bottling module (Mohais et al. 2011) is responsible for generating optimal production schedules for the wineries' bottling operations. Opportunities for optimisation include manipulating the sequencing order, selecting which bottling lines to use, consolidating similar orders within the planning horizon, and suggesting changes to the requested dates to improve the overall schedule. Some of the key objectives are to maximise export and domestic service levels (i.e. DIFOT), maximising production efficiency, and minimising cost. As the module provides decision-support, the user has full control over the optimisation process in that they are able to lock in manual decisions, set the business rules and constraints, re-optimise after making changes, and compare the current solution with an alternative plan. The module also provides a *what-if* tab that can be used to analyse strategic business decisions and events such as capital investment in new equipment, or to look at operational decisions like adding or removing extra shifts, or even for crisis management (what is the impact of a bottling line going down or key staff being ill). Reporting is provided on a number of levels and to suit different stakeholders; for example the daily bottling programs for execution, a report on particular wine blends, or a report on expected production efficiency. Users are also able to generate an alternative to the current production schedule, with the system providing a comparison to help the user evaluate the impact of any changes. The comparison includes performance metrics (KPI's) such as any difference in the number of late orders, and changes to production efficiency and cost (this could include measures such as cost per unit, total production throughput, production line utilisation, etc.). This allows the user to experiment with different schedules before committing to making any changes; for example, trying to incorporate a last minute export order without disrupting existing orders.

Each of these modules (whether predictive modelling for grape, vintage planning, crush scheduling, tank farm optimisation, bottling-line sequencing, and demand planning) represents a significant large-scale optimisation/prediction problem in itself and includes several interesting research aspects (e.g. variable constraints in the bottling module). However, finding the optimal solution in one silo of the operation may have negative impact downstream and/or upstream of the whole operation. Consider, for example, the following example. The bottling plant has to process 700,000 litres of Jacob's Creek[14] to satisfy demand. However, the closest (in terms of volume) tank on

---

[13] The application was developed at SolveIT Software (www.solveitsoftware.com) together with similar applications in other verticals (e.g. for grain handling, pit-to-port mining logistics, and mine-planning activities).

[14] Jacob's Creek is one of the most known, trusted and enjoyed Australian wines around the world.

the tank farm contains 800,000 litres of Jacob's Creek – and what is the optimal decision here? Clearly, from the bottling perspective they should get just 700,000 litres of that wine and process it – but it would be a bad decision from the perspective of another silo: tank farm. They will be left with 100,000 litres leftover with all risks (the quality of these 100,000 litres will go down quickly due to oxygen levels in the tank) and inefficiencies in operations (the tank with the leftover 100,000 litres can't be used in their operation). From tank farm perspective, the 'optimal' decision would be to send 800,000 litres of Jacob's Creek for bottling (no leftovers with clear benefits), but the bottling operation would not be happy with such a solution (note that once a wine is bottled and labelled, the choice for its destination is quite limited as different customers have different requirements for packaging). This is why the global optimum here should consider implications of various decisions across all silos. Further, it is next to impossible to compare the result to some theoretical global optimum – usually the comparisons are made with respect to some metrics from the previous year (i.e. before the system went live). Further, by considering the entire supply chain we can attempt to answer some key (global) questions, as "What is the impact of increasing demand by 10% of Jacob's Creek across all operations"? And the answers for such questions are sought today by businesses and industries.

Darrell Whitley (the former editor-in-chief of Evolutionary Computation Journal who also participated actively in a few real-world projects), wrote:[15] "*Your comments about silos and up-stream and down-stream side effects is spot-on and is something I have worried about for some years now. If you say 'I ran an experiment and I can increase line-loading by 10 percent, what does this do to inventory?' Is that increase in line-loading sustainable or just temporary because if you increase line-loading, you decrease inventory, which might reduce your ability to line-load. When you introduce a change in one silo you move the steady-state in other silos. You almost NEVER see this in academic studies.*"

A decision-support system that optimises multi-silo operational problems is of a great importance for an organisation; it supports what-if analysis for operational and strategic decisions and trade-off analysis to handle multi-objective optimisation problems; it is capable of handling and analysing variances; it is easy to modify – constraints, business rules, and various assumptions can be re-configured by a client. Further, from end-user perspective, such decision-support system must be easy to use, with intuitive interfaces which lead to faster and easier adoption by users with less training.

However, it seems to me that the research done within the EC community diverges further and further from complexities of today's problems mainly because the focus of the research is on issues which are secondary for real-world applications. This is partially due to the reasons identified by Jeff Ullman and discussed in the *Inspiration* section of this article – most researchers follow up of some previous work of other researchers, making progress on some 'open issues', and, in the words of Jeff Ullman: "*it almost guarantees that after a while, the work is driven by what **can** be solved, rather than what **needs** to be solved.*" Thus the influence of many research papers in the EC community, beyond the world of paper-writing, is minimal.

---

[15] Private correspondence.

In my recent correspondence with Hans-Georg Beyer he identified: "... *3 fields where theory should focus on or contribute to in the future: (1) mathematical characterization of evolutionary dynamics (in a sense of a predictive theory); (2) population sizing rules for highly multimodal optimization problems, and (3) development of stopping rules based on evolutionary dynamics.*" However, it is not clear to me how even breakthrough results in these areas would help practitioners for approaching complex real-world problems of growing complexity, as described earlier?

Consider, for example, the number of published papers on EAs in dynamic environments. Most researchers focused at recovery rates in cases there was a change in the shape of a landscape. I think it is of no significance, as in real-world problems (1) the objective is usually fixed (e.g. say, you minimize the cost of some operation and you do not change this objective), (2) constraints are changing (e.g. failure of a truck), and (3) we deal with partially-executed solutions. Thus the concept of recovery is very different. There is an enormous gap between theoretical models and practice. The same is in many other research areas. For example, I remember that I was amazed when (over 20 years ago) I discovered the first textbook on genetic algorithms (Goldberg, 1989). The book was around 400 pages, and it included one paragraph on how to deal with constraints (!). The same is true today – many theoretical results are achieved in constraints-free environments and their applicability to real world situations is quite limited.

Further, as discussed in Section 2, it is unclear what is (and what is not) an Evolutionary Algorithm. On one hand, EA practitioners usually employ hybrid forms of evolutionary algorithms (e.g. extending the system by problem-specific initialisation, problem-specific operators) and a successful application of EA to a complex business problem requires a significant dose of 'art'; on the other hand most of theoretical research concentrates on classic versions of EAs and toy problems. There are many research papers published in the EC community on convergence properties of evolutionary algorithms, diversity, exploration, exploitation, constraint-handling, multi-objective optimisation, parallel EAs, handling noise and robustness, ruggedness of the landscape, deceptiveness, epistasis, pleiotropy – to name just a few areas of research. In most cases some standard versions of EA are studied: binary coded GA with a tournament selection or (1+1) ES. Whatever the results, their applicability to solving complex real-world problem are questionable.

A large portion of the research on Evolutionary Algorithms is experimental – hence the researchers use a variety of benchmark functions and test cases. However, these experiments are usually conducted on simple silo problems. The researchers use some classic sets of functions (from $f_1, ..., f_5$ proposed by Ken De Jong [De Jong,1975] to f_n today, where n approaches 100) for numerical optimisation and classic benchmarks (e.g. on graph colouring, traveling salesman, vehicle routing, job shop scheduling) for combinatorial optimisation. However, these small-scale silo problems are far cry from complexity of real-world problems – consequently these theoretical and experimental results are of little help (if any) to any practitioner who works on EA-based enterprise software applications. There are hundreds (if not thousands) of research papers addressing traveling salesman problems, job shop scheduling problems, transportation problems, inventory problems, stock cutting problems, packing problems, etc. While most of these problems are NP-hard and clearly deserve research efforts, it is not exactly what real-world community needs. Most companies run complex operations and they need solutions for complex multi-silo problems with all their complexities (e.g. many objectives, noise, constraints). In the same time Evolutionary Computation offers various techniques to experiment with, e.g., cooperative coevolution [Potter and De Jong, 1994], where several EAs, each corresponding to a single silo, are run in parallel. Communication between silos may occur during evaluation of individual solutions. Solutions from one silo might be evaluated based on their performance when combined with representative solutions from the other silos.

However, there are very few research efforts which aim in the direction of optimising interdependent multi-silo operational problems with many conflicting objectives, complex constraints and business rules, variability issues and noise. This might be due to the lack of benchmarks or test cases available. It is also much harder to work with a company on such global level as a delivery of successful software solution usually involves many other (apart from optimisation) skills, from understanding the company's internal processes to complex software engineering issues. And it is much harder to report the results, as they may involve revealing company's confidential data. It is also much harder to run significant number of experiments to satisfy requirements of many journals.

Further, in almost in all cases a new method (whether new representation, new set of operators, new selection method, a novel way to incorporate problem-specific knowledge into the algorithm, a novel way to adapt parameters in the algorithm, and so on) is tested against such accepted benchmarks – and the most popular quality measure of a new method is the closeness of the generated solution to the known, global optima in a number of function evaluations. Of course, it is helpful that some silo problems are used for testing, as for many of these the global optima are known (or it is possible to estimate their values). However, for many real-world applications getting to the global optima is secondary. First, the concept of global optima in business is different to that in academia – a global optimum for a silo is referred to as a local optimum solution, as it does not take into account other interacting silos of the business. And the global optimum solution refers to whole multi-silo operation of the organisation. Second, for large-scale (e.g. multi-silo) problems, it would take days (if not more) to generate a global optimum solution, while decision-makers have minutes to react. Third, the multi-silo environment is highly variable (delays, unexpected orders, failures, etc.) and a robust, quality solutions are of higher importance, as the current solution would be modified, anyway, due to changes in the environment. Fourth, due to many, possibly conflicting, objectives, business rules, and soft constraints, the meaning of the term "global optimum" is not that clear – even experienced decision makers often have difficulties in pointing to a better solution out of two available solutions. Finally, the name of the game in industry is not to find an elusive global optimum, but rather to match (and hopefully improve) the results of the human team of experts[16], who have been involved in particular decision-making

---

[16] There is a nice, one sentence summary of how evolution works: "You don't have to outrun the bear, but you just have to outrun the other hunter".

activity for a significant time – however, technical journals reject such comparisons. Not to mention that it would be extremely hard to document such comparisons in a systematic way without revealing sensitive data of an organisation.

Each EA possesses a number of algorithm-specific parameters. Clearly, theory should provide guidelines how to choose those parameters (this is what Dave Davis referred to in his talk from 15 years ago). But theory can only consider simple toy problems. Many researchers would like to believe that if these toy problems cover certain aspects of real-world problem, the results of the theory can be used as a first guideline to choose these parameters. But it seems to me that this is just a wishful thinking – there is no comparison in terms of complexity between real-world problems are and toy problems – and I cannot see any justified transition of results. Further, most research papers focus on one selected aspect of a problem, whether this is constraint-handling method, handling many objectives, dealing with noise and/or uncertain information. In business problems, however, all these aspects are usually present in every problem – and there is hardly a paper which addresses problems of such complexity. Further, real-world applications usually require hybrid approaches – where an 'evolutionary algorithm' is loaded with non-standard features (e.g. decoders, problem-specific variation operators, memory) – but the current theory of evolutionary algorithms does not support such hybrid approaches very well.

5   Conclusions and Recommendations

Let us conclude by returning to the questions from the second paragraph of the abstract: What are the practical contributions coming from the theory of Evolutionary Algorithms? Did we manage to close the gap between the theory and practice? How Evolutionary Algorithms do compare with Operation Research methods in real-world applications? Why do so few papers on Evolutionary Algorithms describe the real-world applications? For what type of problems Evolutionary Algorithm is "the best" method? Let's address these questions in turn.

5.1   What Are the Practical Contributions Coming from the Theory of Evolutionary Algorithms?

It seems that the practical contributions coming from the theory of Evolutionary Algorithms are minimal at the best. When a practitioner faces complex business problem (problem which involved many silos of the operation, significant number of business rules and constraints, many (possibly conflicting) objectives, uncertainties and noise – all of these combined together, there are very few hints available which might be used in the algorithms being developed. None of the results on convergence properties, diversity, exploration, exploitation, ruggedness of the landscape, deceptiveness, epistasis, pleiotropy that I know of would help me directly in developing EA-based enterprise software application. This is partially due to the lack of benchmarks or test cases of appropriate complexity and partially that the EC technical journals are not appropriate for publicizing business successes (see later questions/answers).

5.2   Did We Manage to Close the Gap between the Theory and Practice?

No, we did not – and as a matter of fact, the gap is growing. As discussed in this article, there are two main reasons for this phenomenon (1) Growing complexity of real-world problems and (2) focus of research community on issues which are secondary for real-world applications.

5.3   How Do Evolutionary Algorithms Compare with Operation Research Methods in Real-World Applications?

It seems to me that (after talking to many practitioners at many major corporations) they compare quite poorly... The main reason (I think) is due to the wide spread of standard Operation Research methods which dominated optimisation aspects of business operations for more than 30 years. Further, the Operation Research community has a few standard and powerful tools (e.g. integer programming methods) which are widely in use in many organisations. On the other hand, this is not the case for the EC community – there are no 'plug-in' software tools appropriate to deal with thousands of variables and hundreds of constraints, there are no tools available of comparable power to integer programming. Further, many Operation Research methods are exact – they guarantee the optimum solution, which is not the case with heuristic methods in general and Evolutionary Algorithms in particular.

However, there is one catch here, as every time we solve a problem we must realize that we are in reality only finding the solution to a *model* of the problem. All models are a simplification of the real world – otherwise they would be as complex and unwieldy as the natural setting itself. Thus the process of problem solving consists of two separate general steps: (1) creating a model of the problem, and (2) using that model to generate a solution:

$$\text{Problem} \rightarrow \text{Model} \rightarrow \text{Solution}$$

Note that the "solution" is only a solution in terms of the model. If our model has a high degree of fidelity, we can have more confidence that our solution will be meaningful. In contrast, if the model has too many unfulfilled assumptions and rough approximations, the solution may be meaningless, or worse.

So in solving real-world problem there are at least two ways to proceed:

1.   We can try to simplify the model so that traditional OR-based methods might return better answers.
2.   We can keep the model with all its complexities, and use non-traditional approaches, to find a near-optimum solution.

In either case it's difficult to obtain a precise solution to a problem because we either have to approximate a model or approximate the solution. In other words, neither exact methods nor heuristic methods return optimum solution to *the problem*, as the former methods simplify the problem (by building simplified, usually linear, model of the problem) so the optimum solution to the simplified model does not correspond to the optimum solution of the problem, and the latter methods return near-optimum solutions (but to the more precise models). Further, as discussed in section 4, for many real-world applications the issue of getting to the global optima is secondary as robust, quality (i.e. near-optimum) solutions are of higher importance. And the more complexity in the problem (e.g., size of the search space, conflicting objectives, noise, constraints), the more appropriate it is to use a heuristic method. A large volume of experimental evidence shows that this latter approach can often be used to practical advantage.

5.4 Why Do So Few Papers on Evolutionary Algorithms Describe Real-World Applications?

The journal editors and reviewers of submitted papers are well versed in the standard criteria for acceptance: (a) clearly revealed algorithms, so the reader can at least attempt to replicate you approach, (b) well characterized problems, so the reader can tell if his problem is the right type, (c) rigorous comparison with known results, so the reader can have confidence your results are significant. All this is needed for verifiability – the soul of science. On the other hand, a successful application of EA to complex business problem requires a significant dose of 'art' – and technical journals and (to some lesser extend) conferences generally have difficulties with that. For additional thoughts on this very topic, see (Michalewicz, 2012).

5.5 For What Type of Problems Evolutionary Algorithm Is "The Best" Method?

A submission of a survey article on Evolutionary Algorithms (written with Marc Schoenauer) – for Wiley Encyclopedia of Operations Research and Management Science, 2010 edition – generated editor's comment/request: *"Can the authors provide objective guidance on the types of problems for which evolutionary methods are more appropriate than standard methods? I know a lot of 'hearsay' related to this issue but I am wondering if there is more objective evidence. Is there any solid evidence of EA superiority in a class of problems?"* More and more people question the usefulness and applicability of Evolutionary Computation methods and it is essential that our community would get ready to answer such questions.

And I think that the right answers for above questions are not of the type: "EA techniques are superior for, say, symmetrical TSP" or "EA techniques are superior for, say, such-and-such types of scheduling problems," as the main issue is just in *size* and *complexity* of the problems – for example, multi-silo operational problems with many conflicting objectives, tens of thousands of variables, complex constraints and business rules, variability issues and noise, interdependencies between operational silos – problems, for which standard Operations Research methods are not appropriate. For such problems the business starts looking for optimization support because rational decision making and logical decomposition of the problem are no longer possible. This is a big chance for Evolutionary Computation community, and the time is right to move that direction. However, this would require some fundamental changes in a way the EC community looks at real-world problems…

Many traditional decision-support applications at large corporations worldwide have failed, not realising the promised business value, mainly because small improvements and upgrades of systems created in the 1990s do not suffice any longer for solving 21st century companies' problems. Also, the existing applications often are not flexible enough to cope with exceptions, i.e. it is very difficult, if not impossible, to include problem-specific features – and most businesses have unique features which need to be included in the underlying model, and are not adequately captured by off-the-shelf standard applications. Thus, the results are often not realistic. A new approach is necessary which seamlessly integrates local models and optimisation algorithms for different components of complex business problems with global models and optimisation algorithms for the overall problem. Such decision-support systems should allow also manual adjustments by domain experts, to achieve optimal decisions with the flexibility to be adapted to business rules and unforeseen circumstances. And I believe that this new generation of decision-support systems will be based on Evolutionary Algorithms (understood in a broad sense of this term).

So let's move to the last part of this article and discuss what should be done to remedy the current state of the art with respect to Evolutionary Algorithms? Before we start this discussion, let's go back 20 years...

Most of the readers might be aware that the field of evolvable hardware (EHW) is about 20 years old. Some of the first really serious work was done by John Koza – in fact, EHW was one of the applications Koza used to show the power of Genetic Programming (GP). He used GP to evolve a series of analog filters (both circuit configuration and passive component values). He was able to evolve a series of filters that were essentially identical to filters that were patented in the 1930s for the telephone industry. Based on those results he coined the phrase "human competitive designs" and claimed this was something GP can do. Ten years ago there was the annual (small, but enthusiastic) NASA/DOD sponsored EHW conference. However, that conference is no longer held and a majority of the main players have moved on to other areas. Why? Well, the human competitive designs were just that and little else. GP could duplicate what people had done before, but really didn't (couldn't?) evolve much new and innovative. Consequently people figured all of the low hanging fruit had been picked and EHW had little new to offer. So they moved on to new areas...

This story should serve as a warning to the whole EC community – I believe that the future of EAs would be determined by the applicability of EAs. The excitement connected with terms like 'genetic', 'evolution', 'emergence', if not supported by practice, would wear off (this is, to some extent, already happening in industry). And as long as EAs would produce just 'interesting', 'promising', or 'comparable to an OR method' results on some benchmark problems, and the theory would focus on standard properties of EAs, it would be hard to compete in the real-world environment.

Some theoreticians do not worry: one of them wrote to me: *"Good problem solvers/theoreticians can also work in other fields. I do not see this as a severe problem."* I found such comments quite depressing, as they remind me of a sad story where a group of scientists (astronomers) were studying climate patterns of a planet X which was located millions of light-years from the Earth. One day the planet exploded, and these scientists moved to study another planet, Y...

So, what should we do? What can be done? Well, apart from suggesting that the IEEE should give a post mortem Evolutionary Computation Pioneer Award to Isaac Newton, it seems there are a few things that are worth considering; these include:

1. revisiting the policies of some EC journals – and introduce a different set of criteria to evaluate application papers. After all, the whole purpose of the journal is to disseminate knowledge. It assumes that only the top researchers, the leading edge theory types, are the only ones who might read the journals. Garry Greenwood summarized it nicely: *"The notion that a paper has to have some theoretical component or it doesn't rise to the level of a journal publication is absurd"*. However, due to complexities of evaluating the merit of an application paper, this task is far from trivial. There is need for a framework that would give the reader confidence that: (a) the described methods really are an advance on previous industrial practice (validity); (b) the described methods will often port to other industrial problems (generalisability), and (c) the advances described really are new (novelty). And there are no easy ways to accomplish that. Note that application papers that describe systems which are in daily use (whatever their significance) are routinely rejected as, for example, it is impossible to run millions scenarios, which is usually required for the evaluation of the approach. However, papers which document millions of scenarios run on a meaningless problem (e.g. an unconstrained sphere problem) have much better chance – as this is real science... Whichever way I look at this, it does not seem right.

2. removing 'lip-service' from publications, e.g. the authors of accepted papers should be asked to remove sentences about applicability of EAs to a wide-range of real-world problems in many verticals (as discussed in section 3 of this paper), not to misled the EC community (unless some concrete examples are provided).

3. educating the research community on real-world problems, as it seems that most of the researchers are disconnected from the real-world and its challenges. Some of these challenges include separation of business rules and constraints from the optimisation engine, as no industrial user is going to accept a software that needs to be modified (or worse, redesigned) if the application problem slightly changes. The methods have to be robust in the sense they can work with just about any variation of the application problem domain without redesign or recoding – but rather just by changing environmental settings in the configuration file of the system[17]. Garry Greenwood, summarised this[18] very well: *"If you have to take your EC method offline to change it for every new application problem variation and the user interface keeps changing,*

---

[17] Similar issue was raised in [Goertzel, 1997]: AI-based software applications are quite fragile, often a slight change in problem definition will render them useless. This problem is the main motivation for the establishment of Artificial General Intelligence community.

[18] Private correspondence.

forget it." Additional challenges include "explanatory features" of the optimiser – as usually end-users hate "black-boxes" which return just a recommended result without any "explanations" and interactive components, which are very helpful in influencing the optimiser into particular directions, to name a few.

4. encouraging researchers who deal with theoretical aspects of EAs to pay attention to commercial developments. Theoreticians should understand that the work in vacuum is not that significant. For many people it is clear that a researcher working on internet search engines should not ignore developments at Google – so the same should apply to an EA researcher who works on optimisation of supply chain type of problems and such researcher should be knowledgeable on offerings of SAP (Kallrath and Maindl, 2006).

5. developing artificial problem sets that better reflect real-world difficulties which the research community can use to experience (and appreciate) for themselves what it really means to tackle a real-world problem. This would lead to some meaningful classification of different types of problems to understand the effectiveness of various algorithms. This would require studying the problem representation and modeling issues, as these are the key components in approaching any real-world problem. Finally, more emphasis should be placed on studying the reliability of algorithms versus the frequency of hitting the global optimum, as in the real world setting reliability (in the sense of getting quality solution every run) is more desirable than getting global optimum in 95% of runs (and poor quality solutions in the remaining 5% of runs).

I think the following story nicely concludes this chapter and illustrates the gap between the theory and practice in Evolutionary Algorithms:

A scientist discovered a special dog food with amazing characteristics. He has proved (by a scientific method) that if a dog eats this food on regular basis, its fur would be always shiny, its teeth would be always white, it would never be sick, it would be well-behaved, together with many additional advantages. However, there was just one problem with this invention – when it was commercialized, it was discovered that dogs refused to eat this food...

Acknowledgements. The author is grateful to Manuel Blanco Abello, Brad Alexander, Hans-Georg Beyer, Yuri Bykov, Ken De Jong, Fred Glover, Garry Greenwood, Dennis Hooijmaijers, Pablo Moscato, James Whitacre, and Darrell Whitley for their comments – many of them were incorporated in the final version of this chapter. Also, a few quotes from our private correspondence were included in the text (with permission of the author).

References

[Ackoff, 1979]   Ackoff, R.: The Future of OR is Past. JORS (1979)
[Chow & Yuen, 2011]   Chow, C.K., Yuen, S.Y.: An Evolutionary Algorithm That Makes Decision Based on the Entire Previous Search History. IEEE Transactions on Evolutionary Computation 15(6), 741–769 (2011)

[De Jong, 1975]   De Jong, K.A.: An Analysis of the Behavior of a Class of Genetic Adaptive Systems. Doctoral Dissertation, University of Michigan, Ann Arbor, MI. Dissertation Abstract International 36(10), 5140B (University Microfilms No 76-9381) (1975)

[De Jong, 2002]   De Jong, K.A.: Evolutionary Computation: A unified approach. Bradford Book (2002)

[Gattorna, 2010]   Gattorna, J.: Dynamic Supply Chains. Prentice Hall (2010)

[Goertzel, 1997]   Goertzel, B.: From Complexity to Creativity: Explorations in Evolutionary, Autopoietic, and Cognitive Dynamics. Plenum Press (1997)

[Goldberg, 1989]   Goldberg, D.E.: Genetic Algorithms in Search, Optimization, and Machine Learning. Addison-Wesley (1989)

[Hinterding et al. 1999]   Hinterding, R., Michalewicz, Z.: Your Brains and My Beauty: Parent Matching for Constrained Optimisation. In: Proceedings of the 5th IEEE International Conference on Evolutionary Computation, Anchorage, Alaska, May 4-9, pp. 810–815 (1998)

[Ibrahimov et al. 2011]   Ibrahimov, M., Mohais, A., Schellenberg, S., Michalewicz, Z.: Advanced Planning in Vertically Integrated Supply Chains. In: Bouvry, P., González-Vélez, H., Kołodziej, J. (eds.) Intelligent Decision Systems in Large-Scale Distributed Environments. SCI, vol. 362, pp. 125–148. Springer, Heidelberg (2011)

[Kallrath andMaindl, 2006]   Kallrath, J., Maindl, T.I.: Real Optimization with SAP-APO. Springer (2006)

[Michalewicz, 1992]   Michalewicz, Z.: Genetic Algorithms + Data Structures = Evolution Programs, 1st edn. Springer (1992)

[Michalewicz, 2012]   Michalewicz, Z.: The Emperor is Naked: Evolutionary Algorithms for Real-World Applications. ACM Ubiquity (2012)

[Mohais et al. 2011]   Mohais, A., Ibrahimov, M., Schellenberg, S., Wagner, N., Michale wicz, Z.: An Integrated Evolutionary Approach to Time-Varying Constraints in Real-World Problems. In: Chiong, R., Weise, T., Michalewicz, Z. (eds.) Variants of Evolutionary Algorithms for Real-World Applications. Springer (2011)

[Potter and De Jong, 1994]   Potter, M.A., De Jong, K.A.: A Cooperative Coevolutionary Appro   ach to Function Optimization. In: Davidor, Y., Männer, R., Schwefel, H.-P. (eds.) PPSN 1994. LNCS, vol. 866, pp. 249–257. Springer, Heidelberg (1994)

[Ullman, 2009]   Ullman, J.D.: Advising Students for Success. Communications of the ACM 53(3), 34–37 (2009)

Probabilistic Graphical Approaches for Learning, Modeling, and Sampling in Evolutionary Multi-objective Optimization

Vui Ann Shim and Kay Chen Tan

Department of Electrical and Computer Engineering,
National University of Singapore, 4 Engineering Drive 3, 117576, Singapore
{g0800438,eletankc}@nus.edu.sg

Abstract. Multi-objective optimization is widely found in many fields, such as logistics, economics, engineering, or whenever optimal decisions need to be made in the presence of tradeoff between two or more conflicting objectives. The synergy of probabilistic graphical approaches in evolutionary mechanism may enhance the iterative search process when interrelationships of the archived data has been learned, modeled, and used in the reproduction. This paper presents the implementation of probabilistic graphical approaches in solving multi-objective optimization problems under the evolutionary paradigm. First, the existing work on the synergy between probabilistic graphical models and evolutionary algorithms in the multi-objective framework will be presented. We will then show that the optimization problems can be solved using a restricted Boltzmann machine (RBM). The learning, modeling as well as sampling mechanisms of the RBM will be highlighted. Lastly, five studies that implement the RBM for solving multi-objective optimization problems will be discussed.

Keywords: Evolutionary algorithm, multi-objective optimization, probabilistic graphical model, restricted Boltzmann machine.

1 Introduction

Many real-world problems involve the simultaneous optimization of several competing objectives and constraints that are difficult, if not impossible, to solve without the aid of powerful optimization algorithms. In a multi-objective optimization problem (MOP) [1, 2], no one solution is optimal to all objectives. Therefore, in order to solve the MOP, search methods employed must be capable of finding a number of alternative solutions representing the tradeoff among the various conflicting objectives. An MOP in minimization case can be formulated as follows. Minimize:

$$F(X) = (f_1(X), ..., f_m(X)) \tag{1}$$

subject to $X = \{x_1, ..., x_n\} \in \theta$, $F \in \mathbf{R}^m$, where θ is the decision space and $\mathbf{R}^m$ is the objective space.

J. Liu et al. (Eds.): WCCI 2012 Plenary/Invited Lectures, LNCS 7311, pp. 122–144, 2012.

Evolutionary algorithms (EAs) are a class of stochastic search methods that have been found to be efficient and effective in solving MOPs. The advantage of EAs can be attributed to their capabilities of sampling multiple candidate solutions simultaneously, a task that most conventional optimization techniques fail to work well. Nonetheless, the stochastic recombination in standard EAs may disrupt the building of strong schemas of a population and thus movement towards optimal is extremely difficult to predict. Motivated by the idea of exploiting the linkage information among the decision variables, estimation of distribution algorithm (EDA) has been regarded as a new computing paradigm in the field of evolutionary computation [3–5].

In EDAs, the discovered knowledge of the data is used to predict the location or pattern of the Pareto front or to predict the favorable movement in the search space. By using the discovered correlations of the parameters of a cost function, the search can be regulated to follow the correlated patterns when generating an offspring solution. The correlations as well as the probability distribution of the cardinalities of the parameters can be learned and modeled by a probabilistic model. In order to effectively learn and model that information, the probabilistic graphical approach is one of the well-known and promising techniques [6–8]. In EDAs, the reproduction of children solutions is carried out by building a representative probabilistic model of the parent solutions, and new solutions are then generated through the sampling of the constructed probabilistic model. Therefore, the learning, modeling, and sampling mechanisms are important features of an EDA.

This paper focuses on the implementation of EDAs for solving MOPs. First, a literature review that studies the potential of EDAs in solving MOPs is given. Then, the focus is tailored to introduce an EDA that uses the restricted Boltzmann machine (RBM) as its modeling technique. An insight discussion on how and what information is learned and modeled by the RBM will be covered. Subsequently, five studies that implement the RBM-based EDA (REDA) in solving scalable problems, epistatic problems, noisy problems, combinatorial problems, and global unconstrained continuous optimization problems will be presented.

The rest of the paper is as follows. Section 2 presents a literature review on the implementation of probabilistic graphical models (PGMs) in solving MOPs. Section 3 describes the RBM as well as its training and modeling mechanisms. Five case studies that implement the REDA in solving MOPs are presented in Section 4. The conclusion is drawn in Section 5.

2   Probabilistic Graphical Models in Multi-objective Evolutionary Algorithms

EDAs draw its inspiration from the use of the probability distribution of promising solutions to predict the Pareto optimal front or the favorable movement in the search space. Based on this idea, the linkage information or the regularity patterns that appear quite often in a set of promising solutions can be captured and used to predict the probability distribution of other superior solutions. In the literature, the probability information can be captured using at least three methods depends on how the interactions among the decision variables are taken into consideration. Those methods are the univariate modeling, bivariate modeling, and multivariate modeling [3]. Over the last decade, several attempts have been devoted to developing EDAs in the context of multi-objective optimization (MOEDAs) [9]. The main differences among the MOEDAs are the employment of different modeling and sampling mechanisms.

The first MOEDA was introduced in [10]. The authors proposed a mixture-based multi-objective iterated density-estimation evolutionary algorithm (MIDEA) with both discrete and continuous representations. The mixture probability distribution in MIDEA was constructed using the univariate factorization. MIDEA is well-known for its simplicity, speed, and effectiveness. Furthermore, it can also serve as a baseline algorithm for other MOEDAs. The simulation results indicated that MIDEA is able to generate a set of diverse solutions that is close to the Pareto optimal front.

In [11], Laumanns and Ocenasek examined the effect of incorporating mutual dependencies between the decision variables in approximating the set of Pareto optimal solutions. The mutual dependencies were captured using the Bayesian optimization algorithm with binary decision trees. The experimental results showed that the proposed Bayesian multi-objective optimization algorithm (BMOA) is effective in approximating the Pareto front for simple test instances. In order to deal with harder test instances, an additional computational time is required.

In [12], MOEDA based on the Parzen estimator was introduced. The Parzen estimator, a non-parametric technique, was used to estimate the kernel density through the learning of the multivariate dependencies among the decision variables. The Parzen estimator was also used in the objective space to enhance the diversity preservation of the algorithm. The empirical results indicated that the proposed algorithm has better convergence rate and is able to obtain a set of well spread solutions.

Li et al. [13] suggested a hybrid binary EDA with mixture distribution (MOHEDA) for solving the multi-objective 0/1 knapsack problems. One of the simplest EDA, the univariate marginal distribution algorithm (UMDA), was hybridized with a weighted sum local search method. This hybridization enable MOHEDA took advantage of both local and global information to guide the search towards optimality. Furthermore, the population was clustered into several groups using a proposed stochastic clustering algorithm before the mixture distribution was constructed. In [14], Okabe et al. proposed an EDA that uses Voronoi diagram (VEDA) as its probabilistic modeling method. The Voronoi diagram takes into account the problem structure in estimating the most appropriate probability distribution. Instead of determining the distribution from individuals in the selected population, the implementation also makes use of those that are not selected. The experimental results showed that VEDA performs better than NSGA-II [15] for a limited number of fitness evaluations.

In [16], Pelikan et al. modified the Bayesian optimization algorithm and introduced a hierarchical Bayesian optimization algorithm (hBOA) to solve multi-objective decomposable problems. hBOA adapted the NSGA-II's domination-based framework and used k-mean clustering for modeling purpose. hBOA were used to solve scalable deceptive problems. The simulation results demonstrated that hBOA successes to obtain the optimal solutions and has faster convergence rate. They inferred that the clustering and linkage learning are the main criteria that contribute to the success of the algorithm in solving decomposable multi-objective problems. In [17], Sastry et al. proposed another MOEDA, called extended compact genetic algorithm (ECGA), to solve the scalable deceptive problems. The paper analyzed the characteristics of the algorithm on a class of bounding adversarial problems with scalable decision variables. m-k deceptive trap problems [16] were used to investigate the performance of the proposed algorithm.

In [18], Soh and Kirley proposed a parameter-less MOEA, which combines the ECGA with external ϵ-Pareto archive, clustering, and competent mutation to deal with scalable problems. Two types of scalable problems were studied, including deceptive problems with scalable decision variables and DTLZ problems with scalable objective functions. The proposed algorithm showed promising results in those test problems due to the incorporation of linkage learning, clustering, and local search. In [19], the authors examined the limitation of maximum-likelihood estimators in the problems which may lead to the prematurely vanishing variance. Using the framework of MIDEA, the authors combined the normal mixture distribution with adaptive variance scaling to remedy the vanishing variance problem. Under this scheme, the premature convergence was prevented in the condition that the estimated probability distribution is enlarged beyond its original maximum likelihood estimation. Zhong and Li [20] presented a decision-tree-based multi-objective estimation of distribution algorithm (DT-MEDA) for global optimization in the continuous-valued representation. The conditional dependencies among the decision variables are learned by the decision tree. The children solutions are generated through the sampling of the constructed conditional probability distribution.

Zhang et al. [21] proposed a regularity model-based multi-objective estimation of distribution algorithm (RM-MEDA) for solving continuous multi-objective optimization problems with linkage dependencies. A local principle component analysis (PCA) was used to model the probability distribution of promising individuals. The experimental results showed that RM-MEDA has good scalability in terms of decision variables and less sensitive to algorithmic parameter settings. In order to further improve the algorithm, Zhou et al. [22] generalized the idea of RM-MEDA and proposed a probabilistic model-based multi-objective evolutionary algorithm, named as MMEA, which is able to simultaneously approximate both the Pareto set and Pareto front of an MOP. Martí et al. [23] developed another MOEDA using growing neural gas (GNG) network - multi-objective neural EDA (MONEDA). GNG network is a self-organizing neural network based on

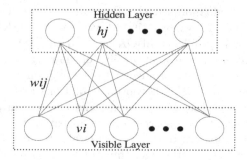

Fig. 1. Architecture of a RBM

neural gas model. This model creates an ordered cluster of input data set; a new cluster will then be inserted based on the topology and cumulative errors. WFG problems [24] were used to evaluate the search capability of the MONEDA.

In [25], an MOEDA, which uses Gaussian model as its modeling technique, was used to optimize the radio frequency identification (RFID) network design. In order to enhance the search capability of the algorithm, a particle swarm optimization (PSO) [26] algorithm was hybridized with the MOEDA. A number of children solutions are generated by the EDA while the rest of them are produced by the velocity-free PSO. The algorithm succeeded to obtain a set of tradeoff solutions in RFID network design.

3   Restricted Boltzmann Machine (RBM)

The RBM [27–29] is an energy-based binary stochastic neural network. The architecture of the network is illustrated in Fig. 1. The network consists of two layers of neurons - a visible layer and a hidden layer. The visible layer, denoted as v_i, is an input layer of the network. The hidden layer, denoted as h_j, is a latent layer that determines the capability of the network in modeling the probability distribution of the input stimuli. The network does not have the output layer. Instead, the output information is represented by the energy values of the network. w_{ij} is the weight that connecting visible unit i and hidden unit j. b_i is the bias of the visible unit i and d_j is the bias of hidden unit j. Both of the layers are fully connected to one another and the weights are symmetric. In this way, the information can flow from one layer to another, increasing the learning capability of the network. Furthermore, there is no interconnection among the neurons within the same layer. Thus, the hidden units are conditionally independent. Besides, the visible units can be updated in parallel given the hidden states. This behavior improves the training speed of the network. The weights and biases of an RBM define the energy function of the network. The energy function is presented as follows.

$$E(v, h) = -\sum_i \sum_j v_i h_j w_{ij} - \sum_i v_i b_i - \sum_j h_j d_j \qquad (2)$$

Using the energy function of the network, the probability distribution of any global state can be derived as follows

$$P(v, h) = \frac{exp(-E(v, h))}{Z = \sum_{x,y} exp(-E(x, y))} \tag{3}$$

where Z is the normalizing constant, which is defined by the energy of all the global states while the numerator of the equation is the energy of a particular state. By summing all the configurations of the hidden units, the probability distribution of a visible state can be clamped to be

$$P(v) = \frac{\sum_h exp(-E(v, h))}{\sum_{x,y} exp(-E(x, y))} \tag{4}$$

3.1 Training

The training is one of the main issues in the RBM. In the literature, the contrastive divergence (CD) training method [30, 31] is the most well-known training mechanism for the RBM. In the CD training, two phases (positive phase and negative phase) are carried out. In the positive phase, the input stimuli or input data are rendered into the visible units of the network. Subsequently, the hidden states, given the visible states, are constructed by performing the Gibbs sampling as follows

$$P(h_j|v) = \varphi(\sum_i w_{ij}v_i - d_j) \tag{5}$$

where $\varphi(x) = \frac{1}{1+e^{-x}}$ is the logistic function. In the negative phase, given the hidden states, the visible states are reconstructed using the same logistic function. The process of these two phases is repeated S times. Next, the weights and biases of the network are updated as follows

$$w'_{ij} = w_{ij} + \epsilon(<v_i h_j>_0 - <v_i h_j>_1) \tag{6}$$

$$b'_i = b_i + \epsilon(<v_i>_0 - <v_i>_1) \tag{7}$$

$$d'_j = d_j + \epsilon(<h_j>_0 - <h_j>_1) \tag{8}$$

where ϵ is the learning rate, $<>_0$ is the original states of the neurons, and $<>_1$ is the states of the neurons after a single step of reconstruction. The overall CD training is repeated until a stopping criterion is met. The process of CD training is further demonstrated in Fig. 2.

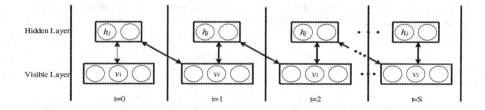

Fig. 2. The contrastive divergence (CD) training mechanism

3.2   Modeling

This paper will only study a restrcited Boltzmann machine-based estimation of distribution algorithm for multi-objective optimization (REDA). In the implementation stage, the alleles of the decision variables in the cost function are the input data that will be rendered to the visible layer of an RBM. Therefore, the RBM has n visible units if an MOP has n decision variables to be optimized. The setting of the number of hidden units is to be determined by users. The complexity of the network is directly proportional to the setting of the number of visible and hidden units. Since the probability distribution of the population needs to be constructed at every generation, it is essential for the model to be kept simple. Therefore, the number of hidden units is set to as small as possible as long as the probability model is representative.

After performing the CD training, a set of trained weights, biases, and hidden states are obtained. Subsequently, in binary representation, the joint probability distribution with n decision variables in generation g is formulated as follows.

$$P_g(v) = \prod_{i=1}^{n} p_g(v_i) \tag{9}$$

where $p_g(v_i)$ is the marginal probability of decision variable i (v_i) at generation g. The marginal probability of each decision variable is obtained through (4). Expanding the equation,

$$p_g(v_i = 1) = \frac{\sum_{l=1}^{N} \delta_l(v_i^+)}{\sum_{l=1}^{N} \delta_l(v_i^+) + \sum_{l=1}^{N} \delta_l(v_i^-)} \tag{10}$$

$$\delta_l(v_i^+) = \sum_{h=1}^{H} e^{-E(v_i=1,h)} \tag{11}$$

$$\delta_l(v_i^-) = \sum_{h=1}^{H} e^{-E(v_i=0,h)} \tag{12}$$

where $\delta_l(v_i^+)$ is the marginal cost of v_i when the cardinality of $v_i = 1$, $\delta_l(v^-)$ is the marginal cost of v_i when the cardinality cost of $v_i = 0$, N is the number of selected solutions or parent solutions, and H is the number of hidden units.

Direct sampling from the above probabilistic model reaches a limit in progress when the probability reaches a maximum value of 1.0 or a minimum value of 0.0. Therefore, the lower and upper bounds are added to the probability distribution based on the average cost of cardinality. The modified version of the marginal probability is given as below

$$p_g(v_i = 1) = \frac{\sum_{l=1}^{N} \delta_l(v_i^+) + avg(\sum_{l=1}^{N} \delta_l(v_i))}{\sum_{l=1}^{N} \delta_l(v_i^+) + \sum_{l=1}^{N} \delta_l(v_i^-) + r_i \times avg(\sum_{l=1}^{N} \delta_l(v_i))} \quad (13)$$

where $avg(\sum_{l=1}^{N} \delta_l(v_i)) = \frac{\sum_{l=1}^{N} \delta(v_i)}{N}$ and r_i is the number of different values that v_i may take. In binary case, r_i is 2.

3.3   Sampling

The children solutions are generated through the sampling of the constructed probabilistic model as follows

$$v_i = \begin{cases} 1 \text{ if random } (0,1) \leq p_g\,(v_i = 1) \\ 0 \qquad\qquad \text{otherwise} \end{cases} \quad (14)$$

where random (0,1) is a randomly generated value between $[0, 1]$.

3.4   Learning Capability of REDA

In this section, a detailed description of the behaviors of the RBM in the evolutionary perspective is presented. Three main issues will be covered: (1) How and what information is captured in an RBM (2) How to effectively train the RBM in the evolutionary perspective (3) What can be elucidated from the energy values of the RBM in the fitness landscape perspective.

How and What Information Is Captured in an RBM. In an RBM, the neurons between two layers are fully connected via weighted synaptic connections, and there is no intra-layer connection. These weight connections are used by the neurons to communicate their activations to one another. The quality of training of the network corresponds directly to the effectiveness at which the algorithm learns the probability distribution. Whenever the number of hidden units is sufficiently large, the network can represent any discrete distribution. This behavior that a sufficient number of hidden units in the network would guarantee improvement in the training error has been proven mathematically in [31]. During the training process, the aim is to minimize the energy equilibrium of the network such that the implicit correlations as well as the probability distribution of the input stimuli is captured and stored in the synaptic weights of the network. This distribution-based model allows the RBM to globally learn the probability distribution of the decision variables by considering the interdependencies of the data.

How to Effectively Train an RBM in the Evolutionary Perspective.
The weight update process in an RBM requires calculating the gradient of log-likelihood of the input data. The gradient is minimal when the reconstructed data is exactly similar to the input stimuli. Contrastive divergence training [31] method aims to minimize the energy level and training error of the network. The primary understanding is that the minimal energy level and training error can be achieved when sufficient number of hidden units and training epochs are applied. This is because the learning capability of the network is determined by the number of hidden units. A larger number of hidden units gives extra flexibility for the network to model the global distribution of the input stimuli, and thus could yield better convergence. On the other hand, CD training will require a large number of training epochs to train the network well. When the RBM is modeled as EDA (REDA), another factor that can reduce the energy level and training error is the number of generations of an optimization process. Over generations, the training error and energy level of the network are reduced. This is most likely due to the reduction in the size of more promising search space when the search converges to near optimal points. By taking this into consideration, the computational time of the algorithm can be improved by eliminating unnecessary training of the network in each generation.

What Can Be Elucidated from the Energy Values of an RBM in the Fitness Landscape Perspective. In EDAs, the two main mechanisms that determine the success of the algorithms are probabilistic model construction and sampling technique. The core purpose of probabilistic modeling is to learn the probability distribution of the candidate solutions and to capture the dependencies among the decision variables. By using the linkage information of known solutions, the characteristics of the unknown solutions can be studied. In EDAs, the sampled solutions are the unknown solutions. If the characteristics of these solutions are known, this additional information can be taken into consideration during the optimization process. In an RBM, the energy-based model captures the dependencies of the decision variables by associating a scalar energy value from the network to each solution. Thus, it can be inferred that sampled solutions may have higher energy if the solutions are outside the boundary modeled by the RBM. In pattern recognition, a lower energy level suggests that a test sample is more likely to belong to a certain class of patterns. However, this is not the case in EDAs as a lower energy level does not mean that the solutions are fitter, and vice versa. The choice of a solution outside the boundary of the modeled energy distribution may imply an increase in the exploration capability, while focusing on the solutions inside the boundary may imply an increase in the exploitation capability of the algorithm.

3.5   Algorithmic Process Flow

The general evolutionary process flow of REDA is presented in Fig. 3.

Begin

    Initialization: At generation $g=0$, randomly generate N solutions as the initial population, $Pop(g = 0)$

    Evaluation: Evaluate all solutions

    Do While ("maximum generation is not reached")

       1. **Fitness Assignment**: Perform Pareto ranking and crowding distance over the population

       2. **Selection**: Select N promising individuals based on the binary tournament selection

       3. **Training**: Train the RBM by using CD training method to obtain the weights, biases, and hidden states

       4. **Probabilistic model**: Compute the probability of the joint configuration $P(v)$ by using the trained weights, biases, and hidden states of the RBM

       5. **Reproduction**: Generate new set of N solutions (P)from $P(v)$

       6. **Evaluation**: Calculate the fitness values of all offspring

       7. **Elitism**: Select N individuals from $P \cup Pop(g)$ to form $Pop(g + 1)$. $g = g + 1$

    End Do

End

Fig. 3. Pseudo-code of REDA

4 Restricted Boltzmann Machine-Based Estimation of Distribution Algorithm for Solving Multi-objective Optimization Problems

4.1 REDA with Clustering for Solving High Dimensional Problems

Many real-world optimization problems are challenged by the different characteristics and difficulties. The problems may be non-linear, restricted to several constraints, has complex relationships within the decision variables, has a large number of variables, and even consists of several conflicting objectives [32, 33]. High dimensional problems with many decision variables and conflicting objective functions to be optimized simultaneously are hard problems which may challenge the algorithm in finding the global optimal solutions. In the problems with many decision variables, the complexity of the problems increase with an increase in the number of decision variables. This is due to the enlargement of the search space and an increase in the number of possible moves towards optimality. In the problems with many conflicting objective functions, the selection pressure in selecting fitter individuals is reduced when the number of conflicting objective functions is increased. This is due to the high rate of non-dominance among individuals during the evolutionary process. This may hinder the search towards optimality or result in the population getting trapped in a local optimal. One of the ideas to overcome these issues is to exploit extra information (e.g. linkage dependencies) from within the selected population. This information is hypothesized to provide valuable guidance in driving the search process.

In order to solve MOPs with scalable number of decision variables, the algorithm presented in Fig. 3 is implemented directly. The simulation results presented in [34] indicated that REDA performs equally well or slightly inferior to NSGA-II in ZDT and DTLZ problems with a smaller number of decision variables. This may be attributed to the fact that, REDA does not directly use location information of the selected solutions in exploiting the new solutions. Furthermore, REDA, which models the global probability of the selected individuals, is not able to effectively escape from local optima. REDA shows superior convergence performances in problems with a larger number of decision variables. Besides, REDA also able to produce more non-dominated solutions compared to NSGA-II. The increase in number of decision variables increases the size of the search space, and thus increases the complexity of the problems. The stochastic behavior of NSGA-II prevents the algorithm from evolving a good set of solutions since the number of possible combinations towards optimality is increased. The incorporation of dependency information as well as the use of the global probability distribution of solutions enables REDA to effectively explore the huge search space.

In problems with many objective functions, clustering is incorporated into REDA in order to build a probabilistic model from different regions in the search space. For sake of simplicity, the number of cluster, k, is determined by the user. In the implementation stage, a probabilistic model is built for each cluster, $L^1, L^2, ..., L^k$. Subsequently, the new population $Pop(g+1)$ is generated by sampling the probabilistic model constructed from each cluster L, where N new solutions are generated and equal number of individuals is sampled from each cluster.

In problems with three objective functions, the performance of REDA is comparable to NSGA-II. In a higher number of objective functions, REDA gives the superior performance in converging to the Pareto optimal front as well as maintaining a set of diverse solutions. The superior performance of REDA may be due to the incorporation of linkage information in driving the search. This information is learned by the network and is clamped into the probability distribution before the sampling takes place. Some flexibility is given to the algorithm in exploring the search space by allowing the training to stop before the energy reaches the minimum. The good performance of REDA in these test instances supports the claim that REDA scales well with the number of objective functions compared to NSGA-II. Clustering is important for problems with solutions that are hard to represent by a single probabilistic model.

Most of the probabilistic modeling techniques for learning the linkage dependencies of the solutions incur additional computational cost and time. In an RBM, the most time consuming part is the network training. Training is conducted at each generation and stops when the maximum number of training epochs is reached. This training process is more complicated than the genetic operators in standard MOEAs, and thus incurs additional simulation time. Even though REDA may spend more simulation time, it has a faster convergence rate compared to NSGA-II. This is one of the strengths of REDA, especially when dealing with real-world problems where the fitness evaluations are computationally expensive. Detailed information of the implementation and experimental results can be referred to [34].

Even though REDA showed promising results in solving scalable problems, it still suffers several flaws which requires further investigation. Firstly, univariate sampling in REDA may limit its ability to generate new solutions. This is because univariate sampling does not consider the correlation between the decision variables when performing the sampling. A more sophisticated sampling mechanism that is able to take into account the multivariate dependencies between the decision variables may enhance the search capability of REDA. Secondly, REDA fails to converge to the global Pareto optimal front in problems with many local optima. This is because REDA in particular and MOEDAs in general will model the probability distribution of the solutions even though they are trapped at local optima, and subsequently use the constructed probabilistic information as a reference model to produce offspring. Hybridization with local search algorithms may be one of the approaches in dealing with problems with many local optima. Thirdly, REDA is sensitive to bias. This is because the modeling in EDAs only estimates the probability distribution of the current best solutions. In other words, only global information is used. Whenever the maintained solutions are biased towards certain search regions, EDAs may consider the maintained solutions are the promising one, thus, construct their probability distribution accordingly. Therefore, it is necessary to enhance the diversity preservation of REDA especially the ability to produce a set of diverse solutions. This can be achieved by combining EDAs with other search algorithms which use location information in producing offspring, including genetic algorithms, differential evolution, particle swarm optimization algorithms, or any other algorithms with similar features.

4.2   An Energy-Based Sampling Mechanism of REDA

In our another study, the sampling mechanism of REDA was investigated [35]. In [34], the simple probability sampling is employed in REDA. This sampling technique may, however, limit the production of appropriate solutions if the decision variables are highly correlated or have a high dimension. This is because, during sampling, marginal probability distribution considers the distribution of the particular decision variable but not the correlation between the decision variables. As a result, the sampled solutions have difficulties following the correlated distribution. One way to tackle this problem is to sample an infinite number of solutions. This may increase the number of possible combinations of the solutions and thus increase the chance of producing fitter individuals. However, sampling of an infinitely large number of solutions may lead to an increase in the number of fitness evaluations and computational time. It is known that some real world problems are very time consuming and such an algorithm would not be practical. To deal with this problem, energy value can be taken into consideration. Firstly, $N \times M$ solutions are generated. Then, the energy value serves as the main criterion for forming new N solutions from the alleles of the $N \times M$ solutions, where $M > 1$ is a multiplier. A lower energy level implies that the solution is in a more stable state while a higher energy level means that the solution is not in energy equilibrium. The energy-based sampling mechanism will, therefore, prefer the alleles of solutions with lower energy levels.

As probabilistic modeling only models the previous best topology, the solutions that are located inside the modeled topology are stable (lower energy levels) in terms of energy equilibrium and are generally fit. On the other hand, the solutions outside the modeled topology (higher energy levels) may be considered unstable but not unfit. This means that the solutions with higher energy levels may be the promising solutions that are not modeled by the network and thus will be worth preserving to the next generation. Therefore, it is required to give the algorithm the flexibility of choosing the alleles of solutions with high energy levels in order to achieve a more explorative search.

By incorporating the above-mentioned approach (energy-based sampling approach) into REDA, the simulation results in [35] indicated that the modified REDA, which models the probability distribution of the solutions by applying the energy information to detect the dependencies, is able to perform well on epistatic problems. The results also showed that the incorporation of the energy-based sampling mechanism enhances the exploration and exploitation capability of REDA. However, the limitation of REDA in escaping from local optima has not been significantly improved through this mechanism. REDA with energy-based sampling mechanism is able to escape the local optima in some simulation runs, however, is trapped in local optima in most of the runs. Therefore, it can be concluded that the REDA with energy-based sampling mechanism is sensitive to different initializations for problem with multi-modality.

4.3 REDA in Noisy Environments

In [36], we have extended the study of REDA in dealing with MOPs with noisy objective functions. In noisy environments, the presence of noise in the cost functions may affect the ability of the algorithms to drive the search process towards optimality. Beyer [37] carried out an investigation and found that the presence of noise may reduce the convergence rate, resulting in suboptimal solutions. In another study by Goh and Tan [38], it was reported that the low level of noise helps an MOEA to produce better solutions for some problems, but a higher noise level may degenerate the optimization process into a random search. Darwen and Pollack [39] concluded that re-sampling can reduce the effect of noise for a small population, but may not be as helpful for a larger population.

EDAs surpass the standard MOEAs in handling noisy information by constructing a noise handling feature in the built probabilistic model. In order to show this advantage, a likelihood correction feature was proposed in [36] in order to tune the error in the constructed probabilistic model. The previous studies showed that REDA has its limitations in exploiting the search space and may be trapped at any local optimum [34]. In order to overcome these limitations, REDA was hybridized with a particle swarm optimization (PSO) algorithm. This hybridization is expected to improve the performance since the particles may now move out of the regions modeled by REDA, and thus provide extra solutions that REDA alone was not able to tap on.

In the likelihood correction feature, the concept of probability dominance, proposed in [40], was employed. This concept is implemented to determine the probability error of each selected individual. This probability error is then used to group the population into several clusters before a probabilistic model is built. Assume that $f_i(A)$ and $f_i(B)$ are two solutions in the objective space with m objective functions ($i = 1, ..., m$). In a noise free situation, $f_i(A)$ is said to strictly dominate $f_i(B)$ if $f_i(A)$ is smaller than $f_i(B)$ in all the objective values in minimization case. On the other hand, $f_i(A)$ and $f_i(B)$ are mutually non-dominated only if not all the objective values in one solution are lower than that of the other. In a noisy domain, the above statements may not correctly represent the true domination behavior. Even through $f_i(A)$ appears to strictly dominate $f_i(B)$, the noise may distort the actual fitness values where $f_i(B)$ is supposed to dominate $f_i(A)$. In the selection process, the selection error occurs when the less fit individual is chosen. Therefore, the probability to make an error in the selection process could be utilized to improve the decision making process.

The likelihood correction feature is based on the heuristic that if the distribution can be approximated as close to the real distribution of the best solutions, the detrimental effect of the noise can then reduced. To approximate the real distribution, the probability of making an error in the selection process is adapted in the probabilistic modeling. In binary tournament selection, if two solutions in the tournament have a huge distinction in their objective values, for example $f_i(A)$ dominates $f_i(B)$ by far, then the selection error for selecting $f_i(A)$ is small. On the other hand, if two individuals in the tournament are near to each other in the objective space, then the selection error is larger. Therefore, if the probabilistic model built by REDA is only based on individuals with small selection error, then, the model may avoid distortions caused by those solutions with a large selection error. However, the probability distribution may not come close to the real distribution if the number of individuals with smaller selection error is too little. Thus, a method to combine the distribution between solutions with small selection error and those with large selection error was designed. This combination was based on the penalty approach where individuals with smaller selection error will be penalized less while solutions with larger selection error will be more heavily penalized. This is because the real distribution is more likely to follow the distribution of the population with smaller selection error than those with larger selection error.

The simulation results in [36] demonstrated that the hybridization between REDA and PSO may slightly deteriorate the search ability of REDA in some noiseless circumstances. However, its performance was outstanding in noisy environments. The results also showed that REDA is more robust than NSGA-II because the performance of REDA is better than NSGA-II in noisy conditions. REDA, which performs the search by modeling the global probability distribution of the population, is more responsive since the reproduction is based on the global information and not individual solutions. Furthermore, likelihood correction is able to tune the probability distribution so that the distribution of the solutions is more likely to follow the one with a smaller selection error. The hybridization has further enhanced the ability of REDA in exploring the search space, especially in noisy conditions. This hybridization is utterly important when REDA fails to model the promising regions in the search space. In that case, the hybridization could provide opportunities to explore those regions, thus, improving the search ability.

The scalability issue was also investigated. The results indicated that the hybrid REDA obtained the most promising results in the noisy test problems with different number of decision variables. This result demonstrated that the hybridization with PSO has enhanced the search ability of REDA. This is important as PSO provides a directional search which may explore the promising regions where probabilistic model may fail to explore. It can be concluded that the hybrid REDA scaled well with the number of decision variables compared to NSGA-II.

In order to study the potential of other hybridizations, REDA was hybridized with a genetic algorithm (GA) and a differential evolution (DE). The simplest and most common GA is applied, where single point crossover and bit-flip mutation are implemented. For DE, the standard recombination proposed in [41] is applied. The results indicated that all hybridizations are able to improve the performance of REDA, in most of the test problems, under both noiseless and noisy environments. Among them, hybridization with PSO gave the best results followed by DE and then GA. The function of hybridization is to provide extra search ability for REDA as REDA performs the search by using only global statistical information. This hybridization therefore enhances the ability for REDA in exploring the search space, especially in the early stage of evolutions where the search space is huge. The search using position information (GA, PSO, DE) is also essential and useful especially to explore and exploit certain promising regions. Thus, hybridization is an important mechanism to improve the search performance of EDAs.

4.4 REDA for Solving the Multi-objective Multiple Traveling Salesman Problem

In [42], REDA was implemented to solve the multi-objective multiple traveling salesman problem (MmTSP). The multiple travelling salesman problem (mTSP) is a generalization of the classical travelling salesman problem (TSP). In the mTSP, Ω salesmen are involved in a routing to visit Ψ cities ($\Omega < \Psi$) in order to achieve a common goal. In the routing, all the salesmen will start from and end at the single depot after visiting the ordered cities. Each city can only be visited once, and the total cost for all salesmen is required to be minimized. The cost can be defined as distance, time, expense, risk, etc. The complexity of the mTSP is higher than the TSP since it is required to allot a set of cities to each salesman in an optimal ordering while minimizing the total travelling cost for all salesmen. Furthermore, the mTSP is more appropriate for real life scheduling or logistic problems than the TSP because more than one salesman is usually involved. Over the past few decades, research on the TSP has attracted a great deal of attention. However, the mTSP has not received the same amount of research effort compared to the TSP.

In an MOP, no single point is an optimal solution. Instead, the optimal solution is a set of non-dominated solutions, which represents the tradeoff between the multiple objectives. In this case, fitness assignment to each solution in the evolutionary multi-objective evolutionary optimization is an important feature for the assurance of the survival of fitter and less crowded solutions to the next generation. Much research has been carried out over the past few decades to address this issue, and fitness assignment based on the domination approach is one of the most popular approaches. However, the fitness assignment in the domination approach is less efficient in solving many-objective problems. This is because the strength of the domination among the solutions in a population is weakened with the increase in the number of objective functions. This phenomenon results in poor decision making in the selection of promising solutions.

Recently, the classical method for multi-objective optimization based on decomposition has been re-formularized into a population-based approach [43, 44]. The decomposition approach decomposes an MOP into several subproblems and subsequently optimizes all the subproblems concurrently. Under this approach, it is not required to differentiate the domination behaviors of the solutions. Instead, the subproblems are constructed using any aggregation approach, and the superiority of the solutions is determined using the aggregated values.

Problem Formulation. The aim of the mTSP is to minimize the total traveling cost of all the salesmen under the condition that each city must be visited strictly once by any salesman, and all the salesmen must return to the starting depot after visiting their final ordered city. The traveling cost could be defined as the traveling distance, traveling time, traveling expense, traveling risk, etc incurred. Each salesman will have his own route and there should be no repeated visit on any city in the route of the salesman.

In the literature, the aim of the mTSP is specified to be either minimizing the total traveling cost of all salesmen or the highest traveling cost incurred by any single salesman [45]. In [42], the focus is tailored specifically for the mTSP with single depot; considering the minimization of the total traveling cost and the balancing of the workload among all salesmen. This is achieved by formularizing the objective function to be the weighted sum of the total traveling cost of all salesmen and the highest traveling cost of any single salesman. In the context of multi-objective optimization (MmTSP), more than one objective is subject to be minimized, which is formulated as follows.

Minimize:

$$F(x) = (F_1(x), ..., F_m(x))$$
$$F_1(x) = \omega_1 \times TC^1 + \omega_2 \times MC^1$$

$$\vdots$$

$$F_m(x) = \omega_1 \times TC^m + \omega_2 \times MC^m$$

where

$$TC^k(x) = \sum_{j=1}^{\Omega} IC_j^k(x)$$

$$MC^k(x) = \max_{1 \le j \le \Omega}(IC_j^k(x))$$

$$IC_j^k(x) = \sum_{i=1}^{n_j-1} D^k(a_{i,j}, a_{i+1,j}) + D^k(a_{n_j,j}, a_{1,j})$$

In the above formulation, $x \in \phi$, ϕ is the decision space, $a_{i,j}$ is the ith visiting city by salesman j, m is the number of objective functions, ω_1 and ω_2 are the weights to balance between total cost and highest cost ($\omega_1 + \omega_2 = 1.0$), TC is the total traveling cost of all salesmen, MC is the highest traveling cost of any single salesman, IC is the individual traveling cost, Ω is the number of salesmen, n_j is the number of cities traveled by salesman j, $D^k(a_{i,j}, a_{i+1,j})$ is the traveling cost (for the kth objective function) between cities at locations i and $i+1$ for salesman j. In a chromosome, two conditions should be met, which are all the cities must be visited exactly once and each salesman must be assigned at least one city in his traveling route.

REDA in this section was developed in the decomposition-based framework of multi-objective optimization. Furthermore, REDA was also hybridized with the evolutionary gradient search [46] (hREDA). In the mTSP, integer number representation is used to represent the permutation of the cities. The modeling and sampling steps of REDA or hREDA is illustrated as follows.

1. Modeling. Decode the integer representation of the cities into the binary representation. Train the network. Compute the $\delta(x_i^j)$ as (11) and (12). Encode the binary representation of $\delta(x_i^j)$ into integer representation. Construct the probabilistic model $P_g(x)$ by computing the marginal probability of each city $(c_1, ..., c_\beta)$, where $\beta = \Psi + \Omega - 1$, in each permutation location as follows.

$$P_g(x) = \begin{bmatrix} p_g(x_1 = c_1) & \cdots & p_g(x_\beta = c_1) \\ \vdots & \ddots & \vdots \\ p_g(x_1 = c_\beta) & \cdots & p_g(x_\beta = c_\beta) \end{bmatrix}$$

$$p_g(x_i = c_j) = \frac{\sum_{l=1}^{N} \delta_l(x_i = c_j) + \frac{Z_i}{\beta \times N}}{Z_i + \frac{Z_i^2}{\beta \times N}}$$

where $p_g(x_i)$ is the probability distribution of the cities at the position x_i of the chromosomes at generation g, $p_g(x_i = c_j)$ is the probability of city j to be located at the ith position of the chromosomes, c_j is the city j ($c_1 = 2 - \Omega, ..., c_\beta = \Psi$) and Z_i is the normalizing constant.

2. Sampling. Sample the constructed probabilistic model, $p_g(x)$, to generate N children solutions as follows.

$$
y_i = \begin{cases}
c_1 & \text{if random}\,(0,1) \leq p_g\,(x_i = c_1) \\
c_2 & \text{if } p_g(x_i = c_1) < \text{random}\,(0,1) \leq \sum_{j=1}^{2} p_g(x_i = c_j)) \\
\vdots & \\
c_\beta & \text{if } \sum_{j=1}^{\beta-1} p_g(x_i = c_j) < \text{random}(0,1) \leq \sum_{j=1}^{\beta} p_g(x_i = c_j)
\end{cases}
$$

where y_i is a newly generated city at i^{th} position of a chromosome.

The formulation of the MmTSP takes into account the weighted sum of total traveling cost of all salesmen and the highest traveling cost of any single salesman. The weight setting is dependent on the preference of the manager whether he wants to achieve the lowest total traveling cost of all salesmen or he wants to achieve the balancing of workload of all salesmen. If the aim is to obtain the lowest total traveling cost, the weights will be set to $\omega_1 = 1.0$, $\omega_2 = 0.0$. On the other hand, if the final objective is to balance the workload of all salesmen, the weights will then be set to $\omega_1 = 0.0$, $\omega_2 = 1.0$. However, if the aim is to achieve tradeoff between the two aims, then different weight settings should be employed.

The simulation results reported in [42] indicated that the hybrid REDA with decomposition (hREDA) is able to produce a set of diverse solutions but it is slightly inferior in terms of proximity to MOEA/D [44] in problems with small number of cities. In problems with a large number of cities, the simulation results showed that the decomposition algorithms (hREDA, REDA, and MOEA/D) achieve better Pareto front than the domination algorithms (NSGA-II). For the decomposition algorithms, hREDA generates a better set of diverse solutions than REDA and MOEA/D. However, the solutions generated by REDA have a better proximity than hREDA. Shim et al. [42] concluded that the decomposition algorithms scale well with the increase in the number of decision variables compared to the algorithms using the concept of domination. REDA uses global distribution of the parent solutions to guide the search process. The results showed that REDA have good proximity results, but poor solution diversity. Introducing local information into the evolutionary process, which helps the algorithm to further explore and exploit the search space, rectifies this limitation of REDA.

The findings also revealed that the total traveling cost increases with the increase in the number of salesmen. This is because when more salesmen are involved, the task gets more difficult since the algorithms need to determine the route for each salesman while maintaining the minimum total traveling cost at the same time. Since all salesmen need to return to the home city and the final assigned city could be far from the depot, additional traveling cost may be incurred. For hREDA, the gradient information weakens with the increase in the number of salesmen, resulting in the algorithm not being able to exploit the information as effectively. However, its performance was the best compared to the other algorithms.

Overall, the performances of algorithms using the decomposition framework (hREDA, REDA and MOEA/D) were superior to those of the algorithms based on the concept of domination (NSGA-II) in most of the problem settings. The superiority of the decomposition algorithms is attributed to the aggregation principle used for fitness assignment. The tournament could be carried out by simply comparing the aggregated fitness values of solutions. Solutions with higher fitness values will always be selected to survive and reproduce. On the other hand, the concept of domination requires that fitness be assigned to each solution based on their rank of domination. In many-objective problems, most of the solutions are non-dominated and are given lower ranks. This may prevent the tournament process from selecting promising solutions to survive. Thus, NSGA-II performed poorly compared to the decomposition algorithms.

4.5 Hybrid Adaptive MOEAs for Solving Continuous MOPs

In our recent work [47], we introduced two versions of hybrid adaptive MOEAs for solving continuous MOPs. The motivation of this study is as follows. Many multi-objective evolutionary algorithms (MOEAs) have been designed to solve MOPs. For example, MOEAs that use genetic algorithms (GAs) as the search technique are NSGA-II [15] and MOEA/D [44], among others. MOEAs that use differential evolutions (DE) as the search technique are Pareto differential evolution (PDE) [48], generalized differential evolution3 (GDE3) [49], and MOEA/D with DE [50], among others. Next, MOEAs that use estimation of distribution algorithms (EDAs) as the search approach are as discussed in Section II. Each of the above-mentioned algorithms is efficient in solving certain MOPs and has their own strengths and weaknesses. Furthermore, no evidence indicates that one of the EAs is superior to the others. Thus, it is possible that the synergy among the EAs can complement their weaknesses while maintaining their strengths.

In [47], an adaptive feature, which determines the proportion of the number of solutions to be produced by each EA in a generation, was proposed. Initially, each EA is given an equal chance to produce the initial solutions. After the reproduction processes, a number of promising solutions are selected and stored in an archive. Then, the proportion of the number of solutions to be generated by each optimizer in the next generation is calculated according to the proposed adaptive mechanism as illustrated in Fig. 4. Let ψ as the solutions in an archive. First, calculate the number of solutions in ψ that are generated by each EA. Afterward, the adaptive proportion rate ($Ar_g^{EA_i}$) at generation g for each EA is calculated according to Step 2. A learning rate ($\epsilon < 0$) is incorporated to the updating rule in Step 2 in order to moderate the influences of the proportion of the number of selected solutions in generation g to the whole evolutionary processes. This is because the optimizers that are able to generate a more number of promising solutions in the current generation may not be the superior optimizers in the next generation. In Step 3, a lower bound is set to the adaptive proportion rate. This is necessary since an optimizer may dominate other EAs and finally the adaptive proportion rate of this optimizer will become 1.0 while the adaptive proportion rate of other EAs will become 0.0. When this happens, all children solutions will only be generated by

%%Given a set of selected solutions that are stored in an archive (ψ)

1. Calculate the number of solutions in ψ that are generated by each EA, denoted as $D_g^{EA_i}$ where $i = 1, ..., M$, M is the number of EAs that are involved in the hybridization. In this paper, three EAs are involved. Thus, the number of solutions in ψ that are generated by each EA are denoted as $D_g^{EA\_1} = D_g^{GA}$, $D_g^{EA\_2} = D_g^{DE}$ and $D_g^{EA\_3} = D_g^{EDA}$.

2. Calculate the adaptive proportion rate for each EA as follows.

For $i = 1: M$

$Ar_g^{EA\_i} = Ar_{g-1}^{EA\_i} + \epsilon \times Pr_g^{EA\_i}$, where $Pr_g^{EA\_i} = D_g^{EA\_i}/N$

End For

where $Ar_g^{EA\_i}$ is the adaptive proportion rate at generation g for i^{th} EA, ϵ is the learning rate, $Pr_g^{EA\_i}$ is the current proportion rate and N is the archive size or the number of solutions in an archive.

3. Check for the lower bound of the adaptive proportion rate

For $i = 1: M$

If $Ar_g^{EA\_i} < l\_bound$

$Ar_g^{EA\_i} = l\_bound$

End For

4. Normalize the adaptive proportion rate so that the sum of the adaptive proportion rates is equal to 1.0

For $i = 1: M$

$Ar_g^{EA\_i} = Ar_g^{EA\_i} / (\sum_{i=1}^{M} Ar_g^{EA\_i})$

End For

Fig. 4. Pseudo-code of the adaptive mechanism

this optimizer till the end of the evolutionary processes. Thus, it is necessary to set a lower bound to the adaptive proportion rate to guarantee that the problem would not exist. Since the summation of all the adaptive proportion rates should be equal to 1.0, the final adaptive proportion rates should be normalized especially when Step 3 is applied (Step 4). Afterward, a typical evolutionary process is continued. Through this hybridization, the hybrid algorithms showed the best results in most of the MOPs.

5 Conclusion

This paper presented our recent works on synergy between PMGs and MOEAs. More specifically, a literature review focused on the existing work on MOEDAs has been outlined. Next, the possibility of constructing an EDA based on RBM has also been demonstrated. Next, five studies that implement REDA for solving different MOPs were given. The case studies are:(1) REDA with clustering in solving scalable MOPs (2) sampling study of REDA and its implementation in solving epistatic MOPs (3) synergy between REDA and PSO in tackling noisy MOPs (4) hybrid REDA with the EGS in evolving a set of permutation of cities in MmTSP problems and (5) hybrid adaptive MOEAs in solving various types of MOPs. The simulation results indicated that REDA is weak in addressing multi-modality problems, inferior to NSGA-II in exploiting the near optimal solutions, and mediocre in generating a set of diverse solutions. The positive aspects are that REDA can perform well in high dimensional problems and many objective problems, REDA is more robust than NSGA-II in noisy MOPs, and REDA has a faster convergence rate than NSGA-II even though it takes a higher computational time. The performance of REDA was improved through the hybridization with local search algorithms and the synergy with other evolutionary algorithms.

References

1. Deb, K.: Multi-objective Optimization using Evolutionary Algorithms. John Wiley & Sons, Chichester (2001)
2. Tan, K.C., Khor, E.F., Lee, L.H.: Multiobjective Evolutionary Algorithms and Applications. Springer (2005)
3. Larrañaga, P., Lozano, J.A.: Estimation of Distribution Algorithms: A New Tool for Evolutionary Computation. Kluwer, Norwell (2001)
4. Lozano, J.A., Larrañaga, P., Bengoetxea, E.: Towards a New Evolutionary Computation: Advances on Estimation of Distribution Algorithms. STUDFUZZ. Springer, New York (2006)
5. Neri, F., Mininno, E.: Memetic Compact Differential Evolution for Cartesian Robot Control. IEEE Computational Intelligence Magazine 5(2), 54–65 (2010)
6. Koller, D., Friedman, N.: Probabilistic Graphical Models: Principles and Techniques. MIT Press (2009)
7. Larrañaga, P.: Probabilistic Graphical Models and Evolutionary Computation. In: Plenary Lecturer in 2010 IEEE World Congress on Computational Intelligence (2010)
8. Ting, C.K., Zeng, W.M., Lin, T.C.: Linkage Discovery Through Data Mining. IEEE Computational Intelligence Magazine 5(1), 10–13 (2010)
9. Pelikan, M., Sastry, K., Goldberg, D.E.: Multiobjective Estimation of Distribution Algorithm. In: Scalable Optimization via Probabilistic Modeling, pp. 223–248. Springer (2006)
10. Bosman, P.A.N., Thierens, D.: Multi-objective Optimization with Diversity Preserving Mixture-based Iterated Density Estimation Evolutionary Algorithms. International Journal of Approximate Reasoning 31(3), 259–289 (2002)
11. Laumanns, M., Ocenasek, J.: Bayesian Optimization Algorithms for Multi-objective Optimization. In: Proceedings of the 7th International Conference on Parallel Problem Solving from Nature, pp. 298–307 (2002)
12. Costa, M., Minisci, E.: MOPED: A Multi-objective Parzen-Based Estimation of Distribution Algorithm for Continuous Problems. In: Fonseca, C.M., Fleming, P.J., Zitzler, E., Deb, K., Thiele, L. (eds.) EMO 2003. LNCS, vol. 2632, pp. 282–294. Springer, Heidelberg (2003)
13. Li, H., Zhang, Q., Tsang, E., Ford, J.A.: Hybrid Estimation of Distribution Algorithm for Multiobjective Knapsack Problem. In: Gottlieb, J., Raidl, G.R. (eds.) EvoCOP 2004. LNCS, vol. 3004, pp. 145–154. Springer, Heidelberg (2004)
14. Okabe, T., Jin, Y., Sendhoff, B., Olhofer, M.: Voronoi-based Estimation of Distribution Algorithm for Multi-objective Optimization. In: IEEE Congress on Evolutionary Computation, pp. 1594–1601 (2004)
15. Deb, K., Pratap, A., Agarwal, S., Meyarivan, T.: A Fast and Elitist Multiobjective Genetic Algorithm: NSGA-II. IEEE Transactions on Evolutionary Computation 6(2), 182–197 (2002)
16. Pelikan, M., Sastry, K., Goldberg, D.E.: Multiobjective hBOA, Clustering, and Scalability. In: Proceedings of the 2005 Conference on Genetic and Evolutionary Computation, pp. 663–670 (2005)

17. Sastry, K., Goldberg, D.E., Pelikan, M.: Limits of Scalability of Multiobjective Estimation of Distribution Algorithms. In: IEEE Congress on Evolutionary Computation, pp. 2217–2224 (2005)
18. Soh, H., Kirley, M.: moPGA: Towards a New Generation of Multi-objective Genetic Algorithms. In: IEEE Congress on Evolutionary Computation, pp. 1702–1709 (2006)
19. Bosman, P.A.N., Thierens, D.: Adaptive Variance Scaling in Continuous Multi-objective Estimation of Distribution Algorithms. In: Proceedings of the 9th Annual Conference on Genetic and Evolutionary Computation, pp. 500–507 (2007)
20. Zhong, X., Li, W.: A Decision-tree-based Multi-objective Estimation of Distribution Algorithm. In: Proceedings of the 2007 International Conference on Computational Intelligence and Security, pp. 114–118 (2007)
21. Zhang, Q., Zhou, Z., Jin, Y.: RM-MEDA: A Regularity Model-based Multiobjective Estimation of Distribution Algorithm. IEEE Transactions on Evolutionary Computation 12(1), 41–63 (2008)
22. Zhou, A., Zhang, Q., Jin, Y.: Approximating the Set of Pareto-optimal Solutions in Both the Decision and Objective Spaces by an Estimation of Distribution Algorithm. IEEE Transactions on Evolutionary Computation 13(5), 1167–1189 (2009)
23. Martí, L., García, J., Berlanga, A., Molina, J.: Solving Complex High-dimensional Problems with the Multi-objective Neural Estimation of Distribution Algorithm. In: Proceedings of the 11th Annual Conference on Genetic and Evolutionary Computation, pp. 619–626 (2009)
24. Huband, S., Hingston, P., Barone, L., While, L.: A Review of Multiobjective Test Problems and a Scalable Test Problem Toolkit. IEEE Transactions on Evolutionary Computation 10(5), 477–506 (2006)
25. Gao, Y., Hu, X., Liu, H., Feng, Y.: Multiobjective Estimation of Distribution Algorithm Combined with PSO for RFID Network Optimization. In: International Conference on Measuring Technology and Mechatronics Automation, pp. 736–739 (2010)
26. Li, X.: Niching Without Niching Parameters: Particle Swarm Optimization Using a Ring Topology. IEEE Transactions on Evolutionary Computation 14(1), 150–169 (2010)
27. Salakhutdinov, R., Mnih, A., Hinton, G.: Restricted Boltzmann Machines for Collaborative Filtering. In: Proceedings of the 24th International Conference on Machine Learning, pp. 791–798 (2007)
28. Hinton, G.E.: Training Products of Experts by Minimizing Contrastive Divergence. Neural Computation 14(8), 1771–1800 (2002)
29. Hinton, G.E., Salakhutdinov, R.: Reducing the Dimensionality of Data with Neural Networks. Science 313(5786), 504–507 (2006)
30. Tieleman, T.: Training Restricted Boltzmann Machines Using Approximations to the Likelihood Gradient. In: Proceedings of the 25th International Conference on Machine Learning, pp. 1064–1071 (2008)
31. Carreira-Perpiñán, M., Hinton, G.: On Contrastive Divergence Learning. In: 10th International Workshop on Artificial Intelligence and Statistics, pp. 59–66 (2005)
32. Lu, K., Yen, G.G.: Rank-density-based multiobjective genetic algorithm and benchmark test function study. IEEE Transactions on Evolutionary Computation 7(4), 325–343 (2003)
33. Deb, K., Sinha, A., Kukkonen, S.: Multi-objective Test Problems, Linkages, and Evolutionary Methodologies. In: Proceedings of the 8th Annual Conference on Genetic and Evolutionary Computation, pp. 1141–1148 (2006)

34. Tang, H., Shim, V.A., Tan, K.C., Chia, J.Y.: Restricted Boltzmann Machine Based Algorithm for Multi-objective Optimization. In: IEEE Congress on Evolutionary Computation, pp. 3958–3965 (2010)
35. Shim, V.A., Tan, K.C., Chia, J.Y.: An Investigation on Sampling Technique for Multi-objective Restricted Boltzmann Machine. In: IEEE Congress on Evolutionary Computation, pp. 1081–1088 (2010)
36. Shim, V.A., Tan, K.C., Chia, J.Y., Al Mamun, A.: Multi-objective Optimization with Estimation of Distribution Algorithm in a Noisy Environment. Evolutionary Computation (accepted)
37. Beyer, H.G.: Evolutionary Algorithms in Noisy Environments: Theoretical Issues and Guidelines for Practice. Computer Methods in Applied Mechanics and Engineering 186(2), 239–267 (2000)
38. Goh, C.K., Tan, K.C.: An Investigation on Noisy Environments in Evolutionary Multiobjective Optimization. IEEE Transactions on Evolutionary Computation 11(3), 354–381
39. Darwen, P., Pollack, J.: Coevolutionary learning on Noisy Tasks. In: IEEE Congress on Evolutionary Computation, pp. 1724–1731 (1999)
40. Hughes, E.J.: Evolutionary Multi-objective Ranking with Uncertainy and Noise. In: Proceedings of the First International Conference on Evolutionary Multi-Criterion Optimization, pp. 329–343 (2001)
41. Storn, R., Price, K.: Differential Evolution - A Simple and Efficient Heuristic for Global Optimization Over Continuous Spaces. Journal of Global Optimization 11(4), 341–359 (1997)
42. Shim, V.A., Tan, K.C., Cheong, C.Y.: A Hybrid Estimation of Distribution Algorithm with Decomposition for Solving the Multi-objective Multiple Traveling Salesmen Problem. IEEE Transactions on Systems, Man, and Cybernetics, Part C: Applications and Reviews (accepted)
43. Ishibuchi, H., Yoshida, T., Murata, T.: A Multi-objective Genetic Local Search Algorithm and Its Application to Flowshop Scheduling. IEEE Transactions on Systems, Man, and Cybernetics, Part C: Applications and Reviews 28(3), 204–223 (1998)
44. Zhang, Q., Li, H.: MOEA/D: A Multiobjective Evolutinoary Algorithm Based On Decomposition. IEEE Transactions on Evolutionary Computation 11(6), 712–731 (2007)
45. Okonjo, C.: An Effective Method of Balancing the Workload Amongst Salesmen. Omega 16(2), 159–163 (1998)
46. Goh, C.K., Ong, Y.S., Tan, K.C., Teoh, E.J.: An Investigation on Evolutionary Gradient Search for Multi-objective Optimization. In: IEEE Congress on Evolutionary Computation, pp. 3741–3746 (2008)
47. Shim, V.A., Tan, K.C.: A Hybrid Adaptive Evolutionary Algorithm in the Domination-based and Decomposition-based Frameworks of Multi-objective Optimization. In: IEEE Congress on Evolutionary Computation (accepted, 2012)
48. Abbass, H.A., Sarker, R.: The Pareto Differential Evolution Algorithm. International Journal of Artificial Intelligence Tools 11(4), 531–552 (2002)
49. Kukkonen, S., Lampinen, J.: GDE3: The Third Evolution Step of Generalized Differential Evolution. In: IEEE Congress on Evolutionary Computation, pp. 443–450 (2005)
50. Li, H., Zhang, Q.: Multiobjective Optimization Problems with Complicated Pareto Sets, MOEA/D and NSGA-II. IEEE Transactions on Evolutionary Computation 13(2), 284–302 (2009)

The Quest for Transitivity, a Showcase of Fuzzy Relational Calculus

Bernard De Baets

Department of Mathematical Modelling, Statistics and Bioinformatics,
Ghent University, Coupure links 653, B-9000 Gent, Belgium

Abstract. We present several relational frameworks for expressing similarities and preferences in a quantitative way. The main focus is on the occurrence of various types of transitivity in these frameworks. The first framework is that of fuzzy relations; the corresponding notion of transitivity is C-transitivity, with C a conjunctor. We discuss two approaches to the measurement of similarity of fuzzy sets: a logical approach based on biresidual operators and a cardinal approach based on fuzzy set cardinalities. The second framework is that of reciprocal relations; the corresponding notion of transitivity is cycle-transitivity. It plays a crucial role in the description of different types of transitivity arising in the comparison of (artificially coupled) random variables in terms of winning probabilities. It also embraces the study of mutual rank probability relations of partially ordered sets.

1 Introduction

Comparing objects in order to group together similar ones or distinguish better from worse is inherent to human activities in general and scientific disciplines in particular. In this overview paper, we present some relational frameworks that allow to express the results of such a comparison in a numerical way, typically by means of numbers in the unit interval. A first framework is that of fuzzy relations and we discuss how it can be used to develop cardinality-based, i.e. based on the counting of features, similarity measurement techniques. A second framework is that of reciprocal relations and we discuss how it can be used to develop methods for comparing random variables. Rationality considerations demand the presence of some kind of transitivity. We therefore review in detail the available notions of transitivity and point out where they occur.

This contribution is organised as follows. In Section 2, we present the two relational frameworks mentioned, the corresponding notions of transitivity and the connections between them. In Section 3, we explore the framework of fuzzy relations and its capacity for expressing the similarity of fuzzy sets. Section 4 is dedicated to the framework of reciprocal relations and its potential for the development of methods for the comparison of random variables. We wrap up in Section 5 with a short conclusion.

J. Liu et al. (Eds.): WCCI 2012 Plenary/Invited Lectures, LNCS 7311, pp. 145–165, 2012.
© Springer-Verlag Berlin Heidelberg 2012

2 Relational Frameworks and Their Transitivity

2.1 Fuzzy Relations

Transitivity is an essential property of relations. A (binary) *relation R* on a universe X (the *universe of discourse* or the *set of alternatives*) is called *transitive* if for any $(a, b, c) \in X^3$ it holds that $(a, b) \in R \land (b, c) \in R$ implies $(a, c) \in R$. Identifying R with its characteristic mapping, *i.e.* defining $R(a, b) = 1$ if $(a, b) \in R$, and $R(a, b) = 0$ if $(a, b) \notin R$, transitivity can be stated equivalently as $R(a, b) = 1 \land R(b, c) = 1$ implies $R(a, c) = 1$. Other equivalent formulations may be devised, such as

$$(R(a, b) \geq \alpha \land R(b, c) \geq \alpha) \Rightarrow R(a, c) \geq \alpha, \tag{1}$$

for any $\alpha \in \,]0, 1]$. Transitivity can also be expressed in the following functional form

$$\min(R(a, b), R(b, c)) \leq R(a, c). \tag{2}$$

Note that on $\{0, 1\}^2$ the minimum operation is nothing else but the Boolean conjunction.

A *fuzzy relation R* on X is an $X^2 \to [0, 1]$ mapping that expresses the degree of relationship between elements of X: $R(a, b) = 0$ means a and b are not related at all, $R(a, b) = 1$ expresses full relationship, while $R(a, b) \in \,]0, 1[$ indicates a partial degree of relationship only. In fuzzy set theory, formulation (2) has led to the popular notion of T-transitivity, where a t-norm is used to generalize Boolean conjunction. A binary operation $T : [0, 1]^2 \to [0, 1]$ is called a *t-norm* if it is increasing in each variable, has neutral element 1 and is commutative and associative. The three main continuous t-norms are the minimum operator $T_{\mathbf{M}}$, the algebraic product $T_{\mathbf{P}}$ and the Łukasiewicz t-norm $T_{\mathbf{L}}$ (defined by $T_{\mathbf{L}}(x, y) = \max(x + y - 1, 0)$). For an excellent monograph on t-norms and t-conorms, we refer to [39].

However, we prefer to work with a more general class of operations called conjunctors. A *conjunctor* is a binary operation $C : [0, 1]^2 \to [0, 1]$ that is increasing in each variable and coincides on $\{0, 1\}^2$ with the Boolean conjunction.

Definition 1. *Let C be a conjunctor. A fuzzy relation R on X is called C-transitive if for any $(a, b, c) \in X^3$ it holds that*

$$C(R(a, b), R(b, c)) \leq R(a, c). \tag{3}$$

Interesting classes of conjunctors are the classes of semi-copulas, quasi-copulas, copulas and t-norms. *Semi-copulas* are nothing else but conjunctors with neutral element 1 [30]. Where t-norms have the additional properties of commutativity and associativity, quasi-copulas are 1-Lipschitz continuous [33,44]. A *quasi-copula* is a semi-copula that is *1-Lipschitz continuous*: for any $(x, y, u, v) \in [0, 1]^4$ it holds that $|C(x, u) - C(y, v)| \leq |x - y| + |u - v|$. If instead of 1-Lipschitz continuity, C satisfies the *moderate growth property* (also called *2-monotonicity*): for any

$(x, y, u, v) \in [0, 1]^4$ such that $x \leq y$ and $u \leq v$ it holds that $C(x, v) + C(y, u) \leq C(x, u) + C(y, v)$, then C is called a *copula*.

Any copula is a quasi-copula, and therefore is 1-Lipschitz continuous; the converse is not true. It is well known that a copula is a t-norm if and only if it is associative; conversely, a t-norm is a copula if and only if it is 1-continuous. The t-norms T_M, T_P and T_L are copulas as well. For any quasi-copula C it holds that $T_L \leq C \leq T_M$. For an excellent monograph on copulas, we refer to [44].

2.2 Reciprocal Relations

Another interesting class of $X^2 \to [0, 1]$ mappings is the class of *reciprocal relations* Q (also called *ipsodual relations* or *probabilistic relations*) satisfying $Q(a, b) + Q(b, a) = 1$, for any $a, b \in X$. For such relations, it holds in particular that $Q(a, a) = 1/2$. Reciprocity is linked with completeness: let R be a complete ($\{0, 1\}$-valued) relation on X, which means that $\max(R(a, b), R(b, a)) = 1$ for any $a, b \in X$, then R has an equivalent $\{0, 1/2, 1\}$-valued reciprocal representation Q given by $Q(a, b) = 1/2(1 + R(a, b) - R(b, a))$.

Stochastic Transitivity. Transitivity properties for reciprocal relations rather have the logical flavor of expression (1). There exist various kinds of stochastic transitivity for reciprocal relations [3,42]. For instance, a reciprocal relation Q on X is called *weakly stochastic transitive* if for any $(a, b, c) \in X^3$ it holds that $Q(a, b) \geq 1/2 \wedge Q(b, c) \geq 1/2$ implies $Q(a, c) \geq 1/2$, which corresponds to the choice of $\alpha = 1/2$ in (1). In [11], the following generalization of stochastic transitivity was proposed.

Definition 2. *Let g be an increasing $[1/2, 1]^2 \to [0, 1]$ mapping such that $g(1/2, 1/2) \leq 1/2$. A reciprocal relation Q on X is called g-stochastic transitive if for any $(a, b, c) \in X^3$ it holds that*

$$(Q(a, b) \geq 1/2 \wedge Q(b, c) \geq 1/2) \Rightarrow Q(a, c) \geq g(Q(a, b), Q(b, c)).$$

Note that the condition $g(1/2, 1/2) \leq 1/2$ ensures that the reciprocal representation Q of any transitive complete relation R is always g-stochastic transitive. In other words, g-stochastic transitivity generalizes transitivity of complete relations. This definition includes the standard types of stochastic transitivity [42]:

  (i) *strong* stochastic transitivity when $g = \max$;
  (ii) *moderate* stochastic transitivity when $g = \min$;
(iii) *weak* stochastic transitivity when $g = 1/2$.

In [11], also a special type of stochastic transitivity has been introduced.

Definition 3. *Let g be an increasing $[1/2, 1]^2 \to [0, 1]$ mapping such that $g(1/2, 1/2) = 1/2$ and $g(1/2, 1) = g(1, 1/2) = 1$. A reciprocal relation Q on X is called g-isostochastic transitive if for any $(a, b, c) \in X^3$ it holds that*

$$(Q(a, b) \geq 1/2 \wedge Q(b, c) \geq 1/2) \Rightarrow Q(a, c) = g(Q(a, b), Q(b, c)).$$

The conditions imposed upon g again ensure that g-isostochastic transitivity generalizes transitivity of complete relations. Note that for a given mapping g, the property of g-isostochastic transitivity is much more restrictive than the property of g-stochastic transitivity.

FG-Transitivity. The framework of FG-transitivity, developed by Switalski [51,52], formally generalizes g-stochastic transitivity in the sense that $Q(a, c)$ is now bounded both from below and above by $[1/2, 1]^2 \to [0, 1]$ mappings.

Definition 4. *Let F and G be two $[1/2, 1]^2 \to [0, 1]$ mappings such that $F(1/2, 1/2) \leq 1/2 \leq G(1/2, 1/2)$, and $G(1/2, 1) = G(1, 1/2) = G(1, 1) = 1$ and $F \leq G$. A reciprocal relation Q on X is called FG-transitive if for any $(a, b, c) \in X^3$ it holds that*

$$(Q(a, b) \geq 1/2 \wedge Q(b, c) \geq 1/2)$$
$$\Downarrow$$
$$F(Q(a, b), Q(b, c)) \leq Q(a, c) \leq G(Q(a, b), Q(b, c)).$$

Cycle-Transitivity. For a reciprocal relation Q, we define for all $(a, b, c) \in X^3$ the following quantities [11]:

$$\alpha_{abc} = \min(Q(a, b), Q(b, c), Q(c, a)),$$
$$\beta_{abc} = \mathrm{median}(Q(a, b), Q(b, c), Q(c, a)),$$
$$\gamma_{abc} = \max(Q(a, b), Q(b, c), Q(c, a)).$$

Let us also denote $\Delta = \{(x, y, z) \in [0, 1]^3 \mid x \leq y \leq z\}$. A function $U : \Delta \to \mathbb{R}$ is called an *upper bound function* if it satisfies:

(i) $U(0, 0, 1) \geq 0$ and $U(0, 1, 1) \geq 1$;
(ii) for any $(\alpha, \beta, \gamma) \in \Delta$:

$$U(\alpha, \beta, \gamma) + U(1 - \gamma, 1 - \beta, 1 - \alpha) \geq 1. \tag{4}$$

The function $L : \Delta \to \mathbb{R}$ defined by $L(\alpha, \beta, \gamma) = 1 - U(1 - \gamma, 1 - \beta, 1 - \alpha)$ is called the *dual lower bound function* of the upper bound function U. Inequality (4) then simply expresses that $L \leq U$. Condition (i) again guarantees that cycle-transitivity generalizes transitivity of complete relations.

Definition 5. *A reciprocal relation Q on X is called cycle-transitive w.r.t. an upper bound function U if for any $(a, b, c) \in X^3$ it holds that*

$$L(\alpha_{abc}, \beta_{abc}, \gamma_{abc}) \leq \alpha_{abc} + \beta_{abc} + \gamma_{abc} - 1 \leq U(\alpha_{abc}, \beta_{abc}, \gamma_{abc}), \tag{5}$$

where L is the dual lower bound function of U.

Due to the built-in duality, it holds that if (5) is true for some (a, b, c), then this is also the case for any permutation of (a, b, c). In practice, it is therefore sufficient to check (5) for a single permutation of any $(a, b, c) \in X^3$. Alternatively, due to the same duality, it is also sufficient to verify the right-hand inequality (or equivalently, the left-hand inequality) for two permutations of any $(a, b, c) \in X^3$ (not being cyclic permutations of one another), e.g. (a, b, c) and (c, b, a). Hence, (5) can be replaced by

$$\alpha_{abc} + \beta_{abc} + \gamma_{abc} - 1 \leq U(\alpha_{abc}, \beta_{abc}, \gamma_{abc}).$$

Note that a value of $U(\alpha, \beta, \gamma)$ equal to 2 is used to express that for the given values there is no restriction at all (as $\alpha + \beta + \gamma - 1$ is always bounded by 2).

Two upper bound functions U_1 and U_2 are called *equivalent* if for any $(\alpha, \beta, \gamma) \in \Delta$ it holds that $\alpha + \beta + \gamma - 1 \leq U_1(\alpha, \beta, \gamma)$ is equivalent to $\alpha + \beta + \gamma - 1 \leq U_2(\alpha, \beta, \gamma)$.

If it happens that in (4) the equality holds for all $(\alpha, \beta, \gamma) \in \Delta$, then the upper bound function U is said to be *self-dual*, since in that case it coincides with its dual lower bound function L. Consequently, also (5) and (2.2) can only hold with equality. Furthermore, it then holds that $U(0, 0, 1) = 0$ and $U(0, 1, 1) = 1$.

Although C-transitivity is not intended to be applied to reciprocal relations, it can be cast quite nicely into the cycle-transitivity framework.

Proposition 1. [11] *Let C be a commutative conjunctor such that $C \leq T_M$. A reciprocal relation Q on X is C-transitive if and only if it is cycle-transitive w.r.t. the upper bound function U_C defined by*

$$U_C(\alpha, \beta, \gamma) = \min(\alpha + \beta - C(\alpha, \beta), \beta + \gamma - C(\beta, \gamma), \gamma + \alpha - C(\gamma, \alpha)).$$

Moreover, if C is 1-Lipschitz continuous, then U_C is given by

$$U_C(\alpha, \beta, \gamma) = \alpha + \beta - C(\alpha, \beta).$$

Consider the three basic t-norms (copulas) T_M, T_P and T_L:

(i) For $C = T_M$, we immediately obtain as upper bound function the median (the simplest self-dual upper bound function):

$$U_{T_M}(\alpha, \beta, \gamma) = \beta.$$

(ii) For $C = T_P$, we find

$$U_{T_P}(\alpha, \beta, \gamma) = \alpha + \beta - \alpha\beta.$$

(iii) For $C = T_L$, we obtain

$$U_{T_L}(\alpha, \beta, \gamma) = \begin{cases} \alpha + \beta & , \text{ if } \alpha + \beta < 1, \\ 1 & , \text{ if } \alpha + \beta \geq 1. \end{cases}$$

An equivalent upper bound function is given by $U'_{T_L}(\alpha, \beta, \gamma) = 1$.

Cycle-transitivity also incorporates stochastic transitivity, although the latter fits more naturally in the FG-transitivity framework; in particular, isostochastic transitivity corresponds to cycle-transitivity w.r.t. particular self-dual upper bound functions [11]. We have shown that the cycle-transitivity and FG-transitivity frameworks cannot easily be translated into one another, which underlines that these are two essentially different approaches [6].

One particular form of stochastic transitivity deserves our attention. A probabilistic relation Q on X is called *partially stochastic transitive* [31] if for any $(a, b, c) \in X^3$ it holds that

$$(Q(a, b) > 1/2 \wedge Q(b, c) > 1/2) \Rightarrow Q(a, c) \geq \min(Q(a, b), Q(b, c)).$$

Clearly, it is a slight weakening of moderate stochastic transitivity. Interestingly, also this type of transitivity can be expressed elegantly in the cycle-transitivity framework [24] by means of a simple upper bound function.

Proposition 2. *Cycle-transitivity w.r.t. the upper bound function U_{ps} defined by*

$$U_{ps}(\alpha, \beta, \gamma) = \gamma$$

is equivalent to partial stochastic transitivity.

A Frequentist Interpretation. Finally, we provide an interesting interpretation of some important types of upper bound functions [23].

Definition 6. *Let C be a conjunctor and Q be a reciprocal relation on X. A permutation $(a, b, c) \in X^3$ is called a C-triplet if*

$$C(R(a, b), R(b, c)) \leq R(a, c).$$

Let $\Delta_C(Q)$ denote the greatest number k such that any subset $\{a, b, c\} \subseteq X$ has k C-triplets. Obviously, Q is C-transitive if and only if $\Delta_C(Q) = 6$.

Proposition 3. *For any conjunctor $C \leq T_M$ and any reciprocal relation Q on X it holds that $3 \leq \Delta_C(Q) \leq 6$. More specifically, it holds that*

(i) $\Delta_{T_M}(Q) \in \{3, 5, 6\}$;
(ii) $\Delta_{T_P}(Q) \in \{3, 4, 5, 6\}$;
(iii) $\Delta_{T_L}(Q) \in \{3, 6\}$.

Proposition 4. *Let C be a commutative quasi-copula. A reciprocal relation Q on X is cycle-transitive w.r.t.*

(i) $U(\alpha, \beta, \gamma) = \beta + \gamma - C(\beta, \gamma)$ *if and only if $\Delta_C(Q) \geq 4$;*
(ii) $U(\alpha, \beta, \gamma) = \alpha + \gamma - C(\alpha, \gamma)$ *if and only if $\Delta_C(Q) \geq 5$;*
(iii) $U(\alpha, \beta, \gamma) = \alpha + \beta - C(\alpha, \beta)$ *if and only if $\Delta_C(Q) = 6$.*

Statement (iii) is nothing else but a rephrasing of Proposition 1. According the above proposition (statement (ii) applied to $C = T_M$), partial stochastic transitivity of a reciprocal relation implies that it is 'at least 5/6' T_M-transitive.

For ease of reference, we will refer to cycle-transitivity w.r.t. $U(\alpha, \beta, \gamma) = \beta + \gamma - C(\beta, \gamma)$ as weak C-transitivity, to cycle-transitivity w.r.t. $U(\alpha, \beta, \gamma) = \alpha + \gamma - C(\alpha, \gamma)$ as moderate C-transitivity, and to cycle-transitivity w.r.t. $U(\alpha, \beta, \gamma) = \alpha + \beta - C(\alpha, \beta)$ as (strong) C-transitivity.

3   Similarity of Fuzzy Sets

3.1   Basic Notions

Recall that an *equivalence relation* E on X is a reflexive, symmetric and transitive relation on X and that there exists a one-to-one correspondence between equivalence relations on X and partitions of X. In fuzzy set theory, the counterpart of an equivalence relation is a T-*equivalence*: given a t-norm T, a T-equivalence E on X is a fuzzy relation on X that is reflexive $(E(x,x) = 1)$, symmetric $(E(x,y) = E(y,x))$ and T-transitive. A T-equivalence is called a T-equality if $E(x,y)$ implies $x = y$.

For the prototypical t-norms, it is interesting to note that (see e.g. [15,17]):

(i) A fuzzy relation E on X is a $T_{\mathbf{L}}$-equivalence if and only if $d = 1 - E$ is a pseudo-metric on X.

(ii) A fuzzy relation E on X is a $T_{\mathbf{P}}$-equivalence if and only if $d = -\log E$ is a pseudo-metric on X.

(iii) A fuzzy relation E on X is a $T_{\mathbf{M}}$-equivalence if and only if $d = 1 - E$ is a pseudo-ultra-metric on X. Another interesting characterization is that a fuzzy relation E on X is a $T_{\mathbf{M}}$-equivalence if and only if for any $\alpha \in [0,1]$ its α-cut $E_\alpha = \{(x,y) \in X^2 \mid E(x,y) \geq \alpha\}$ is an equivalence relation on X. The equivalence classes of E_α become smaller for increasing α leading to the concept of a partition tree (see e.g. [26]).

3.2   A Logical Approach

To any left-continuous t-norm T, there corresponds a residual implicator $I_T : [0,1]^2 \to [0,1]$ defined by

$$I_T(x,y) = \sup\{z \in [0,1] \mid T(x,z) \leq y\},$$

which can be considered as a generalization of the Boolean implication. Note that $I_T(x,y) = 1$ if and only if $x \leq y$. In case $y < x$, one gets for the prototypical t-norms: $I_{\mathbf{M}}(x,y) = y$, $I_{\mathbf{P}}(x,y) = y/x$ and $I_{\mathbf{L}}(x,y) = \min(1-x+y,1)$. An essential property of the residual implicator of a left-continuous t-norm is related to the classical syllogism:

$$T(I_T(x,y), I_T(y,z)) \leq I_T(x,z)),$$

for any $(x,y,z) \in [0,1]^3$. The residual implicator is the main constituent of the biresidual operator $\mathcal{E}_T : [0,1]^2 \to [0,1]$ defined by

$$\mathcal{E}_T(x,y) = \min(I_T(x,y), I_T(y,x)) = I_T(\max(x,y), \min(x,y)),$$

which can be considered as a generalization of the Boolean equivalence. Note that $\mathcal{E}_T(x,y) = 1$ if and only if $x = y$. In case $x \neq y$, one gets for the prototypical t-norms: $\mathcal{E}_{\mathbf{M}}(x,y) = \min(x,y)$, $\mathcal{E}_{\mathbf{P}}(x,y) = \min(x,y)/\max(x,y)$ and $\mathcal{E}_{\mathbf{L}}(x,y) = 1 - |x - y|$.

Of particular importance in this discussion is the fact that $\mathcal{E}_T$ is a T-equality on $[0, 1]$. The biresidual operator obviously serves as a means for measuring equality of membership degrees. Any T-equality E on $[0, 1]$ can be extended in a natural way to $\mathcal{F}(X)$, the class of fuzzy sets in X:

$$E'(A, B) = \inf_{x \in X} E(A(x), B(x)).$$

It then holds that E' is a T-equality on $\mathcal{F}(X)$ if and only if E is a T-equality on $[0, 1]$. Starting from $\mathcal{E}_T$ we obtain the T-equality E^T. A second way of defining a T-equality on $\mathcal{F}(X)$ is by defining

$$E_T(A, B) = T(\inf_{x \in X} I_T(A(x), B(x)), \inf_{x \in X} I_T(B(x), A(x))).$$

The underlying idea is that in order to measure equality of two (fuzzy) sets A and B, one should both measure inclusion of A in B, and of B in A. Note that in general $E_T \subseteq E^T$, while $E_M = E^M$. These T-equivalences can be used as a starting point for building metrics on $\mathcal{F}(X)$. The above ways of measuring equality of fuzzy sets are very strict in the sense that the "worst" element decides upon the value.

Without going into detail, it is worth mentioning that there exist an appropriate notion of fuzzy partition, called T-partition [16], so that there exists a one-to-one correspondence between T-equalities on X and T-partitions of X [17].

3.3   A Cardinal Approach

Classical Cardinality-Based Similarity Measures. A common recipe for comparing objects is to select an appropriate set of features and to construct for each object a binary vector encoding the presence (1) or absence (0) of each of these features. Such a binary vector can be formally identified with the corresponding set of present features. The degree of similarity of two objects is then often expressed in terms of the cardinalities of the latter sets. We focus our attention on a family of $[0, 1]$-valued similarity measures that are rational expressions in the cardinalities of the sets involved [12]:

$$S(A, B) = \frac{x\, \alpha_{A,B} + t\, \omega_{A,B} + y\, \delta_{A,B} + z\, \nu_{A,B}}{x'\, \alpha_{A,B} + t'\, \omega_{A,B} + y'\, \delta_{A,B} + z'\, \nu_{A,B}},$$

with $A, B \in \mathcal{P}(X)$ (the powerset of a finite universe X),

$$\alpha_{A,B} = \min(|A \setminus B|, |B \setminus A|),$$
$$\omega_{A,B} = \max(|A \setminus B|, |B \setminus A|),$$
$$\delta_{A,B} = |A \cap B|,$$
$$\nu_{A,B} = |(A \cup B)^c|,$$

and $x, t, y, z, x', t', y', z' \in \{0, 1\}$. Note that these similarity measures are symmetric, *i.e.* $S(A, B) = S(B, A)$ for any $A, B \in \mathcal{P}(X)$.

Reflexive similarity measures, *i.e.* $S(A, A) = 1$ for any $A \in \mathcal{P}(X)$, are characterized by $y = y'$ and $z = z'$. We restrict our attention to the (still large) subfamily obtained by putting also $t = x$ and $t' = x'$ [5,14], *i.e.*

$$S(A, B) = \frac{x \triangle_{A,B} + y \delta_{A,B} + z \nu_{A,B}}{x' \triangle_{A,B} + y \delta_{A,B} + z \nu_{A,B}}, \tag{6}$$

with $\triangle_{A,B} = |A \triangle B| = |A \setminus B| + |B \setminus A|$. On the other hand, we allow more freedom by letting the parameters x, y, z and x' take positive real values. Note that these parameters can always be scaled to the unit interval by dividing both numerator and denominator of (6) by the greatest among the parameters. In order to guarantee that $S(A, B) \in [0, 1]$, we need to impose the restriction $0 \leq x \leq x'$. Since the case $x = x'$ leads to trivial measures taking value 1 only, we consider from here on $0 \leq x < x'$. The similarity measures gathered in Table 1 all belong to family (6); the corresponding parameter values are indicated in the table.

Table 1. Some well-known cardinality-based similarity measures

Measure	expression	x	x'	y	z	T
Jaccard [34]	$\frac{|A \cap B|}{|A \cup B|}$	0	1	0	1	$T_{\mathbf{L}}$
Simple Matching [50]	$1 - \frac{|A \triangle B|}{n}$	0	1	1	1	$T_{\mathbf{L}}$
Dice [29]	$\frac{2|A \cap B|}{|A \triangle B| + 2|A \cap B|}$	0	1	2	0	–
Rogers and Tanimoto [46]	$\frac{n - |A \triangle B|}{n + |A \triangle B|}$	0	2	1	1	$T_{\mathbf{L}}$
Sneath and Sokal 1 [49]	$\frac{|A \cap B|}{|A \cap B| + 2|A \triangle B|}$	0	2	1	0	$T_{\mathbf{L}}$
Sneath and Sokal 2 [49]	$1 - \frac{|A \triangle B|}{2n - |A \triangle B|}$	0	1	2	2	–

The $T_{\mathbf{L}}$- or $T_{\mathbf{P}}$-transitive members of family (6) are characterized in the following proposition.

Proposition 5. [14]

(i) The $T_{\mathbf{L}}$-*transitive members of family (6) are characterized by the necessary and sufficient condition* $x' \geq \max(y, z)$.

(ii) The $T_{\mathbf{P}}$-*transitive members of family (6) are characterized by the necessary and sufficient condition* $x x' \geq \max(y^2, z^2)$.

Fuzzy Cardinality-Based Similarity Measures. Often, the presence or absence of a feature is not clear-cut and is rather a matter of degree. Hence, if instead of binary vectors we have to compare vectors with components in the real unit interval $[0, 1]$ (the higher the number, the more the feature is present), the need arises to generalize the aforementioned similarity measures. In fact, in the same way as binary vectors can be identified with ordinary subsets of a finite universe X, vectors with components in $[0, 1]$ can be identified with fuzzy sets in X.

In order to generalize a cardinality-based similarity measure to fuzzy sets, we clearly need fuzzification rules that define the cardinality of a fuzzy set and translate the classical set-theoretic operations to fuzzy sets. As to the first, we stick to the following simple way of defining the cardinality of a fuzzy set, also known as the sigma-count of A [55]: $|A| = \sum_{x \in X} A(x)$. As to the second, we define the intersection of two fuzzy sets A and B in X in a pointwise manner by $A \cap B(x) = C(A(x), B(x))$, for any $x \in X$, where C is a commutative conjunctor. In [14], we have argued that commutative quasi-copulas are the most appropriate conjunctors for our purpose. Commutative quasi-copulas not only allow to introduce set-theoretic operations on fuzzy sets, such as $A \setminus B(x) = A(x) - C(A(x), B(x))$ and $A \triangle B(x) = A(x) + B(x) - 2C(A(x), B(x))$, they also preserve classical identities on cardinalities, such as $|A \setminus B| = |A| - |A \cap B|$ and $|A \triangle B| = |A \setminus B| + |B \setminus A| = |A| + |B| - 2|A \cap B|$. These identities allow to rewrite and fuzzify family (6) as

$$S(A, B) = \frac{x(a + b - 2u) + yu + z(n - a - b + u)}{x'(a + b - 2u) + yu + z(n - a - b + u)}, \tag{7}$$

with $a = |A|$, $b = |B|$ and $u = |A \cap B|$.

Bell-Inequalities and Preservation of Transitivity. Studying the transitivity of (fuzzy) cardinality-based similarity measures inevitably leads to the verification of inequalities on (fuzzy) cardinalities. We have established several powerful meta-theorems that provide an efficient and intelligent way of verifying whether a classical inequality on cardinalities carries over to fuzzy cardinalities [13]. These meta-theorems state that certain classical inequalities are preserved under fuzzification when modelling fuzzy set intersection by means of a commutative conjunctor that fulfills a number of Bell-type inequalities.

In [35], we introduced the classical Bell inequalities in the context of fuzzy probability calculus and proved that the following Bell-type inequalities for commutative conjunctors are necessary and sufficient conditions for the corresponding Bell-type inequalities for fuzzy probabilities to hold. The Bell-type inequalities for a commutative conjunctor C read as follows:

$$B_1 : T_{\mathbf{L}}(p, q) \leq C(p, q) \leq T_{\mathbf{M}}(p, q)$$
$$B_2 : 0 \leq p - C(p, q) - C(p, r) + C(q, r)$$
$$B_3 : p + q + r - C(p, q) - C(p, r) - C(q, r) \leq 1$$

for any $p, q, r \in [0, 1]$. Inequality B_2 is fulfilled for any commutative quasi-copula, while inequality B_3 only holds for certain t-norms [36], including the members of the Frank t-norm/copula family $T_\lambda^{\mathbf{F}}$ with $\lambda \leq 9 + 4\sqrt{5}$ [45]. Also note that inequality B_1 follows from inequality B_2.

Theorem 1. [13] *Consider a commutative conjunctor I that satisfies Bell inequalities B_2 and B_3. If for any ordinary subsets A, B and C of an arbitrary finite universe X it holds that*

$$\mathcal{H}(|A|, |B|, |C|, |A \cap B|, |A \cap C|, |B \cap C|, |X|) \geq 0,$$

where $\mathcal{H}$ denotes a continuous function which is homogeneous in its arguments, then it also holds for any fuzzy sets in an arbitrary finite universe Y.

If the function $\mathcal{H}$ does not depend explicitly upon $|X|$, then Bell inequality B_3 can be omitted. This meta-theorem allows us to identify conditions on the parameters of the members of family (7) leading to $T_{\mathbf{L}}$-transitive or $T_{\mathbf{P}}$-transitive fuzzy similarity measures. As our fuzzification is based on a commutative quasi-copula C, condition B_2 holds by default. The following proposition then is an immediate application.

Proposition 6. [13]

(i) *Consider a commutative quasi-copula C that satisfies B_3. The $T_{\mathbf{L}}$-transitive members of family (7) are characterized by $x' \geq \max(y, z)$.*

(ii) *The $T_{\mathbf{L}}$-transitive members of family (7) with $z = 0$ are characterized by $x' \geq y$.*

(iii) *Consider a commutative quasi-copula C that satisfies B_3. The $T_{\mathbf{P}}$-transitive members of family (7) are characterized by $x\,x' \geq \max(y^2, z^2)$.*

(iv) *The $T_{\mathbf{P}}$-transitive members of family (7) with $z = 0$ are characterized by $xx' \geq y^2$.*

However, as our meta-theorem is very general, it does not necessarily always provide the strongest results. For instance, tedious and lengthy direct proofs allow to eliminate condition B_3 from the previous theorem, leading to the following general result.

Proposition 7. [13] *Consider a commutative quasi-copula C.*

(i) *The $T_{\mathbf{L}}$-transitive members of family (7) are characterized by the necessary and sufficient condition $x' \geq \max(y, z)$.*

(ii) *The $T_{\mathbf{P}}$-transitive members of family (7) are characterized by the necessary and sufficient condition $x\,x' \geq \max(y^2, z^2)$.*

4 Comparison of Random Variables

4.1 Dice-Transitivity

Consider three dice A, B and C which, instead of the usual numbers, carry the following integers on their faces:

$$A = \{1, 3, 4, 15, 16, 17\}, \quad B = \{2, 10, 11, 12, 13, 14\}, \quad C = \{5, 6, 7, 8, 9, 18\}.$$

Denoting by $\mathcal{P}(X, Y)$ the probability that dice X wins from dice Y, we have $\mathcal{P}(A, B) = 20/36$, $\mathcal{P}(B, C) = 25/36$ and $\mathcal{P}(C, A) = 21/36$. It is natural to say that dice X is strictly preferred to dice Y if $\mathcal{P}(X, Y) > 1/2$, which reflects that dice X wins from dice Y in the long run (or that X statistically wins from Y, denoted $X >_s Y$). Note that $\mathcal{P}(Y, X) = 1 - \mathcal{P}(X, Y)$ which implies that the relation $>_s$ is asymmetric. In the above example, it holds that $A >_s B$, $B >_s C$

and $C >_s A$: the relation $>_s$ is not transitive and forms a cycle. In other words, if we interpret the probabilities $\mathcal{P}(X, Y)$ as constituents of a reciprocal relation on the set of alternatives $\{A, B, C\}$, then this reciprocal relation is even not weakly stochastic transitive.

This example can be generalized as follows: we allow the dice to possess any number of faces (whether or not this can be materialized) and allow identical numbers on the faces of a single or multiple dice. In other words, a generalized dice can be identified with a multiset of integers. Given a collection of m such generalized dice, we can still build a reciprocal relation Q containing the *winning probabilities* for each pair of dice [28]. For any two such dice A and B, we define

$$Q(A, B) = \mathcal{P}\{A \text{ wins from } B\} + \frac{1}{2}\mathcal{P}\{A \text{ and } B \text{ end in a tie}\}.$$

The dice or integer multisets may be identified with independent discrete random variables that are uniformly distributed on these multisets (i.e. the probability of an integer is proportional to its number of occurrences); the reciprocal relation Q may be regarded as a quantitative description of the pairwise comparison of these random variables.

In the characterization of the transitivity of this reciprocal relation, a type of cycle-transitivity, which can neither be seen as a type of C-transitivity, nor as a type of FG-transitivity, has proven to play a predominant role. For obvious reasons, this new type of transitivity has been called dice-transitivity.

Definition 7. *Cycle-transitivity w.r.t. the upper bound function U_D defined by*

$$U_D(\alpha, \beta, \gamma) = \beta + \gamma - \beta\gamma,$$

is called dice-transitivity.

Dice-transitivity is nothing else but a synonym for weak $T_\mathbf{P}$-product transitivity. According to Proposition 4, dice-transitivity of a reciprocal relation implies that it is 'at least 4/6' $T_\mathbf{P}$-transitive. Dice-transitivity can be situated between $T_\mathbf{L}$-transitivity and $T_\mathbf{P}$-transitivity, and also between $T_\mathbf{L}$-transitivity and moderate stochastic transitivity.

Proposition 8. *[28] The reciprocal relation generated by a collection of generalized dice is dice-transitive.*

4.2   A Method for Comparing Random Variables

Many methods can be established for the comparison of the components (random variables, r.v.) of a random vector $(X_1, \ldots, X_n)$, as there exist many ways to extract useful information from the joint cumulative distribution function (c.d.f.) $F_{X_1,\ldots,X_n}$ that characterizes the random vector. A first simplification consists in comparing the r.v. two by two. It means that a method for comparing r.v. should only use the information contained in the bivariate c.d.f. F_{X_i,X_j}. Therefore, one can very well ignore the existence of a multivariate c.d.f. and just describe mutual dependencies between the r.v. by means of the bivariate c.d.f. Of course one should be aware that not all choices of bivariate c.d.f. are compatible with a multivariate c.d.f. The problem of characterizing those ensembles of bivariate c.d.f. that can be identified with the marginal bivariate c.d.f. of a single multivariate c.d.f., is known as the *compatibility problem* [44].

A second simplifying step often made is to bypass the information contained in the bivariate c.d.f. to devise a comparison method that entirely relies on the one-dimensional marginal c.d.f. In this case there is even not a compatibility problem, as for any set of univariate c.d.f. F_{X_i}, the product $F_{X_1} F_{X_2} \cdots F_{X_n}$ is a valid joint c.d.f., namely the one expressing the independence of the r.v. There are many ways to compare one-dimensional c.d.f., and by far the simplest one is the method that builds a partial order on the set of r.v. using the principle of first order stochastic dominance [40]. It states that a r.v. X is weakly preferred to a r.v. Y if for all $u \in \mathbb{R}$ it holds that $F_X(u) \leq F_Y(u)$. At the extreme end of the chain of simplifications, are the methods that compare r.v. by means of a characteristic or a function of some characteristics derived from the one-dimensional marginal c.d.f. The simplest example is the weak order induced by the expected values of the r.v.

Proceeding along the line of thought of the previous section, a random vector $(X_1, X_2, \ldots, X_m)$ generates a reciprocal relation by means of the following recipe.

Definition 8. *Given a random vector* $(X_1, X_2, \ldots, X_m)$, *the binary relation* Q *defined by*

$$Q(X_i, X_j) = \mathcal{P}\{X_i > X_j\} + \frac{1}{2} \mathcal{P}\{X_i = X_j\}$$

is a reciprocal relation.

For two discrete r.v. X_i and X_j, $Q(X_i, X_j)$ can be computed as

$$Q(X_i, X_j) = \sum_{k > l} p_{X_i, X_j}(k, l) + \frac{1}{2} \sum_{k} p_{X_i, X_j}(k, k),$$

with p_{X_i, X_j} the joint probability mass function (p.m.f.) of (X_i, X_j). For two continuous r.v. X_i and X_j, $Q(X_i, X_j)$ can be computed as:

$$Q(X_i, X_j) = \int_{-\infty}^{+\infty} dx \int_{-\infty}^{x} f_{X_i, X_j}(x, y) \, dy,$$

with f_{X_i, X_j} the joint probability density function (p.d.f.) of (X_i, X_j).

For this pairwise comparison, one needs the two-dimensional marginal distributions. Sklar's theorem [44,48] tells us that if a joint cumulative distribution function F_{X_i, X_j} has marginals F_{X_i} and F_{X_j}, then there exists a copula C_{ij} such that for all x, y:

$$F_{X_i, X_j}(x, y) = C_{ij}(F_{X_i}(x), F_{X_j}(y)).$$

If X_i and X_j are continuous, then C_{ij} is unique; otherwise, C_{ij} is uniquely determined on $\mathrm{Ran}(F_{X_i}) \times \mathrm{Ran}(F_{X_j})$.

As the above comparison method takes into account the bivariate marginal c.d.f. it takes into account the dependence of the components of the random vector. The information contained in the reciprocal relation is therefore much richer than if, for instance, we would have based the comparison of X_i and X_j solely on their expected values. Despite the fact that the dependence structure is entirely captured by the multivariate c.d.f., the pairwise comparison is only apt to take into account pairwise dependence, as only bivariate c.d.f. are involved. Indeed, the bivariate c.d.f. do not fully disclose the dependence structure; the r.v. may even be pairwise independent while not mutually independent.

Since the copulas C_{ij} that couple the univariate marginal c.d.f. into the bivariate marginal c.d.f. can be different from one another, the analysis of the reciprocal relation and in particular the identification of its transitivity properties appear rather cumbersome. It is nonetheless possible to state in general, without making any assumptions on the bivariate c.d.f., that the probabilistic relation Q generated by an arbitrary random vector always shows some minimal form of transitivity.

Proposition 9. [7] *The reciprocal relation Q generated by a random vector is $T_{\mathbf{L}}$-transitive.*

4.3 Artificial Coupling of Random Variables

Our further interest is to study the situation where abstraction is made that the r.v. are components of a random vector, and all bivariate c.d.f. are enforced to depend in the same way upon the univariate c.d.f., in other words, we consider the situation of all copulas being the same, realizing that this might not be possible at all. In fact, this simplification is equivalent to considering instead of a random vector, a collection of r.v. and to artificially compare them, all in the same manner and based upon a same copula. The pairwise comparison then relies upon the knowledge of the one-dimensional marginal c.d.f. solely, as is the case in stochastic dominance methods. Our comparison method, however, is not equivalent to any known kind of stochastic dominance, but should rather be regarded as a graded variant of it (see also [8]).

The case $C = T_{\mathbf{P}}$ generalizes Proposition 8, and applies in particular to a collection of independent r.v. where all copulas effectively equal $T_{\mathbf{P}}$.

Proposition 10. [27,28] *The reciprocal relation Q generated by a collection of r.v. pairwisely coupled by $T_{\mathbf{P}}$ is dice-transitive.*

Next, we discuss the case when using one of the extreme copulas to artificially couple the r.v. In case $C = T_{\mathbf{M}}$, the r.v. are coupled comonotonically. Note that this case is possible in reality. Comparing with Proposition 9, the following proposition expresses that this way of coupling does not lead to a gain in transitivity.

Proposition 11. [24,25] *The reciprocal relation Q generated by a collection of r.v. pairwisely coupled by $T_{\mathbf{M}}$ is $T_{\mathbf{L}}$-transitive.*

In case $C = T_{\mathbf{L}}$, the r.v. are coupled countermonotonically. This assumption can never represent a true dependence structure for more than two r.v., due to the compatibility problem.

Proposition 12. [24,25] *The reciprocal relation Q generated by a collection of r.v. pairwisely coupled by $T_{\mathbf{L}}$ is partially stochastic transitive.*

The proofs of these propositions were first given for discrete uniformly distributed r.v. [25,28]. It allowed for an interpretation of the values $Q(X_i, X_j)$ as winning probabilities in a hypothetical dice game, or equivalently, as a method for the pairwise comparison of ordered lists of numbers. Subsequently, we have shown that as far as transitivity is concerned, this situation is generic and therefore characterizes the type of transitivity observed in general [24,27].

The above results are special cases of a more general result [7,9].

Proposition 13. *Consider a Frank copula $T_{\lambda}^{\mathbf{F}}$, then the reciprocal relation Q generated by a collection of random variables pairwisely coupled by $T_{\lambda}^{\mathbf{F}}$ is cycle-transitive w.r.t. to the upper bound function U^{λ} defined by:*

$$U^{\lambda}(\alpha, \beta, \gamma) = \beta + \gamma - T_{1/\lambda}^{\mathbf{F}}(\beta, \gamma).$$

4.4   Comparison of Special Independent Random Variables

Dice-transitivity is the generic type of transitivity shared by the reciprocal relations generated by a collection of independent r.v. If one considers independent r.v. with densities all belonging to one of the one-parameter families in Table 2, the corresponding reciprocal relation shows the corresponding type of cycle-transitivity listed in Table 3 [27].

Note that all upper bound functions in Table 3 are self-dual. More striking is that the two families of power-law distributions (one-parameter subfamilies of the two-parameter Beta and Pareto families) and the family of Gumbel distributions, all yield the same type of transitivity as exponential distributions, namely cycle-transitivity w.r.t. the self-dual upper bound function U_E defined by:

$$U_E(\alpha, \beta, \gamma) = \alpha\beta + \alpha\gamma + \beta\gamma - 2\alpha\beta\gamma.$$

Table 2. Parametric families of continuous distributions

Name	Density function $f(x)$		
Exponential	$\lambda e^{-\lambda x}$	$\lambda > 0$	$x \in [0, \infty[$
Beta	$\lambda x^{(\lambda-1)}$	$\lambda > 0$	$x \in [0, 1]$
Pareto	$\lambda x^{-(\lambda+1)}$	$\lambda > 0$	$x \in [1, \infty[$
Gumbel	$\mu e^{-\mu(x-\lambda)} e^{-e^{-\mu(x-\lambda)}}$	$\lambda \in \mathbb{R}, \mu > 0$	$x \in]-\infty, \infty[$
Uniform	$1/a$	$\lambda \in \mathbb{R}, a > 0$	$x \in [\lambda, \lambda + a]$
Laplace	$(e^{-|x-\lambda|/\mu})/(2\mu)$	$\lambda \in \mathbb{R}, \mu > 0$	$x \in]-\infty, \infty[$
Normal	$(e^{-(x-\lambda)^2/2\sigma^2})/\sqrt{2\pi\sigma^2}$	$\lambda \in \mathbb{R}, \sigma > 0$	$x \in]-\infty, \infty[$

Table 3. Cycle-transitivity for the continuous distributions in Table 1

Name	Upper bound function $U(\alpha, \beta, \gamma)$
Exponential Beta Pareto Gumbel	$\alpha\beta + \alpha\gamma + \beta\gamma - 2\alpha\beta\gamma$
Uniform	$\begin{cases} \beta + \gamma - 1 + \frac{1}{2}[\max(\sqrt{2(1-\beta)} + \sqrt{2(1-\gamma)} - 1, 0)]^2 & \beta \geq 1/2 \\[2mm] \alpha + \beta - \frac{1}{2}[\max(\sqrt{2\alpha} + \sqrt{2\beta} - 1, 0)]^2 & \beta < 1/2 \end{cases}$
Laplace	$\begin{cases} \beta + \gamma - 1 + f^{-1}(f(1-\beta) + f(1-\gamma)) & \beta \geq 1/2 \\[2mm] \alpha + \beta - f^{-1}(f(\alpha) + f(\beta)) & \beta < 1/2 \end{cases}$ with $f^{-1}(x) = \frac{1}{2}\left(1 + \frac{x}{2}\right)e^{-x}$
Normal	$\begin{cases} \beta + \gamma - 1 + \Phi(\Phi^{-1}(1-\beta) + \Phi(1-\gamma)) & \beta \geq 1/2 \\[2mm] \alpha + \beta - \Phi(\Phi^{-1}(\alpha) + \Phi^{-1}(\beta)) & \beta < 1/2 \end{cases}$ with $\Phi(x) = (\sqrt{2\pi})^{-1}\int_{-\infty}^{x} e^{-t^2/2}dt$

Cycle-transitivity w.r.t. U_E can also be expressed as

$$\alpha_{abc}\beta_{abc}\gamma_{abc} = (1 - \alpha_{abc})(1 - \beta_{abc})(1 - \gamma_{abc}),$$

which is equivalent to the notion of multiplicative transitivity [53]. A reciprocal relation Q on X is called *multiplicatively transitive* if for any $(a, b, c) \in X^3$ it holds that

$$\frac{Q(a,c)}{Q(c,a)} = \frac{Q(a,b)}{Q(b,a)} \cdot \frac{Q(b,c)}{Q(c,b)}.$$

In the cases of the unimodal uniform, Gumbel, Laplace and normal distributions we have fixed one of the two parameters in order to restrict the family to a one-parameter subfamily, mainly because with two free parameters, the formulae become utmost cumbersome. The one exception is the two-dimensional family of normal distributions. In [27], we have shown that the corresponding reciprocal relation is in that case moderately stochastic transitive.

4.5   Mutual Rank Transitivity in Posets

Partially ordered sets, posets for short, are witnessing an increasing interest in various fields of application. They allow for incomparability of elements and can be conveniently visualized by means of a Hasse diagram. Two such fields are environmetrics and chemometrics [1,2]. In these applications, most methods eventually require a linearization of the poset. A standard way of doing so is to rank the elements on the basis of their averaged ranks, i.e. their average position computed over all possible linear extensions of the poset. Although the computation of these averaged ranks has become feasible for posets of reasonable size [19], they suffer from a weak information content as they are based on marginal distributions only, as explained further. For this reason, interest is shifting to mutual rank probabilities instead.

The mutual rank probability relation is an intriguing object that can be associated with any finite poset. For any two elements of the poset, it expresses the probability that the first succeeds the second in a random linear extension of that poset. Its computation is feasible as well for posets of reasonable size [19,21], and approximation methods are available for more extensive posets [18]. However, exploiting the information contained in the mutual rank probability relation to come up with a ranking of the elements is not obvious. Simply ranking one element higher than another when the corresponding mutual rank probability is greater than $1/2$ is not appropriate, as it is prone to generating cycles (called linear extension majority cycles in this context [22,38]). A solution to this problem requires a better understanding, preferably a characterization, of the transitivity of mutual rank probability relations, coined proportional probabilistic transitivity by Fishburn [32], and, for the sake of clarity, renamed mutual rank transitivity here. A weaker type of transitivity (called δ^*-transitivity, expression not shown here) has been identified by Kahn and Yu [37] and Yu [54]. We have identified a weaker type of transitivity, yet enabling us to position mutual rank transitivity within the cycle-transitivity framework.

Consider a finite poset $(P, \leq)$. The discrete random variable X_a denotes the position (rank) of an element $a \in P$ in a random linear extension of P. The *mutual rank probability* $p_{a>b}$ of two different elements $a, b \in P$ is defined as the fraction of linear extensions of P in which a succeeds b (a is ranked higher than b), i.e., $p_{a>b} = \text{Prob}\{X_a > X_b\}$. The $[0,1]$-valued relation $Q_P : P^2 \to [0,1]$ defined by $Q_P(a,b) = p_{a>b}$, for all $a, b \in P$ with $a \neq b$, and $Q_P(a,a) = 1/2$, for all $a \in P$, is a reciprocal relation. Note that in the way described above, with any finite poset $P = \{a_1, \ldots, a_n\}$ we associate a unique discrete random vector $X = (X_{a_1}, \ldots, X_{a_n})$ with joint distribution function $F_{X_{a_1}, \ldots, X_{a_n}}$. The mutual rank probabilities $p_{a_i > a_j}$ are then computed from the bivariate marginal distributions $F_{X_{a_i}, X_{a_j}}$.

Note that, despite the fact that the joint distribution function $F_{X_{a_1}, \ldots, X_{a_n}}$ does not lend itself to an explicit expression, a fair amount of pairwise couplings are of a very simple type. If it holds that $a > b$, then a succeeds b in all linear extensions of P, whence X_a and X_b are comonotone. For pairs of incomparable elements, the bivariate couplings can vary from pair to pair. Certainly, these couplings cannot all be counter-monotone. Despite all this, it is possible to obtain transitivity results on mutual rank probability relations [10].

Definition 9. *The mutual rank probability relation Q_P associated with a finite poset $(P, \leq)$ is cycle-transitive w.r.t. the upper bound function U_P defined by*

$$U_P(\alpha, \beta, \gamma) = \alpha + \gamma - \alpha\gamma.$$

Proposition 4 implies that the mutual rank probability relation of a poset it is 'at least 5/6' $T_\mathbf{P}$-transitive.

5   Conclusion

We have introduced the reader to two relational frameworks and the wide variety of types of transitivity they cover. When considering different types of transitivity, we can try to distinguish weaker or stronger types. Obviously, one type is called weaker than another, if it is implied by the latter. Hence, we can equip a collection of types of transitivity with this natural order relation and depict it graphically by means of a Hasse diagram.

The Hasse diagram containing all types of transitivity of reciprocal relations encountered in this contribution is shown in Figure 1. At the lower end of the diagram, $T_\mathbf{M}$-transitivity and multiplicative transitivity, two types of cycle-transitivity w.r.t. a self-dual upper bound function, are incomparable and can be considered as the strongest types of transitivity. At the upper end of the diagram, also $T_\mathbf{L}$-transitivity and weak stochastic transitivity are incomparable and can be considered as the weakest types of transitivity. Furthermore, note that the subchain consisting of partial stochastic transitivity, moderate product

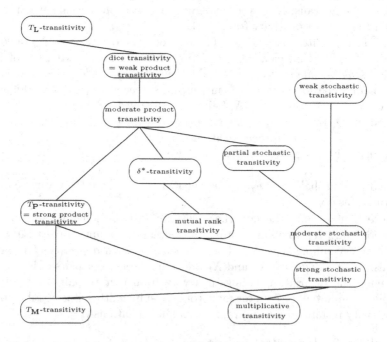

Fig. 1. Hasse diagram with different types of transitivity of reciprocal relations (weakest types at the top, strongest types at the bottom

transitivity and weak product transitivity, bridges the gap between g-stochastic transitivity and T-transitivity.

Anticipating on future work, in particular on applications, we can identify two important directions. The first direction concerns the use of fuzzy similarity measures. Moser [43] has shown recently that the T-equality E^T, with $T = T_{\mathbf{P}}$ or $T = T_{\mathbf{L}}$, is positive semi-definite. This question has not yet been addressed for the fuzzy cardinality-based similarity measures. Results of this type allow to bridge the gap between the fuzzy set community and the machine learning community, making some fuzzy similarity measures available as potential kernels for the popular kernel-based learning methods, either on their own or in combination with existing kernels (see e.g. [41] for an application of this type).

The second direction concerns the further exploitation of the results on the comparison of random variables. As mentioned, the approach followed here can be seen as a graded variant of the increasingly popular notion of stochastic dominance. Future research will have to clarify how these graded variants can be defuzzified in order to come up with meaningful partial orderings of random variables that are more informative than the classical notions of stochastic dominance. Some results into that direction can be found in [8,20].

Acknowledgement. This chapter is a slightly updated version of [4].

References

1. Brüggemann, R., Simon, U., Mey, S.: Estimation of averaged ranks by extended local partial order models. MATCH Communications in Mathematical and in Computer Chemistry 54, 489–518 (2005)
2. Brüggemann, R., Sørensen, P., Lerche, D., Carlsen, L.: Estimation of averaged ranks by a local partial order model. Journal of Chemical Information and Computer Science 4, 618–625 (2004)
3. David, H.A.: The Method of Paired Comparisons, Griffin's Statistical Monographs & Courses, vol. 12. Charles Griffin & Co. Ltd., London (1963)
4. De Baets, B.: Similarity of fuzzy sets and dominance of random variables: a quest for transitivity. In: Della Riccia, G., Dubois, D., Lenz, H.-J., Kruse, R. (eds.) Preferences and Similarities. CISM Courses and Lectures, vol. 504, pp. 1–22. Springer (2008)
5. De Baets, B., De Meyer, H.: Transitivity-preserving fuzzification schemes for cardinality-based similarity measures. European J. Oper. Res. 160, 726–740 (2005)
6. De Baets, B., De Meyer, H.: Transitivity frameworks for reciprocal relations: cycle-transitivity versus FG-transitivity. Fuzzy Sets and Systems 152, 249–270 (2005)
7. De Baets, B., De Meyer, H.: Cycle-transitive comparison of artificially coupled random variables. Internat. J. Approximate Reasoning 96, 352–373 (2005)
8. De Baets, B., De Meyer, H.: Toward graded and nongraded variants of stochastic dominance. In: Batyrshin, I., Kacprzyk, J., Sheremetov, J., Zadeh, L. (eds.) Perception-based Data Mining and Decision Making in Economics and Finance. SCI, vol. 36, pp. 252–267. Springer (2007)
9. De Baets, B., De Meyer, H.: On a conjecture about the Frank copula family. Internat. J. of Approximate Reasoning (submitted)

10. De Baets, B., De Meyer, H., De Loof, K.: On the cycle-transitivity of the mutual rank probability relation of a poset. Fuzzy Sets and Systems 161, 2695–2708 (2010)
11. De Baets, B., De Meyer, H., De Schuymer, B., Jenei, S.: Cyclic evaluation of transitivity of reciprocal relations. Social Choice and Welfare 26, 217–238 (2006)
12. De Baets, B., De Meyer, H., Naessens, H.: A class of rational cardinality-based similarity measures. J. Comput. Appl. Math. 132, 51–69 (2001)
13. De Baets, B., Janssens, S., De Meyer, H.: Meta-theorems on inequalities for scalar fuzzy set cardinalities. Fuzzy Sets and Systems 157, 1463–1476 (2006)
14. De Baets, B., Janssens, S., De Meyer, H.: On the transitivity of a parametric family of cardinality-based similarity measures. Internat. J. Approximate Reasoning 50, 104–116 (2009)
15. De Baets, B., Mesiar, R.: Pseudo-metrics and $\mathcal{T}$-equivalences. J. Fuzzy Math. 5, 471–481 (1997)
16. De Baets, B., Mesiar, R.: T-partitions. Fuzzy Sets and Systems 97, 211–223 (1998)
17. De Baets, B., Mesiar, R.: Metrics and $\mathcal{T}$-equalities. J. Math. Anal. Appl. 267, 531–547 (2002)
18. De Loof, K., De Baets, B., De Meyer, H.: Approximation of average ranks in posets. MATCH - Communications in Mathematical and in Computer Chemistry 66, 219–229 (2011)
19. De Loof, K., De Baets, B., De Meyer, H., Brüggemann, R.: A hitchhiker's guide to poset ranking. Combinatorial Chemistry and High Throughput Screening 11, 734–744 (2008)
20. De Loof, K., De Meyer, H., De Baets, B.: Graded stochastic dominance as a tool for ranking the elements of a poset. In: Lopéz-Díaz, M., Gil, M., Grzegorzewski, P., Hyrniewicz, O., Lawry, J. (eds.) Soft Methods for Integrated Uncertainty Modelling. AISC, pp. 273–280. Springer (2006)
21. De Loof, K., De Baets, B., De Meyer, H.: Exploiting the lattice of ideals representation of a poset. Fundamenta Informaticae 71, 309–321 (2006)
22. De Loof, K., De Baets, B., De Meyer, H.: Cycle-free cuts of mutual rank probability relations. Computers and Mathematics with Applications (submitted)
23. De Loof, K., De Baets, B., De Meyer, H., De Schuymer, B.: A frequentist view on cycle-transitivity w.r.t. commutative dual quasi-copulas. Fuzzy Sets and Systems (submitted)
24. De Meyer, H., De Baets, B., De Schuymer, B.: On the transitivity of the comonotonic and countermonotonic comparison of random variables. J. Multivariate Analysis 98, 177–193 (2007)
25. De Meyer, H., De Baets, B., De Schuymer, B.: Extreme copulas and the comparison of ordered lists. Theory and Decision 62, 195–217 (2007)
26. De Meyer, H., Naessens, H., De Baets, B.: Algorithms for computing the mintransitive closure and associated partition tree of a symmetric fuzzy relation. European J. Oper. Res. 155, 226–238 (2004)
27. De Schuymer, B., De Meyer, H., De Baets, B.: Cycle-transitive comparison of independent random variables. J. Multivariate Analysis 96, 352–373 (2005)
28. De Schuymer, B., De Meyer, H., De Baets, B., Jenei, S.: On the cycle-transitivity of the dice model. Theory and Decision 54, 264–285 (2003)
29. Dice, L.: Measures of the amount of ecologic associations between species. Ecology 26, 297–302 (1945)
30. Durante, F., Sempi, C.: Semicopulæ. Kybernetika 41, 315–328 (2005)
31. Fishburn, P.: Binary choice probabilities: On the varieties of stochastic transitivity. J. Math. Psych. 10, 327–352 (1973)

32. Fishburn, P.: Proportional transitivity in linear extensions of ordered sets. Journal of Combinatorial Theory Series B 41, 48–60 (1986)
33. Genest, C., Quesada-Molina, J.J., Rodríguez-Lallena, J.A., Sempi, C.: A characterization of quasi-copulas. Journal of Multivariate Analysis 69, 193–205 (1999)
34. Jaccard, P.: Nouvelles recherches sur la distribution florale. Bulletin de la Sociétée Vaudoise des Sciences Naturelles 44, 223–270 (1908)
35. Janssens, S., De Baets, B., De Meyer, H.: Bell-type inequalities for quasi-copulas. Fuzzy Sets and Systems 148, 263–278 (2004)
36. Janssens, S., De Baets, B., De Meyer, H.: Bell-type inequalities for parametric families of triangular norms. Kybernetika 40, 89–106 (2004)
37. Kahn, J., Yu, Y.: Log-concave functions and poset probabilities. Combinatorica 18, 85–99 (1998)
38. Kislitsyn, S.: Finite partially ordered sets and their associated sets of permutations. Matematicheskiye Zametki 4, 511–518 (1968)
39. Klement, E., Mesiar, R., Pap, E.: Triangular Norms. In: Trends in Logic. Studia Logica Library, vol. 8. Kluwer Academic Publishers (2000)
40. Levy, H.: Stochastic Dominance. Kluwer Academic Publishers, MA (1998)
41. Maenhout, S., De Baets, B., Haesaert, G., Van Bockstaele, E.: Support Vector Machine regression for the prediction of maize hybrid performance. Theoretical and Applied Genetics 115, 1003–1013 (2007)
42. Monjardet, B.: A generalisation of probabilistic consistency: linearity conditions for valued preference relations. In: Kacprzyk, J., Roubens, M. (eds.) Non-Conventional Preference Relations in Decision Making. LNEMS, vol. 301. Springer, Berlin (1988)
43. Moser, B.: On representing and generating kernels by fuzzy equivalence relations. Journal of Machine Learning Research 7, 2603–2620 (2006)
44. Nelsen, R.: An Introduction to Copulas. Lecture Notes in Statistics, vol. 139. Springer, New York (1998)
45. Pykacz, J., D'Hooghe, B.: Bell-type inequalities in fuzzy probability calculus. Internat. J. of Uncertainty, Fuzziness and Knowledge-Based Systems 9, 263–275 (2001)
46. Rogers, D., Tanimoto, T.: A computer program for classifying plants. Science 132, 1115–1118 (1960)
47. Schweizer, B., Sklar, A.: Probabilistic Metric Spaces. North-Holland (1983)
48. Sklar, A.: Fonctions de répartition à n dimensions et leurs marges. Publ. Inst. Statist. Univ. Paris 8, 229–231 (1959)
49. Sneath, P., Sokal, R.: Numerical Taxonomy. WH Freeman, San Francisco (1973)
50. Sokal, R., Michener, C.: A statistical method for evaluating systematic relationships. Univ. of Kansas Science Bulletin 38, 1409–1438 (1958)
51. Switalski, Z.: Transitivity of fuzzy preference relations – an empirical study. Fuzzy Sets and Systems 118, 503–508 (2001)
52. Switalski, Z.: General transitivity conditions for fuzzy reciprocal preference matrices. Fuzzy Sets and Systems 137, 85–100 (2003)
53. Tanino, T.: Fuzzy preference relations in group decision making. In: Kacprzyk, J., Roubens, M. (eds.) Non-Conventional Preference Relations in Decision Making. LNEMS, vol. 301. Springer, Berlin (1988)
54. Yu, Y.: On proportional transitivity of ordered sets. Order 15, 87–95 (1998)
55. Zadeh, L.: Fuzzy sets. Information and Control 8, 338–353 (1965)

Cognition-Inspired Fuzzy Modelling

Maria Rifqi[1,2,3]

[1] UPMC Univ Paris 06, UMR 7606, LIP6, F-75005, Paris, France
Maria.Rifqi@lip6.fr
[2] CNRS, UMR 7606, LIP6, F-75005, Paris, France
[3] Université Panthéon-Assas - Paris 2, F-75005 Paris, France

Abstract. This chapter presents different notions used for fuzzy modelling that formalize fundamental concepts used in cognitive psychology. From a cognitive point of view, the tasks of categorization, pattern recognition or generalization lie in the notions of similarity, resemblance or prototypes. The same tasks are crucial in Artificial Intelligence to reproduce human behaviors. As most real world concepts are messy and open-textured, fuzzy logic and fuzzy set theory can be the relevant framework to model all these key notions.

On the basis of the essential works of Rosch and Tversky, and on the critics formulated on the inadequacy of fuzzy logic to model cognitive concepts, we study a formal and computational approach of the notions of similarity, typicality and prototype, using fuzzy set theory. We propose a framework to understand the different properties and possible behaviors of various families of similarities. We highlight their semantic specifics and we propose numerical tools to quantify these differences, considering different views. We propose also an algorithm for the construction of fuzzy prototypes that can be extended to a classification method.

Keywords: similarity, typicality, prototype, fuzzy logic, prototype-based learning.

1 Introduction

Artificial Intelligence drew large parts of its inspiration in psychological and cognitive sciences. From a cognitive point of view, the tasks of categorization, pattern recognition, or generalization lie in the notions of similarity [47], resemblance [49] or prototypes [43]. The same tasks are crucial in Artificial Intelligence to reproduce human behaviors, especially in machine learning or in information retrieval.

In this chapter, two fundamental notions will be investigated: the notion of typicality (or prototype) and the notion of similarity. These notions are crucial and linked in cognitive science. In machine learning or in information retrieval, most of the tasks rely on a similarity measure. Then, it is natural to deeply study how assess similarity between objects, on which criteria, for which purpose, etc. Two main approaches can be distinguished: an approach coming from statistical analysis, called geometric model, and another approach coming from cognitive

J. Liu et al. (Eds.): WCCI 2012 Plenary/Invited Lectures, LNCS 7311, pp. 166–184, 2012.
© Springer-Verlag Berlin Heidelberg 2012

science. In the latter point of view, Tversky's contrast model [47] is considered as a reference for the framework of similarity measures we propose. Each approach implies a set of properties that a similarity has to satisfy. Because of the various possible properties and because of the large amount of existing measures, our aim is to propose tools enabling to compare them. We proposed a framework [5] that exhibits various families of similarities. Their semantic specifics are highlighted and numerical tools to quantify these differences, considering different views, are proposed.

The notion of typicality or prototype is less studied in machine learning or in information retrieval. However, the cognitive principle as categorization principle as Rosch [43] explained can be applied for a classification task, as we will show it in the following. The proposed construction of fuzzy prototype fully rely on her vision.

Moreover, in cognitive science, concepts are considered as graded, messy, open-textured, imprecise, uncertain. All these terms evoke the fuzzy set theory which aims at representing such concepts. Hence, fuzzy set theory is our formal framework for dealing with imprecision and graduality, inherent in natural categories.

This chapter is organised as follows: the first section is devoted to present the two main approaches for the issue of similarity judgments by means of a similarity measure: geometric approach and cognitive approach, with a particular attention to Tversky's model. The second section addresses the comparison of measures of similarity: a proposed framework aiming at classifying measures in different families of similarity measures is presented. Besides, numerical tools are given in order to quantify the differences existing between similarity measures, even from a same family. The third section presents the prototype theory as well as the link between graduality, conceptual combination and fuzzy logic. In the last section, an automatic method for constructing fuzzy prototypes is presented as well as its extension to complex classes and its use in machine learning tasks.

2 Similarity Judgments

Similarity playing a key role in the human learning ability to constitute knowledge, many studies, theoretical and experimental, in cognitive psychology [17] have been dedicated to better understand how it can be modeled to be as close as possible to human similarity judgments. In this section, two main approaches are presented: first, the geometric one, and then the cognitive one. Lastly, Tversky's model is described.

2.1 Geometric Approach: Metric Properties

Similarity is fundamental in statistics and data analysis especially in classification that lies in the hypothesis that objects that are similar belong to a same cluster. Distance measure is often used to evaluate the proximity among objects. Its semantics is nothing than a reversed notion of the similarity measure: similarity value can be deduced from the distance value. A distance d is a mathematical concept and follows precise laws: X denoting universe, for all $x, y, z \in X$:

- Reflexivity: $d(x,x) = 0$
- Separation: $d(x,y) = 0$ implies $x = y$
- Symmetry: $d(x,y) = d(y,x)$
- Triangle inequality: $d(x,z) \leq d(x,y) + d(y,z)$

2.2   Cognitive Approach: Experimental Critics

Metric models have been the main approaches for analyzing similarity [46]. Each property satisfied by a distance can appear reasonable. However, Tversky showed [47] that all of them are empirically violated by human similarity judgments. Reflexivity and separation properties are experimentally not observed: an object is identified as another object more frequently than it is identified as itself. The violation of symmetry may appear as a surprising observation. Tversky argued that it is because we forget that the meaning behind the similarity implies a directional comparison. A statement of the form "x is like y" implies that x is a subject and y the referent. Metaphors are typical examples illustrating this asymmetry of similarity: we say "Turks fight like tigers" and not "tigers fight like Turks" or "my love is as deep as the ocean", not "the ocean is as deep as my love" [47]. In all these examples, the subject is compared to a prototype and not the converse. Lastly, the triangular inequality of a distance or the transitivity of a similarity measure "should not be accepted as a cornerstone of similarity models" as many examples do not satisfy this constraint. Later, Hampton [18] demonstrated intransitivities in human similarity judgments.

2.3   Feature-Based Approach: Cognitive Properties

Hence, as "minimality is somewhat problematic, symmetry is apparently false, and the triangle inequality is hardly compelling", Tversky rejected the idea that a similarity measure should lie on a distance measure.

Contrast Model. Against the *geometric model*, he proposed the *contrast model*. In broad outline, this model is usually presented as follows. Considering an object x described by means of a set of features X, the similarity between x and y, $S(x,y)$, is expressed as a linear combination (contrast) of the measure of the common features, ie $X \cap Y$, and distinctive features, ie $X - Y$ and $Y - X$:

$$S(x,y) = \theta f(X \cap Y) - \alpha f(X-Y) - \beta f(Y-X), \; \theta, \alpha, \beta \geq 0 \; \text{ and } f \text{ an interval scale}$$

The contrast model lies on restrictive axioms, coming from decision theory, like the properties of independence, solvability and invariance. In particular, the independence axiom is difficult to interpret in terms of human similarity judgments, and thus, is difficult to see cognitive justification. But it implies useful mathematical properties on the similarity measure: it can not be expressed as a ratio function.

Ratio Model. However, the axioms required in the contrast model are often omitted. This has an important consequence on the generality of the contrast model because the existing similarity measures are often also a combination of common and distinctive features but they are generally not linear. So, it is a mistake that several authors present the contrast model as:

$$S(x,y) = \frac{f(X \cap Y)}{f(X \cap Y) + \alpha f(X - Y) + \beta f(Y - X)}, \quad \alpha, \beta \geq 0$$

This formulation exists in Tversky's paper and he called it the *ratio model*. This model, of course, violates the independence axiom as it is defined as a ratio function, but encompasses many existing measures. For instance, if $\alpha = \beta = 1$, $S(x,y)) = f(X \cap Y)/f(X \cup Y)$ was a measure proposed by Gregson [17], if $\alpha = \beta = 1/2$, then $S(x,y) = 2f(X \cap Y)/(f(X) + f(Y))$ was proposed by Eisler and Ekman [10] and for $\alpha = 1$ and $\beta = 0$, $S(x,y) = f(X \cap Y)/f(X)$, which is not symmetric, was proposed by Bush and Mosteller [6].

3   Comparison of Measures of Similarity

The contrast model being a reference as a model of similarity measures, several authors [45,35,44,4] proposed to extend it for the comparison of fuzzy sets. Basically, the generalization omits the axioms of the contrast model (except in [4]), and focuses on how the final resulting measure of similarity can be generalized to fuzzy sets: definition of intersection, difference, interval scale f of the contrast model,...

3.1   Families of Measures of Similarity

When a similarity measure is needed, and we have seen this situation often occurs, it is difficult to make a choice among the numerous and various existing measures (and the number of measures keeps increasing). First, is it not trivial to see the particular properties of each measure and to understand the impact of the properties in the behaviors of the similarity. Second, it is important to characterize the application where the similarity is needed to match the expected properties with the existing similarities in literature.

In [5], we have proposed a general framework that organises the main families of measures of comparison according to the properties they satisfy and the purpose of their utilisation.

Formally, for any set Ω of elements, let $F(\Omega)$ denote the set of fuzzy subsets of Ω, equipped with a fuzzy set measure $M : F(\Omega) \rightarrow \mathbb{R}^+$ such that $M(\emptyset) = 0$ and $Y \subseteq X$ implies $M(Y) \leq M(X)$. The general form of a measure of comparison is:

$$S(X,Y) = F_S(M(X \cap Y), M(Y - X), M(X - Y))$$

with $F_S : \mathbb{R}^3 \rightarrow [0,1]$.

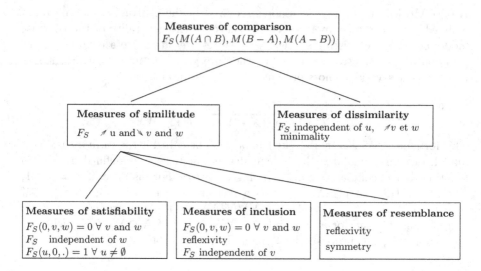

Fig. 1. General framework of measures of comparisons of fuzzy sets [5]

Either we want to evaluate the extent to which two objects are similar and in this case, the *measures of similitude* are devoted to this evaluation, either we want to measure to which extent they are different and then we have to look at the *measures of dissimilarity*.

Among measures of similitude, all the measures do not have the same behaviors. They are measures that are not symmetric, as Tversky's study showed the reality of this property: these measures suppose a subject and a referent when comparing two objects. Inside the similitude family, the *measures of satisfiability* evaluate to which extent Y is compatible with X and it can be used in decision tree algorithms or case-based reasoning for instance. The *measures of inclusion* are also measures of similitude and enable to evaluate to which extent a description can be considered as a particular case of another description, and it is useful when working on databases, semantic networks or relations between properties for instance.

Lastly, we consider that the property of symmetry can be desirable for some situations like in clustering. That is the reason that measures of similitude can be symmetric. They are called *measures of resemblance*. Figure 1 gives the complete hierarchy of comparison measures with the particular properties satisfied by each class of measures.

This general framework gives a way to distinguish measures regarding their properties and their purpose of use, but within a given class of measures, the problem of the choice of a measure is still present, even if it is reduced.

Two main points of view can be adopted to select a measure among a family of measures: either the values of similarity provided are important, either the order induced by a measure is important.

3.2 Value-Based Comparison

In the case where a value is in itself important, we have proposed to consider the power of discrimination of a measure [38,40]. It evaluates the sensitivity of the similarity measures with respect to the values of their arguments: the question is whether small variations of the input values, i.e. small variations in the configurations of the two objects to compare, lead to small differences in the similarity values or large ones. Moreover, this question is considered locally, i.e. the discrimination power studies whether such variations occur for high similarity values or for small ones.

In order to control the discrimination power, we have proposed [38,40] a new parametrized measure of similarity: the Fermi-Dirac measure. This measure generates different behaviors regarding its power of discrimination thanks to its parameter.

3.3 Order-Based Comparison

The case where the values provided by a similarity measure are less important than the order it induces, is always the situation in information retrieval: the results of a search engine take the form of a list of documents ranked by decreasing relevance, often calculated as the similarity between the document and the request. If two search engines give the same ordered list of documents, they are considered as equivalent, even if the relevance values are not the same: the search engines are compared on the basis of their precision and recall which also only depend on the order of the retrieved list.

The theoretical comparison of measures of comparison has been studied by several authors [22,3,2,32]. It leads to the notion of *equivalence*, defined as:

$$\text{two measures } m_1 \text{ and } m_2 \text{ are } equivalent \text{ iff}$$
$$\forall x, y, z, t \quad \begin{cases} m_1(x,y) < m_1(z,t) \Longleftrightarrow m_2(x,y) < m_2(z,t) \\ m_1(x,y) = m_1(z,t) \Longleftrightarrow m_2(x,y) = m_2(z,t) \end{cases}$$

This definition can also be formulated as follows [2,32].

$$\text{two measures } m_1 \text{ and } m_2 \text{ are } equivalent \text{ iff}$$
$$\exists f : Im(m_1) \to Im(m_2) \text{ strictly increasing such that } m_2 = f \circ m_1$$

where $Im(m) \subset \mathbb{R}$ is the set of the values taken by m.

Denoting $a = |X \cap Y|$, $b = |X - Y|$, $c = |Y - X|$, and $d = |\bar{X} \cap \bar{Y}|$ and considering the measures of Table 1, the following classes of equivalence can be established:

- {Jaccard, Dice, all the symmetric Tversky's ratio measures}
- {Sokal and Sneath 1, Rogers et Tanimoto, Simple Matching},
- {Yule Q, Yule Y},
- each remaining measure is a class in itself.

Table 1. Classical measures of similarity, normalised to $[0, 1]$

Similarity measure	Notation	Definition
Jaccard	Jac	$\frac{a}{a+b+c}$
Dice	Dic	$\frac{2a}{2a+b+c}$
Kulczynski 2	Kul	$\frac{1}{2}\left(\frac{a}{a+b} + \frac{a}{a+c}\right)$
Ochiai	Och	$\frac{a}{\sqrt{a+b}\sqrt{a+c}}$
Rogers and Tanimoto	RT	$\frac{a+d}{a+2(b+c)+d}$
Russel and Rao	RR	$\frac{a}{a+b+c+d}$
Simple Matching	SM	$\frac{a+d}{a+b+c+d}$
Sokal and Sneath 1	$SS1$	$\frac{a+d}{a+\frac{1}{2}(b+c)+d}$
Yule Q	YuQ	$\frac{ad}{ad+bc}$
Yule Y	YuY	$\frac{\sqrt{ad}}{\sqrt{ad}+\sqrt{bc}}$

In order to make distinctions among the measures that are not equivalent, a *degree of equivalence* has been proposed [41] to quantify to which extent two measures are in agreement regarding the order they induce. This degree is based on the Kendall coefficient generalized to top-k lists and to ties [12,11].

More formally, for two orders induced by two similarity measures r_1 and r_2: $r(i)$ indicates the rank of the i-th object according to r.

The Kendall coefficient associates to each pair of objects a penalty $P_{r_1,r_2}(i,j)$, and then is computed as the sum of the penalties normalised by the total number of comparisons. The generalized Kendall coefficient distinguishes two particular pairs: the tied pairs and the missing pairs (because of the top-k lists). For the first kind of pairs, a penalty p in $[0,1]$ is associated, and for the second pairs, a penalty p' in $[0,1]$ is associated.

$$K_{p,p'}(r_1, r_2) = \frac{2}{n(n-1)} \sum_{i \neq j} P_{r_1,r_2}(i,j)$$

The degree of equivalence is then defined as:

$$d_k(m_1, m_2) = 1 - K_{p,p'}(r_1^k, r_2^k).$$

Usually, $p = 0.5$ meaning that for tied pairs, the penalty is equal to the probability to have the same rank by forcing arbitrarily the measure to a strict order; and $p' = 1$ for the missing pairs, they are considered as a discordant pair.

The experiments detailed in [28] showed that:

- when considering the totality of the objects, some measures, although not satisfying the definition of equivalence, have very high equivalence degrees: Jaccard/Ochiai, Kulczynski 2/Ochiai, and Jaccard/Kulczynski 2. Actually,

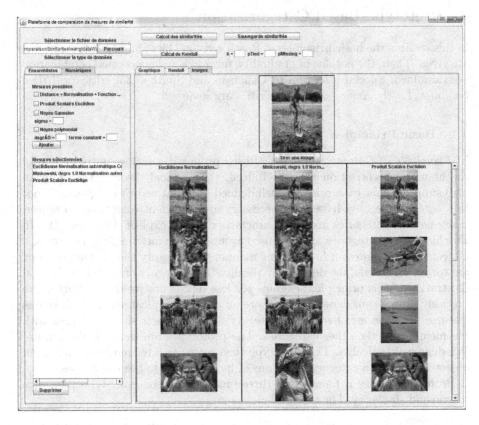

Fig. 2. Screen shot of the platform for the comparison of different types of similarity measures

measures expressed by means of a, b, c lead to very few differences and can thus be considered as quasi-equivalent and then redundant.

– when considering partial lists, the degrees of equivalence appear globally lower than for the full rank comparison: this decrease indicates that the global agreement observed when comparing the full rankings is actually mainly due to the last ranked data. This underlines that a study of the inversion positions, besides their number, is necessary, especially when it comes to selecting non equivalent measures in an information retrieval framework. Still, this decrease does not occur for all measures.

We have developed a platform, for the French project INFOM@GIC, under the umbrella of the business cluster for digital content, CAP DIGITAL, enabling the comparison of different types of similarity measures. Figure 2 is a screen shot of this platform: the top image is the request and the top-4 lists of 3 similarity measures are displayed vertically. It is then possible to see the differences in the rankings of the 3 chosen measures.

4 Prototype-Based Categorization

In this section, the basic principles of Rosch's vision [43] of human categorization are given. Then, the notions of graduality, fuzziness and conceptual combination are examined. Special attention to Osherson and Smith's [33] critics will then be about. Lastly, answerss to their critics are reviewed.

4.1 Basic Principles

In *Principles of Categorization* [43], Rosch proposed a new cognitive vision of the human categorization process. Indeed, the classical theory (Aristotelician one) supposes that categories are well defined and that an object belongs or not to a category because it satisfies necessary and sufficient conditions. However, many natural categories are not characterized by an explicit definition [19]. In Rosch's theory, categories are organised around a core meaning, the prototypes. Objects do not represent in the same manner the category they belong to: they are spread on a scale, the degree of typicality. The category of birds is a classical illustration of this principle: penguins are less representative than robins. More precisely, Rosch considers that the more an object is typical, the more it shares features with the members of its category and the less it shares features with the members of the other categories. The prototypes are those that maximize this degree of typicality. The prototype then serves as a reference to judge if an object belongs to a category by means of its similarity to the prototypes.

Prototype theory is founded on three fundamental concepts: typicality, prototype and similarity. The essential claims are:

1. all members of a category do not represent it equally: they differ by their *degree of typicality* regarding the category.
2. a category has internal structure based on "focal examples" [42] or *prototypes*. They best represent their category and maximize the degree of typicality.
3. a category possesses one or more prototypes.
4. an object is categorized by assessing its *similarity* to the prototypes.

4.2 Typicality, Graduality, Conceptual Combination and Fuzzy Logic

As Rosch's approach naturally mentions the notion of graduality by means of typicality, fuzzy logic can naturally be the theory to formalize the concept of prototype.

But, as many authors have noticed [50,1,33,21,20], the notion of typicality differs from the notion of membership: even if penguins are less typical than robins to represent birds (typicality(penguin) < typicality(robin)), they are all birds (membership(penguin) = membership(robin) = 1). This difference of information between typicality and membership is even the core of the principle of categorisation proposed by Rosch: a prototype of a category can be described by typical properties but these properties are not necessary conditions to categorize objects in this category. Then, it is more relevant and even necessary to distinguish these two degrees, degree of typicality and degree of membership.

However, Osherson and Smith [33] raised several problems both for prototype theory to adequately model the theory of concepts and for fuzzy logic to adequately model concepts. These problems are all related to the problem of conceptual combination. Tversky and Kahneman [48] also have devoted experimental study to the problem of "conjunction fallacy": human thinking apparently violates logical and probabilistic laws. These debates both concerns the prototype theory and the similarity judgments as we have seen in Introduction.

4.3   Osherson and Smith's Critics

In the following, the four points on which Osherson and Smith rejected prototype theory formalised by fuzzy logic are examined. They suppose that the basic operations in fuzzy set theory are formalised by the minimum for the intersection, the maximum for the union and $1 - x$ for the complement.

Conjunctive Concepts. The first problem, and certainly the problem that sparked off the most debate, is the problem known as "the striped apple", that illustrates the fact that the typicality of an object for a category that is a conjunction of categories, can be higher than its typicality to one of the elementary category of this combined category. In the example considered by Osherson and Smith (see Figure 3), the apple (a) is more typical of the concept "striped apple" than of the concept "apple".

If typicality is modeled by fuzzy logic and denoted T, then one has to expect that:

$$\forall x, \ T_{\text{striped apple}}(x) = \min(T_{\text{striped fruit}}(x), T_{\text{apple}}(x))$$

which contradicts the example shown in Figure 3, where:

$$T_{\text{striped apple}}(a) > T_{\text{apple}}(a)$$

(a)   (b)

Fig. 3. Conceptual combination: the striped apple case [33]

Logically Empty and Logically Universal Concepts. The second problem lies in the fact that the concept "an apple that is not an apple" should be logically empty, and the concept "a fruit that is or is not an apple" should be logically universal. The properties required by Osherson and Smith are known as "law of contradiction" and "law of excluded middle" in the classical logic. These concepts can not be modeled by fuzzy logic, because, in general:

$$\forall x, \; T_{\text{apple that is not an apple'}}(x) = \min(T_{\text{apple}}(x), 1 - T_{\text{apple}}(x)) \neq 0$$

and

$$\forall x, \; T_{\text{fruit that is or is not an apple}}(x) = \max(T_{\text{apple}}(x), 1 - T_{\text{apple}}(x)) \neq 1$$

Disjunction of Concepts. The third problem is due to the fuzzy union: a prototype of a category C is necessarily a prototype of the union of the category C and a category C'. Osherson and Smith gave an example but we do not take it up because it doesn't illustrate correctly the problem as noticed by [7,8].

Inclusion of Concepts. The last problem evoked by Osherson and Smith concerns what they call "truth conditions of thoughts". The considered proposition is of the form "All A's are B's" like "All grizzly bears are inhabitants of North America". The authors interpret this proposition as an inclusion of concepts: the concept of "grizzly bears" is included in the concept of "inhabitant of North America". In this case, they use the classical binary inclusion to state that it leads to a contradiction with their intuitions.

4.4   Answers to Osherson and Smith's Critics

Zadeh [50] has answered Osherson and Smith's critics and proposed a new representation of prototypes. Unfortunately, this paper had a limited impact in the community of cognitivists. Remarkable critics are of two kinds: point by point protest or global spirit protest.

Point by Point Protest. Bělohlávek et al. [7,8] proposed point by point answer to Osherson and Smith's critics. The main spirit of their response lies in that fact that Osherson and Smith's mistakes are due to their ignorance of the richness of the theory of fuzzy set and fuzzy logic. For instance, they demonstrated that it is possible to choose the right operators to be in agreement with Osherson and Smith intuitions or just by adequately interpret some propositions, like the ones dealing with the "truth conditions of thoughts". Or, for instance, for the inclusion of concepts, a wrong interpretation is the cause of the conclusion. Instead of using inclusion to formalize the proposition "All A's are B's", it should be understood as a rule R: "for all objects x in the universe, if x is A then x is B" and hence, in a fuzzy logic setting, the formalisation is:

$$f_R(x, y) = \phi(A(x), B(x))$$

where the fuzzy implication ϕ is a function which generalizes the classical implication, in the case where A and B are crisp sets. With this formalisation, the result is the one expected by Osherson and Smith.

Global Spirit Protest. Maybe the most convincing answer comes from Fuhrmann [16]. The problem of conceptual combination is an intricate issue in prototype theory, according to the author: "The meanings of compounds cannot be uniquely derived from that of their constituents by any universal operation". Also, a large amount of papers deals with this problem especially in linguistics where an adjective combined with a noun (like "striped apple") is not always the conjonction of the properties of the adjective with those of the noun (an adjective can be intersective, subsective or intensional [20]). Fuhrman claimed that one "has to build on (element-to-element) similarity and on (category-to-category) representativeness and only on the basis may infer any category-to-category consequence".

The main criticism of Furhman against Osherson and Smith's paper concerns their method for proving the inadequacy of fuzzy set theory for prototype theory: each criteria is translated by choosing one mathematical formula of fuzzy set theory and is then rejected mostly because of being "not compatible with strong intuitions", intuitions that are never really clear. Actually, as noticed by Furhman, the intuitions they refer to are defined on the basis of the Aristotelician logic. So, it is normal that fuzzy logic is not appropriate for their intuitions.

5   Fuzzy Prototypes Construction and Categorisation

Prototypes playing a key role in categorisation, Rifqi [37] proposed a computational model of Rosch's notions of prototype and typicality. First, the basic principles of the fuzzy prototype construction algorithm are given. Then, its use for a classification task is detailed. Lastly, the extended algorithm of the construction of fuzzy prototype is examined.

5.1   Basic Principles of Construction

To construct a fuzzy prototype in agreement with the previous cognitivist prototype view, we consider that the degree of typicality of an object increases with its total resemblance to other objects of its class (*internal resemblance*) and with its total dissimilarity to objects of other classes (*external dissimilarity*). This makes it possible to consider both the common features of the category members, and their distinctive features as opposed to other categories. More precisely, the fuzzy prototype construction principle consists in three steps [37]:

Step 1 Compute the internal resemblance degree of an object with the other members of its category and its external dissimilarity degree with the members of the outside categories.

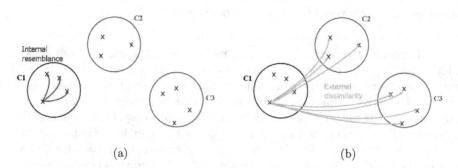

(a)   (b)

Fig. 4. (a) Computation of the internal resemblance, as the resemblance to the other members of the category, (b) computation of the external dissimilarity, as the dissimilarity to members of the other categories

Step 2 Aggregate the internal resemblance and the external dissimilarity degrees to obtain the typicality degree of the considered object.

Step 3 Aggregate the objects that are typical "enough", i.e. with a typicality degree higher than a predefined threshold to obtain the fuzzy prototype.

Figure 4 [29] illustrates the first step of the fuzzy prototype construction. This step requires the choice of a resemblance measure and a dissimilarity measure as defined in Section 3.1. Formally, denoting them r and d respectively, and denoting x an object belonging to a category C, x's internal resemblance with respect to C, $R(x, C)$, and its external dissimilarity, $D(x, C)$, are computed as:

$$R(x, C) = \otimes(r(x, y), y \in C) \qquad D(x, C) = \oplus(d(x, y), y \notin C)$$

i.e. as the aggregated resemblance (by an aggregation operator $\otimes$) to other members of the category and the aggregated dissimilarity (by an aggregation operator $\oplus$) to members of other categories.

Step 2 requires the choice of an aggregation operator to express the dependence of typicality on internal resemblance and external dissimilarity, that is formally written as:

$$T(x, C) = \varphi(R(x, C), D(x, C))$$

where φ denotes an aggregation operator. Lesot et al. [27] showed how the aggregation operator can rule the semantics of typicality and thus that of the prototypes, determining the extent to which the prototype should be a central or a discriminative element.

Originally, this fuzzy prototype construction method was proposed for fuzzy data, described by means of attributes with fuzzy values. It was applied [39] for the characterization and classification of microcalcifications in mammographies: each microcalcification is described by means of 5 fuzzy attributes computed from its fuzzy contour [36] like the perimeter or the elongation. Experts have classified microcalcifications into 2 classes: "round" microcalcifications and "not round" ones, because this property is important to qualify the malignancy of microcalcifications. The aim is then to give the fuzzy prototypes of the classes "round" and "not round".

In the case where the data are described by numerical values but not fuzzy (and it usually the case), Lesot [23] proposed to generalize the previous method to numerical data.

It is to be noticed that there exist other definitions for typicality, as for instance the one used by [34] or the one underlying the Most Typical Value [15]. Yet these definitions are only based on the notion of a deviation with respect to a center: they can be interpreted as internal resemblance measures that describe only one aspect of the Rosch's typicality.

5.2   Fuzzy Prototypes and Conceptual Combination

It is worth noticing that the principles of the construction of a fuzzy prototype described previously can lead to a situation where the typicality of an object for a combined concept (like "striped apple") is higher than the minimum of its typicalities to the constituents of the composite concept. Indeed, the prototype of the combined concept will consider all the objects satisfying it and hence the typicalities of all objectss are recalculated and not deduced from the typicalities of basic concepts.

5.3   Fuzzy Prototypes for Classification

The typicality degrees as well as the fuzzy prototype have been exploited to perform a supervised learning task of classification. It is true that the main interest of a prototype comes from its power of description thanks to its synthetic view of the database. But, as Zadeh underlined [50], a fuzzy prototype can be seen as a schema for generating a set of objects. Thus, in a classification task, when a new object has to be classified, it can be compared to each fuzzy prototype and classified in the class of the nearest prototype (a sort of nearest neighbor algorithm where the considered neighbors are only the prototypes of the classes). Three classification methods based on typicality and fuzzy prototypes can be envisaged:

1. The class of the object to be classified is given by the class of the nearest prototype. The prototype is constructed with the fuzzy value maximizing the typicality degree.
2. As for the first method, the class of the object to be classified is given by the class of the nearest prototype, but the fuzzy prototype is obtained aggregating by the union of the values with a high typicality degree whereas the first one considers only one value.
3. A new object is compared to each object of the database. The comparison is the aggregation of the attribute by attribute comparisons weighted by the degree of typicality of the attribute value of the object in the learning

database. The class given to the unknown object is then the class of the most similar object in the learning database.

The typicality degree framework was also used to perform clustering: the typicality-based clustering algorithm [24] looks for a decomposition such that each point is most typical of the cluster it is assigned to, and aims at maximizing the typicality degrees. It was adapted [25,26] to take into account specific data (non-vectorial data, such as sequences, trees or graphs) or cluster constraints (ellipsoidal clusters are discovered instead of spherical clusters).

5.4   Extended Principle

In Section 4.1, the essential points for the construction of fuzzy prototypes were given. Each of them was respected for the computational model of fuzzy prototype construction, except one: "a category possesses one or more prototypes". In the presented method, this reality was not exploited: essentially one prototype per category (or class) is constructed.

In order to take into account the fact that a category can be represented by prototypes with different descriptions, Forest et al. [13,14] proposed a pre-processing step for the construction of fuzzy prototypes. This step consists in identifying distinct sub-classes inside a same class.

To perform such a task, an algorithm automatically constructs weighted graphs from the data. A graph of an example provides its friends, i.e. examples from the same class that are close in such a way that there is no closer example from another class (enemy). It means that the closest enemy of an example defines

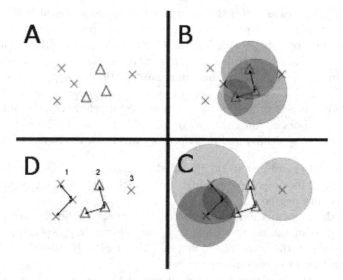

Fig. 5. Identification of subclasses: **A** Database, **B** Hyperspheres of class of red triangles, **C** Hyperspheres of class of blue crosses, **D** Graphs of subclasses [13]

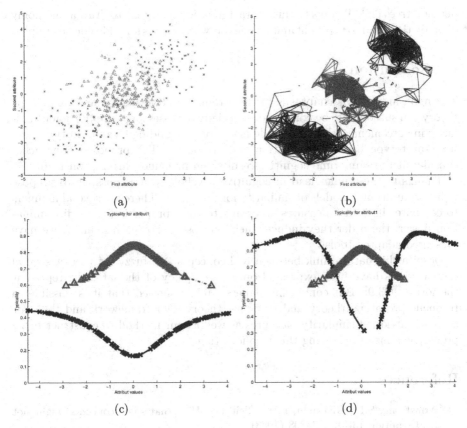

Fig. 6. (a) Artificial two dimensions database, (b) Graphs obtained by the segmentation process, (c) Typicality degrees for the first attribute without the class segmentation, (d) Typicality degrees for the first attribute with a segmentation [13]

the limit of its friends neighbourhood. Edges, representing the proximity, are weighted by a distance between the two nodes (examples). The connex components of the graph (see Figure 5 D), associated with the union of each graph obtained from each example, define the subclasses.

To obtain these graphs, we let a hypersphere grow from an example of a class until it encounters an enemy (Figure 5 B and 5 C). All the examples in this hypersphere are linked to the central example. This process (construction of a hypersphere) is repeated for all the examples. Therefore, this stage of our method allows to discover subclasses identifying the specific behaviours of the class.

Once the sub-classes identified, the algorithm of the construction of fuzzy prototypes is performed on each subclass whose homogeneity is guaranteed and justifies the unique prototype per subclass. The results on artificial data are illustrated in Figure 6.

This segmentation algorithm previously described was used by Marsala and Rifqi [31] to characterize sub-classes of ambiguous areas obtained by a forest of fuzzy decision trees [30]. Thanks to this characterization, it is easy to draw a taxonomy of predictions and to highlight ambiguities between classes for undecided examples.

6   Conclusion

The main aim of this chapter is to show that cognitive studies and fuzzy logic theory can successfully be combined, especially for assessing computational similarity in agreement with human judgments, or for building fuzzy prototypes and fuzzy-prototype based classification and clustering. The proposed approaches consider at the same time cognitive results and machine learning constraints.

Thanks to the studies lead in cognitive psychology, it is possible to propose rich computational models of similarity for fuzzy sets. There are several manners to compare similarity measures according to their properties, their discrimination power, the order they induce, their purpose,... For each manner, we have presented adapted tools.

Despite this natural link between real concepts and fuzzy set theory, several critical views have been expressed on the adequacy of the latter to represent the former [33,9]. But convincing papers [16,7,8] showed that it is possible to reconcile prototype theory and fuzzy set theory. Our framework and tools for the comparison of similarity measures as well as our method to construct fuzzy prototype aims at achieving this reconciliation.

References

1. Armstrong, S.L., Gleitman, L.R., Gleitman, H.: What some concepts might not be. Cognition 13(3), 263–308 (1983)
2. Batagelj, V., Bren, M.: Comparing resemblance measures. Journal of Classification 12, 73–90 (1995)
3. Baulieu, F.B.: A classification of presence/absence based dissimilarity coefficients. Journal of Classification 6, 233–246 (1989)
4. Bouchon-Meunier, B., Coletti, G., Lesot, M.-J., Rifqi, M.: Towards a Conscious Choice of a Fuzzy Similarity Measure: A Qualitative Point of View. In: Hüllermeier, E., Kruse, R., Hoffmann, F. (eds.) IPMU 2010. LNCS, vol. 6178, pp. 1–10. Springer, Heidelberg (2010)
5. Bouchon-Meunier, B., Rifqi, M., Bothorel, S.: Towards general measures of comparison of objects. Fuzzy Sets and Systems 84(2), 143–153 (1996)
6. Bush, R.R., Mosteller, F.: A model for stimulus generalization and discrimination. Psychological Review 58, 413–423 (1951)
7. Bělohlávek, R., Klir, G.J., Lewis, H.W., Way, E.C.: On the capability of fuzzy set theory to represent concepts. International Journal of General Systems 31(6), 569–585 (2002)
8. Bělohlávek, R., Klir, G.J., Lewis, H.W., Way, E.C.: Concepts and fuzzy sets: Misunderstandings, misconceptions, and oversights. International Journal of Approximate Reasoning 51(1), 23–34 (2009)
9. Cohen, B., Murphy, G.L.: Models of concepts. Cognitive Science 8, 27–58 (1984)
10. Eisler, H., Ekman, G.: A mechanism of subjective similarity. Acta Psychologica 16, 1–10 (1959)

11. Fagin, R., Kumar, R., Mahdian, M., Sivakumar, D., Vee, E.: Comparing and aggregating rankings with ties. In: Symposium on Principles of Database Systems, pp. 47–58 (2004)

12. Fagin, R., Kumar, R., Sivakumar, D.: Comparing top k lists. SIAM Journal on Discrete Mathematics 17(1), 134–160 (2003)

13. Forest, J., Rifqi, M., Bouchon-Meunier, B.: Class segmentation to improve fuzzy prototype construction: Visualization and characterization of non homogeneous classes. In: IEEE World Congress on Computational Intelligence (WCCI 2006), Vancouver, pp. 555–559 (2006)

14. Forest, J., Rifqi, M., Bouchon-Meunier, B.: Segmentation de classes pour l'amélioration de la construction de prototypes flous: visualisation et caractérisation de classes non homogénes. In: Rencontres francophones sur la Logique Floue et ses Applications (LFA 2006), Toulouse, pp. 29–36 (2006)

15. Friedman, M., Ming, M., Kandel, A.: On the theory of typicality. International Journal of Uncertainty, Fuzziness and Knowledge-Based Systems 3(2), 127–142 (1995)

16. Fuhrmann, G.: Note on the integration of prototype theory and fuzzy-set theory. Synthese 86, 1–27 (1991)

17. Gregson, R.: Psychometrics of similarity. Academic Press, New York (1975)

18. Hampton, J.A.: A demonstration of intransitivity in natural categories. Cognition 12(2), 151–164 (1982)

19. Hampton, J.A.: The role of similarity in natural categorization. In: Hahn, U., Ramscar, M. (eds.) Similarity and Categorization, pp. 13–28. Oxford University Press (2001)

20. Kamp, H., Partee, B.: Prototype theory and compositionality. Cognition 57(2), 129–191 (1995)

21. Kleiber, G.: Prototype et prototypes. In: Sémantique et Cognition. Editions du C.N.R.S, Paris (1991)

22. Lerman, I.C.: Indice de similarité et préordonnance associée. In: Séminaire Sur Les Ordres Totaux Finis, pp. 233–243. Aix-en-Provence (1967)

23. Lesot, M.J.: Similarity, typicality and fuzzy prototypes for numerical data. Res-Systemica 5 (2005)

24. Lesot, M.J.: Typicality-based clustering. International Journal of Information Technology and Intelligent Computing 1(2), 279–292 (2006)

25. Lesot, M.J., Kruse, R.: Data summarisation by typicality-based clustering for vectorial data and nonvectorial data. In: IEEE International Conference on Fuzzy Systems (Fuzz-IEEE 2006), Vancouver, pp. 3011–3018 (2006)

26. Lesot, M.J., Kruse, R.: Gustafson-Kessel-like clustering algorithm based on typicality degrees. In: International Conference on Information Processing and Management of Uncertainty in Knowledge-Based Systems, IPMU 2006, Paris, pp. 1300–1307 (2006)

27. Lesot, M.J., Mouillet, L., Bouchon-Meunier, B.: Fuzzy prototypes based on typicality degrees. In: Fuzzy Days 2004, pp. 125–138. Springer, Dortmund (2004)

28. Lesot, M.-J., Rifqi, M.: Order-Based Equivalence Degrees for Similarity and Distance Measures. In: Hüllermeier, E., Kruse, R., Hoffmann, F. (eds.) IPMU 2010. LNCS, vol. 6178, pp. 19–28. Springer, Heidelberg (2010)

29. Lesot, M.J., Rifqi, M., Bouchon-Meunier, B.: Fuzzy prototypes: From a cognitive view to a machine learning principle. In: Bustince, H., Herrera, F., Montero, J. (eds.) Fuzzy Sets and Their Extensions: Representation, Aggregation and Models, pp. 431–452. Springer (2007)

30. Marsala, C., Bouchon-Meunier, B.: An adaptable system to construct fuzzy decision trees. In: North American Fuzzy Information Processing Society Annual Conference, NAFIPS 1999, New York, pp. 223–227 (1999)

31. Marsala, C., Rifqi, M.: Characterizing forest of fuzzy decision trees errors. In: 4th International Conference of the ERCIM Working Group on Compting & Statistics (ERCIM 2011), London (2011)

32. Omhover, J.-F., Rifqi, M., Detyniecki, M.: Ranking Invariance Based on Similarity Measures in Document Retrieval. In: Detyniecki, M., Jose, J.M., Nürnberger, A., van Rijsbergen, C.J.'. (eds.) AMR 2005. LNCS, vol. 3877, pp. 55–64. Springer, Heidelberg (2006)

33. Osherson, D.N., Smith, E.E.: On the adequacy of prototype theory as a theory of concepts. Cognition 9, 35–58 (1981)

34. Pal, N., Pal, K., Bezdek, J.: A mixed c-means clustering model. In: IEEE International Conference on Fuzzy Systems, Fuzz-IEEE 1997, Barcelona, pp. 11–21 (1997)

35. Pappis, C.P., Karacapilidis, N.: A comparative assessment of measures of similarity of fuzzy values. Fuzzy Sets and Systems 56, 171–174 (1993)

36. Rick, A., Bothorel, S., Bouchon-Meunier, B., Muller, S., Rifqi, M.: Fuzzy techniques in mammographic image processing. In: Kerre, E., Nachtegael, M. (eds.) Fuzzy Techniques in Image Processing. STUDFUZZ, pp. 308–336. Springer (2000)

37. Rifqi, M.: Constructing prototypes from large databases. In: International Conference on Information Processing and Management of Uncertainty in Knowledge-Based Systems, IPMU 1996, Granada, pp. 301–306 (1996)

38. Rifqi, M., Berger, V., Bouchon-Meunier, B.: Discrimination power of measures of comparison. Fuzzy Sets and Systems 110, 189–196 (2000)

39. Rifqi, M., Bothorel, S., Bouchon-Meunier, B., Muller, S.: Similarity and prototype based approach for classification of microcalcifications. In: IFSA 1997, Prague, pp. 123–128 (1997)

40. Rifqi, M., Detyniecki, M., Bouchon-Meunier, B.: Discrimination power of measures of resemblance. In: IFSA 2003, Istanbul (2003)

41. Rifqi, M., Lesot, M.J., Detyniecki, M.: Fuzzy order-equivalence for similarity measures. In: 27th North American Fuzzy Information Processing Society Annual Conference (NAFIPS 2008), New York (2008)

42. Rosch, E.: Cognitive development and the acquisition of language. In: On the Internal Structure of Perceptual and Semantic Categories, pp. 111–141. Academic Press, Oxford (1973)

43. Rosch, E.: Principles of categorization. In: Rosch, E., Lloyd, B.B. (eds.) Cognition and Categorization, pp. 27–48. Laurence Erlbaum Associates, Hillsdale (1978)

44. Santini, S., Jain, R.: Similarity measures. IEEE Transactions on Pattern Analysis and Machine Intelligence 21(9) (1999)

45. Shiina, K.: A fuzzy-set-theoretic feature model and its application to asymmetric data analysis. Japanese Psychological Research 30(3), 95–104 (1988)

46. Torgerson, W.S.: Multidimensional scaling of similarity. Psychometrika 30, 379–393 (1965)

47. Tversky, A.: Features of similarity. Psychological Review 84, 327–352 (1977)

48. Tversky, A., Kahneman, D.: Extensional versus intuitive reasoning: The conjunction fallacy in probability judgment. Psychological Review 90(4), 293–315 (1983)

49. Wittgenstein, L.: Philosophical investigations. Macmillan, NewYork (1953)

50. Zadeh, L.A.: A note on prototype theory and fuzzy sets. Cognition 12, 291–297 (1982)

A Unified Fuzzy Model-Based Framework for Modeling and Control of Complex Systems: From Flying Vehicle Control to Brain-Machine Cooperative Control[*]

Kazuo Tanaka

Department of Mechanical Engineering and Intelligent Systems,
The University of Electro-Communications,
1-5-1 Chofugaoka, Chofu, Tokyo 182-8585, Japan
ktanaka@mce.uec.ac.jp

Abstract. The invited lecture in 2012 IEEE World Congress on Computational Intelligence (WCCI 2012) presents an overview of a unified fuzzy model-based framework for modeling and control of complex systems. A number of practical applications, ranging from flying vehicles control (including micro helicopter control) to brain-machine cooperative control, are provided in the lecture. The theory and applications have been developed in our laboratory [1] at the University of Electro-Communications (UEC), Tokyo, Japan, in collaboration with Prof. Hua O. Wang and his laboratory [2] at Boston University, Boston, USA. Due to lack of space, this chapter focuses on a unified fuzzy model-based framework for modeling and control of a micro helicopter that is a key application in our research.

1 Introduction

Unmanned aerial vehicles (UAVs) have been an active area of research in recent years. A large number of studies [3] on helicopter control have been conducted as a typical application of UAVs over the last two decades. It is well known that helicopter control is a difficult and challenging problem due to its properties like instability, nonlinearity and coupling, etc. As be mentioned in [3], Sugeno and his group have presented several pioneer and excellent works [4,5] in 1980s and 1990s.

In recent years, in parallel with researches on micro air vehicles (MAVs), autonomous control of micro (small) helicopter (like palm-size helicopters) [6]-[23] has been paid great attention. Due to its restricted payload, the studies [6]-[9] utilize external sensors like CCD camera-type vision sensors. However, use of external vision sensors makes autonomous flight so difficult. The main disadvantage is that micro helicopters can not be controlled in outside vision

---

[*] This work has been done in our laboratory [1] at the University of Electro-Communications (UEC), Tokyo, Japan, in collaboration with Prof. Hua O. Wang and his laboratory [2] at Boston University, Boston, USA.

J. Liu et al. (Eds.): WCCI 2012 Plenary/Invited Lectures, LNCS 7311, pp. 185–208, 2012.
© Springer-Verlag Berlin Heidelberg 2012

sensing area. The study [24] deals with hovering of a micro helicopter carrying vision sensors. However, to accomplish the hovering control, they set up external markers outside the helicopter. As well as the external sensor problem, use of external markers makes autonomous flight so difficult. In addition, the studies [24]-[27] provide no theoretical guarantees of the stability of control system.

Our research targets are to achieve autonomous control of a micro helicopter without any external sensors and markers and to design a controller (theoretically) guaranteeing some kinds of control performance in addition to global and asymptotical stability of control system. The former and latter parts of this chapter focus on the achievement of the second and first targets, respectively.

The pioneer and excellent works by Sugeno and co-workers [4,5] applied traditional model-free fuzzy control to a (large-size) RC helicopter. The author [6] also applied model-free fuzzy control to a micro (palm-size) helicopter. Though the works [4,5] by Sugeno and co-workers particularly achieve great flight control performance such as hovering, turning, taking off and landing by using fuzzy control rules obtained from expert's knowledge and operation manuals, the model-free approaches provide no theoretical guarantees of the stability of control system. In this chapter, to guarantee some kinds of control performance in addition to global and asymptotical stability, we apply two innovative fuzzy model-based control approaches to a micro helicopter. One is a linear matrix inequality (LMI) approach [28] that is a well-known approach and has been widely used in control system design and analysis over the last decade. The other is a sum of squares (SOS) approach [29]-[33] recently presented by the author and co-workers. These [29]-[33] are completely different approaches from the existing LMI approaches. To the best of our knowledge, the paper [29] presented the first attempt at applying an SOS to fuzzy systems. Our SOS approach [29]-[33] provided more extensive results for the existing LMI approaches to Takagi-Sugeno fuzzy model and control. SOS design conditions can be symbolically and numerically solved via the SOSTOOLS [34] and the SeDuMi[35].

Section 2 presents our experimental system and micro helicopter dynamics. In Section 3, we summarize a recent developed SOS design approach for polynomial fuzzy control systems based on polynomial Lyapunov functions. Due to lack of space, we will omit the explanation of LMI-based design approach to Takagi-Sugeno fuzzy systems. For more details of the approach, see [28]. In Section 4, we present a comparison result of a micro helicopter via the LMI and SOS approaches. The simulation result shows that the SOS approach provides better results than the existing LMI approach. Finally, in Section 5, we apply one of the control approaches discussed here to vision-based control of the micro helicopter in real environments [36].

2   Micro Helicopter Dynamics

Fig. 2 shows the micro helicopter that we consider in this study. The helicopter with coaxial counter-rotating blades has two features. One is that rotating torque of yaw-direction of the main body can be canceled by rotating torques between the upper and lower rotors. In other words, the body-turn can be achieved by generating a difference of rotating torques between the upper and lower rotors. The other is that a mechanical stabilizer attached above the upper rotor has a function of keeping the upper rotor horizontally. The two features will be considered in the dynamic model construction. The dynamics of the helicopter can be described as (1)-(6).

$$m(\dot{u}(t) + q(t)w(t) - r(t)v(t)) = F_X(t) \tag{1}$$
$$m(\dot{v}(t) + r(t)u(t) - p(t)w(t)) = F_Y(t) \tag{2}$$
$$m(\dot{w}(t) + p(t)v(t) - q(t)u(t)) = F_Z(t) \tag{3}$$
$$\dot{p}(t)I_X + q(t)r(t)(I_Z - I_Y) = M_X(t) \tag{4}$$
$$\dot{q}(t)I_Y + r(t)p(t)(I_X - I_Z) = M_Y(t) \tag{5}$$
$$\dot{r}(t)I_Z + p(t)q(t)(I_Y - I_X) = M_Z(t) \tag{6}$$

Table 1 shows the definition of variables used in the dynamic models. By considering its co-axial counter structure, the restitutive force (generated by a mechanical stabilizer attached on the helicopter) and gravity compensation, the dynamics can be rewritten as

$$\dot{u}(t) = r(t)v(t) + \frac{1}{m}U_X(t), \tag{7}$$

$$\dot{v}(t) = -r(t)u(t) + \frac{1}{m}U_Y(t), \tag{8}$$

$$\dot{w}(t) = \frac{1}{m}U_Z(t), \tag{9}$$

$$\dot{\psi}(t) = \frac{1}{I_Z}U_\psi(t), \tag{10}$$

Table 1. Definition of variables

x, u	position and velocity (X-axis)
y, v	position and velocity (Y-axis)
z, w	position and velocity (Z-axis)
ϕ, p	angle and angle velocity (X-axis)
θ, q	angle and angle velocity (Y-axis)
ψ, r	angle and angle velocity (Z-axis)
m	mass
I_X, I_Y, I_Z	moments of inertia with respect to X, Y and Z axes
F_X, F_Y, F_Z	translational forces to X, Y and Z axes
M_X, M_Y, M_Z	rotational forces around X, Y and Z axes

where $m = 0.2$ and $I_z = 0.2857$. $U_X(t)$, $U_Y(t)$, $U_Z(t)$ and $U_\psi(t)$ denote new control input variables. We can obtain the original control inputs (to the real helicopter) from $U_X(t)$, $U_Y(t)$, $U_Z(t)$ and $U_\psi(t)$.

3   Polynomial Fuzzy Model and SOS-Based Designs

In general, LMI conditions can be solved numerically and efficiently by interior point algorithms, e.g., by the Robust Control Toolbox of MATLAB[1]. On the other hand, stability [29], stabilization conditions [30,31], guaranteed cost control [32,33] for polynomial fuzzy systems and polynomial Lyapunov functions reduce to SOS problems. Clearly, the problem is never solved by LMI solvers and can be solved via the SOSTOOLS [34] and the SeDuMi[35].

SOSTOOLS [34] is a free, third party MATLAB toolbox for solving sum of squares problems. The techniques behind it are based on the sum of squares decomposition for multivariate polynomials, which can be efficiently computed using semidefinite programming. SOSTOOLS is developed as a consequence of the recent interest in sum of squares polynomials, partly due to the fact that these techniques provide convex relaxations for many hard problems such as global, constrained, and boolean optimization. For more details, see the manual of SOSTOOLS [34].

3.1   Polynomial Fuzzy Model and Controller

In [29], we proposed a new type of fuzzy model with polynomial model consequence, i.e., fuzzy model whose consequent parts are represented by polynomials. Consider the following nonlinear system:

$$\dot{\boldsymbol{x}}(t) = \boldsymbol{f}(\boldsymbol{x}(t), \boldsymbol{u}(t)), \tag{11}$$

where $\boldsymbol{f}$ is a nonlinear function. $\boldsymbol{x}(t) = [x_1(t)\ x_2(t)\ \cdots\ x_n(t)]^T$ is the state vector and $\boldsymbol{u}(t) = [u_1(t)\ u_2(t)\ \cdots\ u_m(t)]^T$ is the input vector. Using the sector nonlinearity concept [28], we exactly represent (11) with the following polynomial fuzzy model (12). The main difference between the Takagi-Sugeno fuzzy model [37] and the polynomial fuzzy model (12) is consequent part representation. The fuzzy model (12) has a polynomial model consequence.

Model Rule i:

$$If\ z_1(t)\ is\ M_{i1}\ and\cdots\ and\ \ z_p(t)\ is\ M_{ip}$$
$$then\ \dot{\boldsymbol{x}}(t) = \boldsymbol{A}_i(\boldsymbol{x}(t))\hat{\boldsymbol{x}}(\boldsymbol{x}(t)) + \boldsymbol{B}_i(\boldsymbol{x}(t))\boldsymbol{u}(t), \tag{12}$$

where $i = 1, 2, \cdots, r$. r denotes the number of *Model Rules*. $z_j(t)$ $(j = 1, 2, \cdots, p)$ is the premise variable. The membership function associated with the ith *Model Rule* and jth premise variable component is denoted by M_{ij}. Each $z_j(t)$ is a measurable time-varying quantity that may be states, measurable external variables and/or time. $\boldsymbol{A}_i(\boldsymbol{x}(t))$ and $\boldsymbol{B}_i(\boldsymbol{x}(t))$ are polynomial matrices in $\boldsymbol{x}(t)$. $\hat{\boldsymbol{x}}(\boldsymbol{x}(t))$ is a column vector whose entries are all monomials in $\boldsymbol{x}(t)$. That is, $\hat{\boldsymbol{x}}(\boldsymbol{x}(t)) \in \boldsymbol{R}^N$ is an $N \times 1$ vector of monomials in $\boldsymbol{x}(t)$. Therefore, $\boldsymbol{A}_i(\boldsymbol{x}(t))\hat{\boldsymbol{x}}(\boldsymbol{x}(t)) + \boldsymbol{B}_i(\boldsymbol{x}(t))\boldsymbol{u}(t)$ is a polynomial vector. Thus, the polynomial fuzzy model (12)

---

[1] A registered trademark of MathWorks, Inc.

has a polynomial in each consequent part. The details of $\hat{x}(x(t))$ will be given in Proposition 1. We assume that

$$\hat{x}(x(t)) = 0 \text{ iff } x(t) = 0$$

throughout this chapter.

The computational method used in this chapter relies on the sum of squares decomposition of multivariate polynomials. A multivariate polynomial $f(x(t))$ (where $x(t) \in R^n$) is a sum of squares (SOS) if there exist polynomials $f_1(x(t))$, $\cdots$, $f_m(x(t))$ such that $f(x(t)) = \sum_{i=1}^{m} f_i^2(x(t))$. It is clear that $f(x(t))$ being an SOS naturally implies $f(x(t)) > 0$ for all $x(t) \in R^n$. This can be shown equivalent to the existence of a special quadric form stated in the following proposition.

Proposition 1. *[38] Let $f(x(t))$ be a polynomial in $x(t) \in R^n$ of degree 2d. In addition, let $\hat{x}(x(t))$ be a column vector whose entries are all monomials in $x(t)$ with degree no greater than d. Then $f(x(t))$ is a sum of squares iff there exists a positive semidefinite matrix P such that*

$$f(x(t)) = \hat{x}^T(x(t))P\hat{x}(x(t)). \tag{13}$$

Expressing an SOS polynomial using a quadratic form as in (13) has also been referred to as the Gram matrix method.

A monomial in $x(t)$ is a function of the form $x_1^{\alpha_1} x_2^{\alpha_2} \cdots x_n^{\alpha_n}$, where α_1, α_2, $\cdots$, α_n are nonnegative integers. In this case, the degree of the monomial is given by $\alpha_1 + \alpha_1 + \cdots + \alpha_n$.

The defuzzification process of the model (12) can be represented as

$$\dot{x}(t) = \sum_{i=1}^{r} h_i(z(t))\{A_i(x(t))\hat{x}(x(t)) + B_i(x(t))u(t)\}, \tag{14}$$

where

$$h_i(z(t)) = \frac{\prod_{j=1}^{p} M_{ij}(z_j(t))}{\sum_{k=1}^{r} \prod_{j=1}^{p} M_{kj}(z_j(t))}.$$

It should be noted from the properties of membership functions that $h_i(z(t)) \geq 0$ for all i and $\sum_{i=1}^{r} h_i(z(t)) = 1$. Thus, the overall fuzzy model is achieved by fuzzy blending of the polynomial system models. As shown in [29]-[31], the number of rules in polynomial fuzzy model generally becomes fewer than that in T-S fuzzy model, and our SOS approach to polynomial fuzzy models provides much more relaxed stability and stabilization results than the existing LMI approaches to T-S fuzzy model and control.

Since the parallel distributed compensation (PDC) [28] mirrors the structure of the fuzzy model of a system, a fuzzy controller with polynomial rule consequence can be constructed from the given polynomial fuzzy model (12).

Control Rule i:

$$If \ z_1(t) \ is \ M_{i1} \ and \cdots \ and \ z_p(t) \ is \ M_{ip}$$
$$then \ \boldsymbol{u}(t) = -\boldsymbol{F}_i(\boldsymbol{x}(t))\hat{\boldsymbol{x}}(\boldsymbol{x}(t)) \quad i = 1, 2, \cdots, r \tag{15}$$

The overall fuzzy controller can be calculated by

$$\boldsymbol{u}(t) = -\sum_{i=1}^{r} h_i(\boldsymbol{z}(t))\boldsymbol{F}_i(\boldsymbol{x}(t))\hat{\boldsymbol{x}}(\boldsymbol{x}(t)). \tag{16}$$

From (14) and (16), the closed-loop system can be represented as

$$\dot{\boldsymbol{x}}(t) = \sum_{i=1}^{r} \sum_{j=1}^{r} h_i(\boldsymbol{z}(t))h_j(\boldsymbol{z}(t))$$
$$\times \{\boldsymbol{A}_i(\boldsymbol{x}(t)) - \boldsymbol{B}_i(\boldsymbol{x}(t))\boldsymbol{F}_j(\boldsymbol{x}(t))\}\hat{\boldsymbol{x}}(\boldsymbol{x}(t)). \tag{17}$$

If $\hat{\boldsymbol{x}}(\boldsymbol{x}(t)) = \boldsymbol{x}(t)$ and $\boldsymbol{A}_i(\boldsymbol{x}(t))$, $\boldsymbol{B}_i(\boldsymbol{x}(t))$ and $\boldsymbol{F}_j(\boldsymbol{x}(t))$ are constant matrices for all i and j, then (14) and (16) reduce to the Takagi-Sugeno fuzzy model and controller, respectively. Therefore, (14) and (16) are more general representation.

3.2   Stable Control

To obtain more relaxed stability results, we employ a polynomial Lyapunov function [29] represented by

$$\hat{\boldsymbol{x}}^T(\boldsymbol{x}(t))\boldsymbol{P}(\tilde{\boldsymbol{x}}(t))\hat{\boldsymbol{x}}(\boldsymbol{x}(t)), \tag{18}$$

where $\boldsymbol{P}(\tilde{\boldsymbol{x}}(t))$ is a polynomial matrix in $\boldsymbol{x}(t)$. If $\hat{\boldsymbol{x}}(t) = \boldsymbol{x}(t)$ and $\boldsymbol{P}(\tilde{\boldsymbol{x}}(t))$ is a constant matrix, then (18) reduces to the quadratic Lyapunov function $\boldsymbol{x}^T(t)\boldsymbol{P}\boldsymbol{x}(t)$. Therefore, (18) is a more general representation.

From now, to lighten the notation, we will drop the notation with respect to time t. For instance, we will employ $\boldsymbol{x}$, $\hat{\boldsymbol{x}}(\boldsymbol{x})$ instead of $\boldsymbol{x}(t)$, $\hat{\boldsymbol{x}}(\boldsymbol{x}(t))$, respectively. Thus, we drop the notation with respect to time t, but it should be kept in mind that $\boldsymbol{x}$ means $\boldsymbol{x}(t)$.

Let $\boldsymbol{A}_i^k(\boldsymbol{x})$ denotes the k-th row of $\boldsymbol{A}_i(\boldsymbol{x})$, $\boldsymbol{K} = \{k_1, k_2, \cdots k_m\}$ denote the row indices of $\boldsymbol{B}_i(\boldsymbol{x})$ whose corresponding row is equal to zero, and define $\tilde{\boldsymbol{x}} = (x_{k_1}, x_{k_2}, \cdots x_{k_m})$.

Theorem 1. *[30] The control system consisting of (14) and (16) is stable if there exist a symmetric polynomial matrix* $\boldsymbol{X}(\tilde{\boldsymbol{x}}) \in \boldsymbol{R}^{N \times N}$ *and a polynomial matrix* $\boldsymbol{M}_i(\boldsymbol{x}) \in \boldsymbol{R}^{m \times N}$ *such that (19) and (20) are satisfied, where* $\epsilon_1(\boldsymbol{x})$ *and* $\epsilon_{2ij}(\boldsymbol{x})$ *are non negative polynomials such that* $\epsilon_1(\boldsymbol{x}) > 0$ $(\boldsymbol{x} \neq 0)$ *and* $\epsilon_{2ij}(\boldsymbol{x}) \geq 0$ *for all* $\boldsymbol{x}$.

$$v^T (X(\tilde{x}) - \epsilon_1(x)I)v \ is \ SOS \qquad (19)$$

$$-v^T \Big(T(x)A_i(x)X(\tilde{x}) - T(x)B_i(x)M_j(x)$$
$$+X(\tilde{x})A_i^T(x)T^T(x) - M_j^T(x)B_i^T(x)T^T(x)$$
$$+T(x)A_j(x)X(\tilde{x}) - T(x)B_j(x)M_i(x)$$
$$+X(\tilde{x})A_j^T(x)T^T(x) - M_i^T(x)B_j^T(x)T^T(x)$$
$$-\sum_{k\in K} \frac{\partial X}{\partial x_k}(\tilde{x})A_i^k(x)\hat{x}(x)$$
$$-\sum_{k\in K} \frac{\partial X}{\partial x_k}(\tilde{x})A_j^k(x)\hat{x}(x) + \epsilon_{2ij}(x)I \Big)v \ is \ SOS \quad i \le j, \qquad (20)$$

where $v \in R^N$ is a vector that is independent of x. $T(x) \in R^{N\times n}$ is a polynomial matrix whose (i, j)-th entry is given by $T^{ij}(x) = \frac{\partial \tilde{x}_i}{\partial x_j}(x)$. In addition,
if (20) holds with $\epsilon_{2ij}(x) > 0$ for $x \ne 0$, then the zero equilibrium is asymp-
totically stable. If $X(\tilde{x})$ is a constant matrix, then the stability holds globally.
A stabilizing feedback gain $F_i(x)$ can be obtained from $X(\tilde{x})$ and $M_i(x)$ as
$F_i(x) = M_i(x)X^{-1}(\tilde{x})$.

3.3   Guaranteed Cost Control

For the polynomial fuzzy model (14) and controller (16), we define the polynomial fuzzy model output as

$$y = \sum_{i=1}^{r} h_i(z)C_i(x)\hat{x}(x), \qquad (21)$$

where $C_i(x)$ is also a polynomial matrix. Let us consider the following performance function to be optimized.

$$J = \int_0^\infty \hat{y}^T \begin{bmatrix} Q & 0 \\ 0 & R \end{bmatrix} \hat{y} dt, \qquad (22)$$

where

$$\hat{y} = \begin{bmatrix} y \\ u \end{bmatrix} = \sum_{i=1}^{r}\sum_{j=1}^{r} h_i(z)h_j(z) \begin{bmatrix} C_i(x) \\ -F_j(x) \end{bmatrix} \hat{x}(x), \qquad (23)$$

Q and R are positive definite matrices.

Theorem 2 provides the SOS design condition that minimizes the upper bound
of the given performance function (22).

Theorem 2. *[32] If there exist a symmetric polynomial matrix $X(\tilde{x}) \in R^{N\times N}$
and a polynomial matrix $M_i(x) \in R^{m\times N}$ such that (24), (25), (26) and (27)
hold, the guaranteed cost controller that minimizes the upper bound of the given
performance function (22) can be designed as $F_i(x) = M_i(x)X^{-1}(\tilde{x})$.*

$$\underset{X(\tilde{x}),M_i(x)}{minimize} \qquad \lambda$$

subject to

$$v_1^T(X(\tilde{x}) - \epsilon_1(x)I)v_1 \text{ is SOS} \tag{24}$$

$$v_2^T \begin{bmatrix} \lambda & \hat{x}^T(0) \\ \hat{x}(0) & X(\tilde{x}(0)) \end{bmatrix} v_2 \text{ is SOS} \tag{25}$$

$$-v_3^T \begin{bmatrix} N_{ii}(x) + \epsilon_{2ii}(x)I & * & * \\ C_i(x)X(\tilde{x}) & -Q^{-1} & * \\ -M_i(x) & 0 & -R^{-1} \end{bmatrix} v_3 \text{ is SOS}, \tag{26}$$

$$-v_4^T \begin{bmatrix} N_{ij}(x) + N_{ji}(x) & * & * \\ \begin{pmatrix} C_i(x)X(\tilde{x}) \\ +C_j(x)X(\tilde{x}) \end{pmatrix} & -2Q^{-1} & 0 \\ -M_i(x) - M_j(x) & 0 & -2R^{-1} \end{bmatrix} v_4 \text{ is SOS}, \quad i < j, \tag{27}$$

where * denotes the transposed elements (matrices) for symmetric positions.

$$\begin{aligned} N_{ij}(x) =& T(x)A_i(x)X(\tilde{x}) - T(x)B_i(x)M_j(x) \\ &+ X(\tilde{x})A_i^T(x)T^T(x) - M_j^T(x)B_i^T(x)T^T(x) \\ &- \sum_{k \in K} \frac{\partial X(\tilde{x})}{\partial x_k} A_i^k(x)\hat{x}. \end{aligned}$$

v_1, v_2, v_3 and v_4 are vectors that are independent of x. $\epsilon_1(x)$ and $\epsilon_{2ii}(x)$ are non negative polynomials such that $\epsilon_1(x) > 0$ and $\epsilon_{2ii}(x) > 0$ at $x \neq 0$, and $\epsilon_1(x) = 0$ and $\epsilon_{2ii}(x) = 0$ at $x = 0$.

Remark 1. Note that v_1, v_2, v_3 and v_4 are vectors that are independent of x, because $L(x)$ is not always a positive semi-definite matrix for all x even if $x^T(x)L(x)x(x)$ is an SOS, where $L(x)$ is a symmetric polynomial matrix in $x(t)$. However, it is guaranteed from Proposition 2 in [32] that if $v^T L(x)v$ is an SOS, then $L(x) \geq 0$ for all x.

Remark 2. To avoid introducing non-convex condition, we assume that $X(\tilde{x})$ only depends on states $\tilde{x}$ whose dynamics is not directly affected by the control input, namely states whose corresponding rows in $B_i(x)$ are zero. In relation to this, it may be advantageous to employ an initial state transformation to introduce as many zero rows as possible in $B_i(x)$.

Remark 3. When $X(\tilde{x})$ is a constant matrix and $\hat{x}(x) = x$, the system representation is the same as the Takagi-Sugeno fuzzy model and control used in many of the references, e.g., [28,39]. Thus, our SOS approach to fuzzy model and control with polynomial rule consequence contains the existing LMI approaches to Takagi-Sugeno fuzzy model and control as a special case. Therefore, our SOS approach provides much more relaxed results than the existing approaches to Takagi-Sugeno fuzzy model and control.

4   Controller Designs

For the dynamics of the helicopter (7)-(10), we consider the local linear feedback control with respect to the yaw angle $\psi(t)$. From the practical control points of view, we design a local stable feedback controller $U_\psi(t) = -a \cdot \psi(t)$, where a is a positive value. Clearly, the yaw dynamics can be stabilized by the local feedback controller. As a result, we can focus on the remaining $x(t)$, $y(t)$ and $z(t)$ position control. Then, the dynamics can be rewritten as

$$\dot{u}(t) = -\frac{a}{I_z}\psi(t)v(t) + \frac{1}{m}U_X(t), \tag{28}$$

$$\dot{v}(t) = \frac{a}{I_z}\psi(t)u(t) + \frac{1}{m}U_Y(t), \tag{29}$$

$$\dot{w}(t) = \frac{1}{m}U_Z(t). \tag{30}$$

Based on the concept of sector nonlinearity [28], the nonlinear system can be exactly represented by a Takagi-Sugeno fuzzy model for $\psi(t) \in [-\pi \ \pi]$. The Takagi-Sugeno fuzzy model is obtained as

$$\dot{\boldsymbol{x}}(t) = \sum_{i=1}^{2} h_i(\boldsymbol{z}(t))\{\boldsymbol{A}_i\boldsymbol{x}(t) + \boldsymbol{B}_i\boldsymbol{u}(t)\}, \tag{31}$$

$$\dot{\boldsymbol{y}}(t) = \sum_{i=1}^{2} h_i(\boldsymbol{z}(t))\boldsymbol{C}_i\boldsymbol{x}(t), \tag{32}$$

where $\boldsymbol{z}(t) = \psi(t)$ and

$$\boldsymbol{x}(t) = [u(t) \ v(t) \ w(t) \ e_x(t) \ e_y(t) \ e_z(t)]^T,$$
$$\boldsymbol{u}(t) = [U_X(t) \ U_Y(t) \ U_Z(t)]^T.$$

The elements $e_x(t)$, $e_y(t)$ and $e_z(t)$ are defined as $e_x(t) = x(t) - x_{ref}$, $e_y(t) = y(t) - y_{ref}$, $e_z(t) = z(t) - z_{ref}$, where x_{ref}, y_{ref} and z_{ref} are constant target positions. $\boldsymbol{A}_i$, $\boldsymbol{B}_i$ and $\boldsymbol{C}_i$ matrices and the membership functions are given as follows.

$$\boldsymbol{A}_1 = \begin{bmatrix} 0 & -\frac{a\pi}{I_z} & 0 & 0 & 0 & 0 \\ \frac{a\pi}{I_z} & 0 & 0 & 0 & 0 & 0 \\ 0 & 0 & 0 & 0 & 0 & 0 \\ 1 & 0 & 0 & 0 & 0 & 0 \\ 0 & 1 & 0 & 0 & 0 & 0 \\ 0 & 0 & 1 & 0 & 0 & 0 \end{bmatrix}, \quad \boldsymbol{A}_2 = \begin{bmatrix} 0 & \frac{a\pi}{I_z} & 0 & 0 & 0 & 0 \\ -\frac{a\pi}{I_z} & 0 & 0 & 0 & 0 & 0 \\ 0 & 0 & 0 & 0 & 0 & 0 \\ 1 & 0 & 0 & 0 & 0 & 0 \\ 0 & 1 & 0 & 0 & 0 & 0 \\ 0 & 0 & 1 & 0 & 0 & 0 \end{bmatrix},$$

$$B_1 = B_2 = \begin{bmatrix} \frac{1}{m} & 0 & 0 \\ 0 & \frac{1}{m} & 0 \\ 0 & 0 & \frac{1}{m} \\ 0 & 0 & 0 \\ 0 & 0 & 0 \\ 0 & 0 & 0 \end{bmatrix}, \quad C_1 = C_2 = \begin{bmatrix} 0 & 0 & 0 & 1 & 0 & 0 \\ 0 & 0 & 0 & 0 & 1 & 0 \\ 0 & 0 & 0 & 0 & 0 & 1 \end{bmatrix},$$

$$h_1(\psi(t)) = \frac{\psi(t) + \pi}{2\pi}, \quad h_2(\psi(t)) = \frac{\pi - \psi(t)}{2\pi}.$$

Note that the Takagi-Sugeno fuzzy model exactly represents the dynamics (28) - (30) for the range $\psi(t) \in [-\pi \; \pi]$. In addition, the local stable controller $U_\psi(t) = -a \cdot \psi(t)$ guarantees $\psi(t_1) > \psi(t_2)$ for $t_1 < t_2$. The asymptotic stability property means that the helicopter describing by the dynamics (7) - (10) can be stabilized if we can design a stable controller for (28) - (30).

4.1 LMI Design Approach

Consider the performance index (22) again. We can find feedback gains that minimizes the upper bound of (22) by solving the following LMIs [28]. From the solutions X and M_i, the feedback gains can be obtained as $F_i = M_i X^{-1}$. Then, the controller satisfies $J < x^T(0) X x(0) < \lambda$.

$$\underset{X, M_i,}{minimize} \; \lambda$$
$$subject \; to$$

$$X > 0, \quad \begin{bmatrix} \lambda & x^T(0) \\ x(0) & X \end{bmatrix} > 0, \tag{33}$$

$$\hat{U}_{ii} < 0 \tag{34}$$

$$\hat{V}_{ij} < 0 \qquad i < j, \tag{35}$$

where

$$\hat{U}_{ii} = \begin{bmatrix} H_{ii} & X C_i^T & -M_i^T \\ C_i X & -Q^{-1} & 0 \\ -M_i & 0 & -R^{-1} \end{bmatrix},$$

$$\hat{V}_{ij} = \begin{bmatrix} H_{ij} + H_{ji} & X C_i^T & -M_j^T & X C_j^T & -M_i^T \\ C_i X & -Q^{-1} & 0 & 0 & 0 \\ -M_j & 0 & -R^{-1} & 0 & 0 \\ C_j X & 0 & 0 & -Q^{-1} & 0 \\ -M_i & 0 & 0 & 0 & -R^{-1} \end{bmatrix},$$

$$H_{ij} = X A_i^T + A_i X - B_i M_j - M_j^T B_i^T.$$

4.2 Simulation Results

The above LMI conditions are feasible. Both SOS design conditions in Theorems 1 and 2 are also feasible. We compare the LMI-based guaranteed-cost controller (designed by solving the (33) - (35)) with the controller (designed by the SOS conditions in Theorem 2), that is, with the SOS-based guaranteed-cost controller. Table 2 shows comparison results of performance function values J for the LMI controller and the SOS controller, where the initial positions are $u(0) = 0.5$, $v(0) = 0.5$, $w(0) = 0.5$ $e_x(0) = -0.6$, $e_y(0) = -0.4$ and $e_z(0) = -1$. In Table 2, Cases I, II and III denote three cases of selecting the weighting matrices $(Q, R) = (I, 0.1I)$, $(Q, R) = (I, I)$, and $(Q, R) = (I, 10I)$, respectively. In the SOS controller design, the order of $M(x)$ is one, i.e., all the elements of $M(x)$ are permitted to be a linear combination of one order with respect to state variables (namely affine), and the order of $X(\tilde{x})$ is zero, i.e., $X(\tilde{x})$ is a constant matrix.

Table 2. Comparison of performance function values J

	Case I	Case II	Case III
LMI controller	0.57724	1.392	5.0064
SOS controller (Order of M is 1)	0.49951	0.84659	2.8677
Reduction rate of J [%]	13.4658	39.1818	42.7193

It is found from Table 2 that the performance index values of the SOS based guaranteed-cost control (Theorem 2) are better than those of the LMI based guaranteed-cost control ((33) - (35)) in all the cases. When the orders of $X(\tilde{x})$ and $M(x)$ are zero, that is, when $X(\tilde{x})$ and $M(x)$ are constant matrices instead of polynomial matrices in x, the design conditions in Theorems 1 and 2 reduce to the existing LMI design conditions. In other words, when $X(\tilde{x})$ and $M(x)$ are constant matrices, the polynomial fuzzy controller reduces to the Takagi-Sugeno fuzzy controller. Thus, the SOS approach provides more relaxed results than the existing LMI approach.

Fig. 1 shows the SOS control result in the following target trajectory: $[x_{ref}\ y_{ref}\ z_{ref}]$ given as $[0\ 0\ 0]$ at $t = 0$, $[0\ 0\ 1]$ at $0 < t < 60$, $[1\ 0\ 1]$ at $60 \le t < 120$, $[1\ 1\ 1]$ at $120 \le t < 180$, $[0\ 1\ 1]$ at $180 \le t < 240$, and $[0\ 1\ 0]$ at $240 \le t \le 300$, where $\psi = 0$ for all t. The designed SOS controller perfectly works even for the trajectory task since $x(t) \to 0$ implies $e_x(t) \to 0$, $e_y(t) \to 0$ and $e_z(t) \to 0$. Table 3 shows comparison results of performance function values J in the above trajectory control. The SOS control result is better than the LMI control result also in the trajectory control.

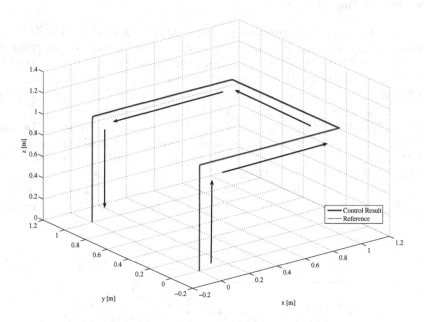

Fig. 1. Trajectory control via SOS control (Case I)

Table 3. Comparison of performance function values J

	Case I	Case II	Case III
LMI controller	0.0479	0.4733	8.0186
SOS controller (Order of M is 1)	0.0274	0.1614	5.1378

5   Vision-Based Micro Helicopter Control in Real Environments

In the previous sections, we have mainly discussed LMI and SOS based fuzzy control system design approaches for the micro helicopter. In this section, we apply one of the approaches to vision-based control of the micro helicopter in real environments [36].

Autonomous control (e.g., [4]-[16]) of helicopters has been conducted for outdoor helicopters with expensive inertial measurement units (IMU) including GPS, 3D acceleration sensors, 3D gyro sensors, etc. On the other hand, it is possible to control a small helicopter with inexpensive (reasonable price) sensors. However, it is, in general, harder to stabilize smaller helicopters due to their smaller moment of inertia in addition to the properties of unstable, nonlinear and coupling.

Fig. 2 shows the micro (palm-size) helicopter that is a co-axial counter rotating helicopter produced by HIROBO. Table 4 shows the specification of helicopter. The weight of the helicopter itself is 200g. It should be noted that the payload is only 60 g due to micro (palm-size) helicopter. Thus, due to the payload restriction, it is difficult to put a 3D acceleration sensor, a 3D gyro sensor and a 3D geomagnetic sensor, etc., on the micro helicopter.

Fig. 2. R/C Micro Helicopter

Table 4. Specification of helicopter

Mass	0.20[kg]	Length	0.40[m]
Width	0.23[m]	Height	0.20 [m]
Blade diameter	0.35 [m]	Payload	60 [g]

We put only a small-light wireless camera on the micro-helicopter. The camera is employed for detecting the position and attitude and for gathering flight visual information. The first point is accomplished by the so-called parallel tracking and mapping (PTAM) [40]. Thus, the PTAM technique using a small single wireless camera on the helicopter is utilized to detect the position and attitude of the helicopter. We construct the measurement system that is able to calibrate the mapping between local coordinate system in the PTAM and world coordinate system and is able to realize noise detection and elimination. In addition, we design the guaranteed cost (stable) controller for the dynamics of the helicopter via an LMI approach. Although path tracking control only via the small single wireless vision sensor is a quite difficult task, the control results demonstrate the utility of our approach. We have verified in the previous sections that the proposed SOS approach is better than the existing LMI approaches. The SOS design for the micro helicopter is currently ongoing and is expected to be presented elsewhere.

Subsection 5.1 presents our experimental system and micro helicopter with a wireless camera. We also discuss the PTAM as a visual SLAM to detect the position and attitude of the helicopter. In addition, we construct the measurement system that is able to calibrate the mapping between local coordinate system in the PTAM and world coordinate system and is able to realize noise detection and elimination. In Subsection 5.2, we design the guaranteed cost (stable) controller for the dynamics of the helicopter via an LMI approach. Subsection 5.3 demonstrates that the constructed system with the guaranteed cost controller achieves path tracking control well even though stabilization of the indoor micro helicopter only via the small single wireless vision sensor is a quite difficult task.

5.1 Experimental System

Fig. 3 shows the experimental system using the small-light wireless camera (TINY-3H) produced by RFSystem Co.,Ltd. The weight is 55 g and is within the payload limitation (60 g). Table 5 summarizes the specification of the wireless camera. The six degree of freedom of the helicopter is calculated by the PTAM based on vision obtained from the wireless camera. After the computation using the PTAM, the control input determined by a stable controller (that will be discussed later) is sent to the helicopter via an R/C transmitter. The sampling rate is 30 [Hz].

In this research, an open-source software, parallel tracking and mapping (PTAM) developed by Klein and Murray [40], is employed to detect the position and attitude of the indoor micro helicopter. The PTAM is a method of estimating camera pose in an unknown scene. They proposed a system specifically designed to track a hand-held camera in a small augmented reality (AR) workspace. For more details of the PTAM, see [40]-[45].

As mentioned before, the PTAM is supposed to use for tracking a hand-held camera in a small AR workspace. Hence, we need to add two functions to achieve stabilization of the micro helicopter.

- Accurate calibration of the mapping between world coordinate system and PTAM coordinate system.
- Compensation of the vision noise contaminated by wireless vision transmission, electromagnetic devices, or body vibrations of the helicopter, etc.

Fig. 4 shows an example of the vision noise contaminated by transmission, electromagnetic devices, or body vibrations of the helicopter, etc., where i-th frame

Table 5. Specification of TINY-3H

Weight	55g
Image Sensor	270000 pixels 1/4 inch color CCD
Unobstructed Effective Range	100m (Transmission distance)
Size	117 18 75

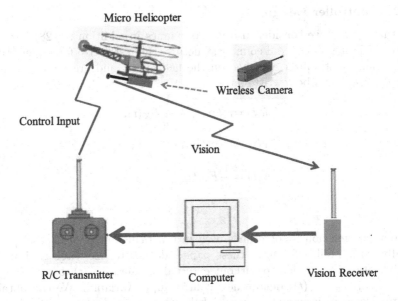

Fig. 3. Experimental System

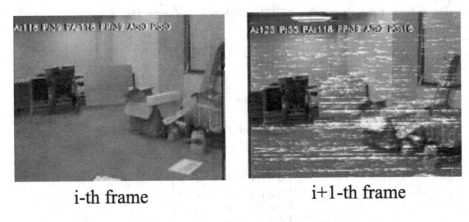

Fig. 4. Vision contaminated by noise

is clear, but i+1-th frame is contaminated by noise. In the i+1-th frame, the position and attitude of the helicopter calculated using the PTAM for noise vision suddenly change for the calculation result from the previous i-th frame. In this case, we ignore the calculation result for the i+1-th frame and still utilizes the calculation result for the previous i-th frame. The threshold of judging the noise frame is ϵ % change of at least one of six variables in the position and attitude. The value of ϵ is adjusted through experiments.

5.2   Controller Design

In this section, we consider also yaw dynamics in addition to (28)-(30). Then, by considering its co-axial counter structure, the restitutive force (generated by a mechanical stabilizer attached on the helicopter) and gravity compensation, the dynamics can be rewritten as

$$\dot{u}(t) = r(t)v(t) + \frac{1}{m}F_X(t), \tag{36}$$

$$\dot{v}(t) = -r(t)u(t) + \frac{1}{m}F_Y(t), \tag{37}$$

$$\dot{w}(t) = \frac{1}{m}F_Z(t), \tag{38}$$

$$\dot{\psi}(t) = \frac{1}{I_Z}U_\psi(t). \tag{39}$$

The approximation is sometimes used in practical control field [46,47] and actually works well. Of course, these papers do not realize wireless vision-based stabilization in addition to without external markers. $U_X(t)$, $U_Y(t)$, $U_Z(t)$ and $U_\psi(t)(\dot{U}_\psi(t) = M_z(t))$ denote new control input variables. We can obtain the original control inputs (to the real helicopter) from $F_X(t)$, $F_Y(t)$, $F_Z(t)$ and $U_\psi(t)$.

By taking time derivative of (36)-(39) and defining the state and control vectors as

$$\boldsymbol{x}(t) = [x(t)\ y(t)\ z(t)\ \psi(t)\ u(t)\ v(t)\ w(t)]^T,$$
$$\boldsymbol{u}(t) = [F_X(t)\ F_Y(t)\ F_Z(t)\ U_\psi(t)]^T,$$

we arrive at the following state equation

$$\frac{d}{dt}\dot{\boldsymbol{x}}(t) = \boldsymbol{A}\dot{\boldsymbol{x}}(t) + \boldsymbol{B}\dot{\boldsymbol{u}}(t) \tag{40}$$

where

$$\boldsymbol{A} = \begin{bmatrix} 0 & 0 & 0 & 0 & 1 & 0 & 0 \\ 0 & 0 & 0 & 0 & 0 & 1 & 0 \\ 0 & 0 & 0 & 0 & 0 & 0 & 1 \\ 0 & 0 & 0 & 0 & 0 & 0 & 0 \\ 0 & 0 & 0 & 0 & 0 & r(t) & 0 \\ 0 & 0 & 0 & 0 & -r(t) & 0 & 0 \\ 0 & 0 & 0 & 0 & 0 & 0 & 0 \end{bmatrix},$$

$$\boldsymbol{B} = \begin{bmatrix} 0 & 0 & 0 & 0 \\ 0 & 0 & 0 & 0 \\ 0 & 0 & 0 & 0 \\ 0 & 0 & 0 & \frac{1}{I_Z} \\ \frac{1}{m} & 0 & 0 & \frac{1}{I_Z}v(t) \\ 0 & \frac{1}{m} & 0 & -\frac{1}{I_Z}u(t) \\ 0 & 0 & \frac{1}{m} & 0 \end{bmatrix},$$

$$C = \begin{bmatrix} 1 & 0 & 0 & 0 & 0 & 0 & 0 \\ 0 & 1 & 0 & 0 & 0 & 0 & 0 \\ 0 & 0 & 1 & 0 & 0 & 0 & 0 \\ 0 & 0 & 0 & 1 & 0 & 0 & 0 \end{bmatrix}.$$

We consider reasonable assumption such that $-u_d \leq u(t) \leq u_d$, $-v_d \leq v(t) \leq v_d$ and $-r_d \leq r(t) \leq r_d$. Applying the sector nonlinearity procedure [28] to the augmented system, we obtain the fuzzy model (41) that exactly represent the dynamics (40) under the assumption. In this case, we set $u_d = 2$, $v_d = 2$ and $r_d = \pi$.

The Takagi-Sugeno fuzzy model for micro helicopter dynamics considering the above two features can be represented as

$$\frac{dt}{d}\dot{x}(t) = \sum_{i=1}^{8} h_i(u(t), v(t), r(t))\{A_i\dot{x}(t) + B_i\dot{u}(t)\}, \tag{41}$$

where

$$x(t) = [x(t)\ y(t)\ z(t)\ \psi(t)\ u(t)\ v(t)\ w(t)]^T,$$
$$u(t) = [F_X(t)\ F_Y(t)\ F_Z(t)\ U_\psi(t)]^T,$$

$$A_1 = A_3 = A_5 = A_7 = \begin{bmatrix} 0 & 0 & 0 & 0 & 1 & 0 & 0 \\ 0 & 0 & 0 & 0 & 0 & 1 & 0 \\ 0 & 0 & 0 & 0 & 0 & 0 & 1 \\ 0 & 0 & 0 & 0 & 0 & 0 & 0 \\ 0 & 0 & 0 & 0 & 0 & r_d & 0 \\ 0 & 0 & 0 & 0 & -r_d & 0 & 0 \\ 0 & 0 & 0 & 0 & 0 & 0 & 0 \end{bmatrix},$$

$$A_2 = A_4 = A_6 = A_8 = \begin{bmatrix} 0 & 0 & 0 & 0 & 1 & 0 & 0 \\ 0 & 0 & 0 & 0 & 0 & 1 & 0 \\ 0 & 0 & 0 & 0 & 0 & 0 & 1 \\ 0 & 0 & 0 & 0 & 0 & 0 & 0 \\ 0 & 0 & 0 & 0 & 0 & -r_d & 0 \\ 0 & 0 & 0 & 0 & r_d & 0 & 0 \\ 0 & 0 & 0 & 0 & 0 & 0 & 0 \end{bmatrix},$$

$$B_1 = B_2 = \begin{bmatrix} 0 & 0 & 0 & 0 \\ 0 & 0 & 0 & 0 \\ 0 & 0 & 0 & 0 \\ 0 & 0 & 0 & \frac{1}{I_Z} \\ \frac{1}{m} & 0 & 0 & \frac{v_d}{I_Z} \\ 0 & \frac{1}{m} & 0 & -\frac{u_d}{I_Z} \\ 0 & 0 & \frac{1}{m} & 0 \end{bmatrix}, B_3 = B_4 = \begin{bmatrix} 0 & 0 & 0 & 0 \\ 0 & 0 & 0 & 0 \\ 0 & 0 & 0 & 0 \\ 0 & 0 & 0 & \frac{1}{I_Z} \\ \frac{1}{m} & 0 & 0 & -\frac{v_d}{I_Z} \\ 0 & \frac{1}{m} & 0 & -\frac{u_d}{I_Z} \\ 0 & 0 & \frac{1}{m} & 0 \end{bmatrix},$$

$$B_5 = B_6 = \begin{bmatrix} 0 & 0 & 0 & 0 \\ 0 & 0 & 0 & 0 \\ 0 & 0 & 0 & 0 \\ 0 & 0 & 0 & \frac{1}{I_Z} \\ \frac{1}{m} & 0 & 0 & \frac{v_d}{I_Z} \\ 0 & \frac{1}{m} & 0 & \frac{u_d}{I_Z} \\ 0 & 0 & \frac{1}{m} & 0 \end{bmatrix}, B_7 = B_8 = \begin{bmatrix} 0 & 0 & 0 & 0 \\ 0 & 0 & 0 & 0 \\ 0 & 0 & 0 & 0 \\ 0 & 0 & 0 & \frac{1}{I_Z} \\ \frac{1}{m} & 0 & 0 & -\frac{v_d}{I_Z} \\ 0 & \frac{1}{m} & 0 & \frac{u_d}{I_Z} \\ 0 & 0 & \frac{1}{m} & 0 \end{bmatrix},$$

$$h_1(u(t), v(t), r(t)) = \frac{u(t) + u_d}{2u_d} \cdot \frac{v(t) + v_d}{2v_d} \cdot \frac{r(t) + r_d}{2r_d},$$

$$h_2(u(t), v(t), r(t)) = \frac{u(t) + u_d}{2u_d} \cdot \frac{v(t) + v_d}{2v_d} \cdot \frac{r_d - r(t)}{2r_d},$$

$$h_3(u(t), v(t), r(t)) = \frac{u(t) + u_d}{2u_d} \cdot \frac{v_d - v(t)}{2v_d} \cdot \frac{r(t) + r_d}{2r_d},$$

$$h_4(u(t), v(t), r(t)) = \frac{u(t) + u_d}{2u_d} \cdot \frac{v_d - v(t)}{2v_d} \cdot \frac{r_d - r(t)}{2r_d},$$

$$h_5(u(t), v(t), r(t)) = \frac{u_d - u(t)}{2u_d} \cdot \frac{v(t) + v_d}{2v_d} \cdot \frac{r(t) + r_d}{2r_d},$$

$$h_6(u(t), v(t), r(t)) = \frac{u_d - u(t)}{2u_d} \cdot \frac{v(t) + v_d}{2v_d} \cdot \frac{r_d - r(t)}{2r_d},$$

$$h_7(u(t), v(t), r(t)) = \frac{u_d - u(t)}{2u_d} \cdot \frac{v_d - v(t)}{2v_d} \cdot \frac{r(t) + r_d}{2r_d},$$

$$h_8(u(t), v(t), r(t)) = \frac{u_d - u(t)}{2u_d} \cdot \frac{v_d - v(t)}{2v_d} \cdot \frac{r_d - r(t)}{2r_d}.$$

By defining the error $e(t) = r - y(t)$, we have the following augmented system.

$$\frac{d}{dt}\hat{x}(t) = \sum_{i=1}^{8} h_i(u(t), v(t), r(t))\{\hat{A}_i\hat{x}(t) + \hat{B}_i\hat{u}(t)\} \tag{42}$$

$$y(t) = \sum_{i=1}^{8} h_i(u(t), v(t), r(t))\hat{C}_i\hat{x}(t), \tag{43}$$

where $\hat{u}(t) = \dot{u}(t)$,

$$\hat{x}(t) = \begin{bmatrix} \dot{x}(t) \\ e(t) \end{bmatrix}, \quad \hat{A}_i = \begin{bmatrix} A_i & 0 \\ \hat{C}_i & 0 \end{bmatrix}, \quad \hat{B}_i = \begin{bmatrix} B_i \\ 0 \end{bmatrix},$$

$$\hat{C}_i = \begin{bmatrix} 1 & 0 & 0 & 0 & 0 & 0 & 0 \\ 0 & 1 & 0 & 0 & 0 & 0 & 0 \\ 0 & 0 & 1 & 0 & 0 & 0 & 0 \\ 0 & 0 & 0 & 1 & 0 & 0 & 0 \end{bmatrix}.$$

We design the following dynamic fuzzy controller to stabilize the augmented system.

$$\hat{u}(t) = - \sum_{i=1}^{8} h_i(u(t), v(t), r(t)) F_i \hat{x}(t) \tag{44}$$

Let us consider the following performance function to be optimized.

$$J = \int_0^{\infty} \hat{y}^T(t) \begin{bmatrix} Q & 0 \\ 0 & R \end{bmatrix} \hat{y}(t) dt, \tag{45}$$

where

$$\hat{y}(t) = \begin{bmatrix} y(t) \\ \hat{u}(t) \end{bmatrix} = \sum_{i=1}^{r} \sum_{j=1}^{r} \hat{h}_i(t) \hat{h}_j(t) \begin{bmatrix} \hat{C}_i \\ -F_j \end{bmatrix} \hat{x}(t), \tag{46}$$

Q and R are positive definite matrices, and $\hat{h}_i(t) = h_i(u(t), v(t), r(t))$.

We can find feedback gains that minimizes the upper bound of (45) by solving the following linear matrix inequalities (LMIs) (47)-(49) [28]. From the solutions X and M_i, the feedback gains can be obtained as $F_i = M_i X^{-1}$. Then, the controller satisfies $J < x^T(0) X x(0) < \lambda$.

$\underset{X, M_i,}{minimize}\ \lambda$

$subject\ to$

$$X > 0, \quad \begin{bmatrix} \lambda & x^T(0) \\ x(0) & X \end{bmatrix} > 0, \tag{47}$$

$$\hat{U}_{ii} < 0 \tag{48}$$

$$\hat{V}_{ij} < 0 \qquad i < j, \tag{49}$$

where

$$\hat{U}_{ii} = \begin{bmatrix} H_{ii} & X\hat{C}_i^T & -M_i^T \\ \hat{C}_i X & -Q^{-1} & 0 \\ -M_i & 0 & -R^{-1} \end{bmatrix},$$

$$\hat{V}_{ij} = \begin{bmatrix} H_{ij} + H_{ji} & X\hat{C}_i^T & -M_j^T & X\hat{C}_j^T & -M_i^T \\ \hat{C}_i X & -Q^{-1} & 0 & 0 & 0 \\ -M_j & 0 & -R^{-1} & 0 & 0 \\ \hat{C}_j X & 0 & 0 & -Q^{-1} & 0 \\ -M_i & 0 & 0 & 0 & -R^{-1} \end{bmatrix},$$

$$H_{ij} = XA_i^T + A_i X - B_i M_j - M_j^T B_i^T.$$

It should be noted that the controller satisfying the LMIs (47)-(49) is a stable controller.

5.3   Experimental Results

Control experiment is performed from the take-off on the floor at the origin $(x(0), y(0), z(0)) = (0[mm], 0[mm], 0[mm])$ and rectangular trajectory flight during keeping z(t)=1000 [mm]. The vertex points of the rectangular trajectory are Point A (0 [mm], 0 [mm], 1000 [mm]), Point B (2000 [mm], 0 [mm], 1000 [mm]), Point C (2000 [mm], 1500 [mm], 1000 [mm]) and Point D (0 [mm], 1500 [mm], 1000 [mm]). The flight task is to make two circles around Points A, B, C and D during keeping the altitude z(t)=1000 [mm].

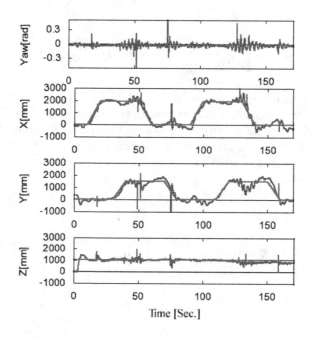

Fig. 5. Experimental result of path tracking flight

Fig. 5 shows the trajectory control result (yaw, x, y and z) via the fuzzy controller, where the target (green lines) and control result (red lines) are plotted. It can be seen that the helicopter can follow the trajectory well even though the impulse noises are sometimes caused by wireless vision transmission, electromagnetic devices, or body vibrations of the helicopter, etc. Thus, the control result shows the utility of our wireless vision-based control system. Fig 6 shows photographs of the experimental result, where the red boxes indicate the positions of the micro helicopter and the small window in each the photograph shows vision views from the wireless camera on the micro helicopter. It should be noted that the trajectory of the micro helicopter is stabilized only by using the wireless vision sensor without 3D acceleration sensors, 3D gyro sensors and 3D geomagnetic sensors, etc., in addition to without external markers.

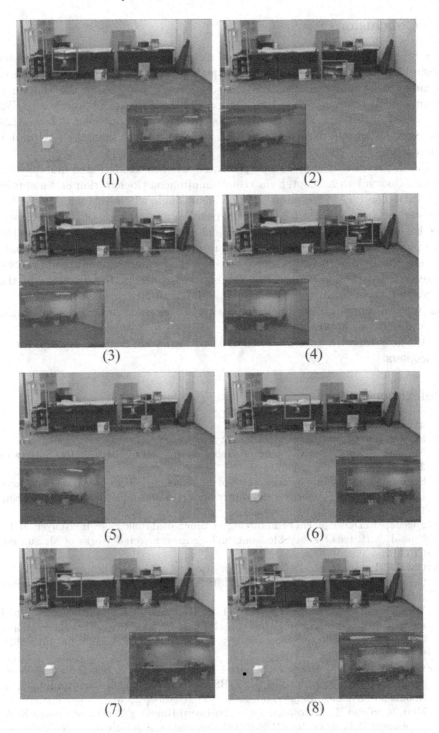

Fig. 6. Experimental results (photos)

6   Conclusions

The former of this chapter has presented a comparison result of micro helicopter control via a typical linear matrix inequality (LMI) approach and a sum of squares (SOS) approach. The SOS design approach discussed in this chapter is more general than that based on the existing LMI design approaches to T-S fuzzy control systems. The control results of a micro helicopter have shown that the SOS design approach provides better control results than the LMI design approach.

The latter of this chapter has presented wireless vision-based stabilization of an indoor micro helicopter via visual simultaneous localization and mapping (SLAM). The PTAM technique using a small single wireless camera on the helicopter has been utilized to detect the position and attitude of the helicopter. We have also constructed the measurement system that is able to calibrate the mapping between local coordinate system in the PTAM and world coordinate system and is able to realize noise detection and elimination. In addition, we have designed the guaranteed cost (stable) controller for the dynamics of the helicopter via a linear matrix inequality (LMI) approach. The path tracking control results only via the small single wireless vision sensor have demonstrated the utility of our approach.

Our next target is to realize vision-based formation control of plural micro helicopters.

References

1. http://www.rc.mce.uec.ac.jp/
2. http://www.bu.edu/me/people/faculty/pz/wang/
3. Nonami, K.: Rotating Wing Aerial Robotics. Journal of the Robotics Society of Japan 24(8), 890–896 (2006)
4. Sugeno, M., et al.: Intelligent Control of an Unmanned Helicopter Based on Fuzzy Logic. In: Proc. of American Helicopter Society 51st Annual Forum, Texas (May 1995)
5. Sugeno, M.: Development of an Intelligent Unmanned Helicopter. In: Nguyen, H.T., Prasad, N.R. (eds.) Fuzzy Modeling and Control: Selected Works of M. Sugeno. CRC Press (1999)
6. Ohtake, H., Iimura, K., Tanaka, K.: Fuzzy Control of Micro Helicopter with Coaxial Counter-rotating Blades. In: SCIS and ISIS 2006, Tokyo, pp. 1955–1958 (September 2006)
7. Yoshihata, Y., Watanabe, K., Iwatani, Y., Hashimoto, K.: Multi-camera visual servoing of a micro helicopter under occlusions. In: 2007 IEEE/RSJ International Conference on Intelligent Robots and Systems, San Diego, CA, USA, pp. 2615–2620 (October 2007)
8. Wang, W., et al.: Autonomous Control for Micro-Flying Robot and Small Wireless Helicopter X.R.B. In: 2006 IEEE/RSJ International Conference on Intelligent Robots and Systems, Beijing, pp. 2906–2911 (October 2006)
9. Mori, R., Kubo, T., Kinoshita, T.: Vision-Based Hovering Control of a Small-Scale Unmanned Helicopter. In: SICE-ICASE International Joint Conference 2006, pp. 1274–1278 (2006)

10. Pastor, E., Lopez, J., Royo, P.: UAV Payload and Mission Control Hardware/Software Architecture. IEEE Aerospace and Electronic Systems Magazine 22(6), 3–8 (2007)
11. Antsev, G.V., et al.: UAV landing system simulation model software system. IEEE Aerospace and Electronic Systems Magazine 26(3), 26–29 (2011)
12. Lin, Y., Hyyppa, J., Jaakkola, A.: Mini-UAV-Borne LIDAR for Fine-Scale Mapping. IEEE Geoscience and Remote Sensing Letters 8(3), 426–430 (2011)
13. Kumon, M., et al.: Autopilot System for Kiteplane. IEEE/ASME Transactions on Mechatronics 11(5), 615–624 (2006)
14. Tanaka, K., et al.: Development of a Cyclogyro-Based Flying Robot With Variable Attack Angle Mechanisms. IEEE/ASME Transactions on Mechatronics 12(5), 565–570 (2007)
15. Jinok, S., et al.: Attitude Control and Hovering Control of Radio-Controlled Helicopter. Journal of the JSME, Part C 68(675), 3284–3291 (2002)
16. Kanade, T., Amidi, O., Ke, Q.: Real-time and 3D vision for autonomous small and micro air vehicles. In: 43rd IEEE Conference on Decision and Control, pp. 1655–1662 (2004)
17. Wu, H., Sun, D., Zhou, Z.: Micro air vehicle: configuration, analysis, fabrication, and test. IEEE/ASME Transactions on Mechatronics 9(1), 108–117 (2004)
18. Madangopal, R., Khan, Z.A., Agrawal, S.K.: Energetics-based design of small flapping-wing micro air vehicles. IEEE/ASME Transactions on Mechatronics 11(4), 433–438 (2006)
19. Gebre-Egziabher, D., Elkaim, G.H.: MAV attitude determination by vector matching. IEEE Transactions on Aerospace and Electronic Systems 44(3), 1012–1028 (2008)
20. McIntosh, S.H., et al.: Design of a mechanism for biaxial rotation of a wing for a hovering vehicle. IEEE/ASME Transactions on Mechatronics 11(2), 145–153 (2006)
21. Valenti, M., Bethke, B., How, J.P., de Farias, D.P., Vian, J.: Embedding Health Management into Mission Tasking for UAV Teams. In: Proceedings of American Control Conference 2007, pp. 5777–5783 (2007)
22. Ohtake, H., et al.: Fuzzy Control of Micro RC Helicopter with Coaxial Counter-rotating Blades. Journal of Society for Fuzzy Theory and Intelligent Informatics 21(1), 100–106 (2009)
23. Wang, W., et al.: Autonomous Control for Micro-Flying Robot and Small Wireless Helicopter X.R.B. In: 2006 IEEE/RSJ International Conference on Intelligent Robots and Systems, Beijing, pp. 2906–2911 (October 2006)
24. Mori, R., Kubo, T., Kinoshita, T.: Vision-Based Hovering Control of a Small-Scale Unmanned Helicopter. In: SICE-ICASE International Joint Conference 2006, pp. 1274–1278 (2006)
25. Hirose, H., et al.: Visual Feedback Control of a Small RC Helicopter Using a Vehicle-Mounted Camera. In: 49th Administration Committee of Japan Joint Automatic Control Conference, pp. 332–335 (2006)
26. Mori, R., et al.: Vision-Based Guidance Control of a Small-Scale Unmanned Helicopter. In: Proceedings of the 2007 IEEE/RSJ International Conference on Intelligent Robots and Systems, pp. 2648–2653 (2007)
27. Iimura, K., et al.: Position Estimation and Control of an Indoor Micro Helicopter using an Attached Single Camera. In: The Proceedings of RoboMec Symposium 2009, Fukuoka, pp. 215–218 (May 2009)
28. Tanaka, K., Wang, H.O.: Fuzzy Control Systems Analysis and Design: A Linear Matrix Inequality Approach. John Wiley and Sons Publisher, New York (2001)

29. Tanaka, K., Yoshida, H., Ohtake, H., Wang, H.O.: A Sum of Squares Approach to Stability Analysis of Polynomial Fuzzy Systems. In: 2007 American Control Conference, New York, pp. 4071–4076 (July 2007)

30. Tanaka, K., Ohtake, H., Wang, H.O.: A Sum of Squares Approach to Modeling and Control of Nonlinear Dynamical Systems with Polynomial Fuzzy Systems. IEEE Transactions on Fuzzy Systems 17(4), 911–922 (2009)

31. Tanaka, K., Ohtake, H., Wang, H.O.: An SOS-based Stable Control of Polynomial Discrete Fuzzy Systems. In: 2008 American Control Conference, Seattle, Washington, pp. 4881–4886 (June 2008)

32. Tanaka, K., Ohtake, H., Wang, H.O.: Guaranteed Cost Control of Polynomial Fuzzy Systems via a Sum of Squares Approach. IEEE Transactions on Systems, Man and Cybernetics Part B 39(2), 561–567 (2009)

33. Tanaka, K., Ohtake, H., Wang, H.O.: A Sum of Squares Approach to Guaranteed Cost Control of Polynomial Discrete Fuzzy Systems. In: 17th IFAC World Congress, Seoul, Korea, pp. 6850–6854 (July 2008)

34. Prajna, S., Papachristodoulou, A., Seiler, P., Parrilo, P.A.: SOSTOOLS:Sum of Squares Optimization Toolbox for MATLAB, Version 2.00 (2004)

35. Sturm, J.F.: Using SeDuMi 1.02:a MATLAB toolbox for optimization over symmetric cones. Optimization Methods and Software 11(1), 625–653 (1999)

36. Tanaka, K., Ohtake, H., Tanaka, M., Wang, H.O.: Wireless Vision-based Stabilization of Indoor Micro Helicopter. IEEE/ASME Transactions on Mechatronics (accepted)

37. Takagi, T., Sugeno, M.: Fuzzy Identification of Systems and Its Applications to Modeling and Control. IEEE Trans. on SMC 15(1), 116–132 (1985)

38. Parrilo, P.A.: Structured Semidefinite Programs and Semialgebraic Geometly Methods in Robustness and Optimization. PhD thesis, California Institute of Technology, Pasadena, CA (2000)

39. Feng, G.: A Survey on Analysis and Design of Model-Based Fuzzy Control Systems. IEEE Trans. on Fuzzy Systems 14(5), 676–697 (2006)

40. Klein, G., Murray, D.: Parallel Tracking and Mapping for Small AR Workspaces. In: Proc. International Symposium on Mixed and Augmented Reality (ISMAR 2007), Nara (2007)

41. Davison, A., Reid, I., Molton, N.D., Stasse, O.: MonoSLAM: Realtime single camera SLAM. IEEE Trans. Pattern Analysis and Machine Intelligence 29(6), 1052–1067 (2007)

42. Eade, E., Drummond, T.: Edge landmarks in monocular slam. In: Proc. British Machine Vision Conference (BMVC 2006), Edinburgh (September 2006); BMVA

43. Eade, E., Drummond, T.: Scalable monocular slam. In: Proc. IEEE Intl. Conference on Computer Vision and Pattern Recognition (CVPR 2006), New York, NY, pp. 469–476 (2006)

44. Smith, R.C., et al.: On the representation and estimation of spatial uncertainty. International Journal of Robotics Research 5(4), 56–68 (1986)

45. Stewénius, H., Engels, C., Nistér, D.: Recent developments on direct relative orientation. ISPRS Journal of Photogrammetry and Remote Sensing 60, 284–294 (2006)

46. Komatsu, T., et al.: Model-based Control of a Small Indoor-type Helicopter. In: The Proceedings of RoboMec Symposium 2008, Nagano, pp. 101–104 (May 2008)

47. Tanaka, K., et al.: Micro Helicopter Control. In: IEEE World Congress on Computational Intelligence, Hong Kong, pp. 347–353 (June 2008)

Predictive Learning, Knowledge Discovery
and Philosophy of Science

Vladimir Cherkassky

Electrical & Computer Engineering, University of Minnesota, Minneapolis MN 55455
`cherk001@umn.edu`

Abstract. Various disciplines, such as machine learning, statistics, data mining and artificial neural networks, are concerned with estimation of data-analytic models. A common theme among all these methodologies is estimation of *predictive* models from data. In our digital age, an abundance of data and cheap computing power offers hope of knowledge discovery via application of statistical and machine learning algorithms to empirical data. This data-analytic knowledge has similarities and differences with classical scientific knowledge. For example, any scientific theory can be viewed as an inductive theory because it generalizes over a finite number of observations (or experiments). The philosophical aspects of induction and knowledge discovery have been thoroughly explored in Western philosophy of science. This philosophical analysis dates back to Kant and Hume. Any knowledge involves a combination of hypotheses/ideas and empirical data. In the modern digital age, the balance between ideas (mental constructs) and observed data (facts) has completely shifted. Classical scientific knowledge was produced mainly by a stroke of genius (e.g., Newton, Maxwell, and Einstein). In contrast, much of modern knowledge in life sciences and social sciences is derived via data-analytic modeling. We argue that such data-driven knowledge can be properly described following the methodology of *predictive learning* originally developed in VC-theory. This paper presents a brief survey of the philosophical concepts related to inductive inference, and then extends these ideas to predictive data-analytic knowledge discovery. We contrast the differences between classical first-principle knowledge, data-analytic knowledge and beliefs. Several application examples are used to illustrate the differences between classical statistical and predictive learning approaches to data-analytic modeling. Finally, we discuss interpretation of data-analytic models under predictive learning framework.

1   Introduction

We live in a world surrounded by data. With the advent of computer technology, most information now is digital, and its amount doubles every few years. This information, however, is useful for decision making only if there are associations and stable relationships present within the data. For example, it is easy to memorize a song or a poem, because it rhymes and has meaning, but it is difficult to remember 200

J. Liu et al. (Eds.): WCCI 2012 Plenary/Invited Lectures, LNCS 7311, pp. 209–233, 2012.

randomly chosen unrelated words. Learning, or making sense of observed data, is central to human intelligence.

Much of human knowledge is based on observations of repeatable events. For example, we 'know' that the Sun rises in the East every morning. People knew this fact thousands of years ago, before the advent of astronomy and physics. In essence, such knowledge is a result of generalization from many observed instances (of the Sun rising in the East every morning). This process of making a general statement from many regular observations is called 'induction' or 'inductive inference'. It is also an example of 'empirical knowledge' that is purely data-driven. In contrast, 'scientific knowledge' provides much deeper insights into observed empirical data. For example, Kepler's laws can be used to predict planets' movement, and Newton's law of gravity can be used to derive Kepler's laws. Scientific laws *explain and predict* many seemingly unrelated events, such as the motion of planetary bodies, and the motion of a falling object, in the case of Newton's law.

For humans, it is not sufficient just to detect regularities from observations of repeatable events; these regularities need to be 'explained' in terms of a small number of basic concepts and causal relationships. This human desire for logical and deterministic explanations is evident in the following quotations:

- All men by nature desire knowledge (Aristotle);
- Man has an intense desire for assured knowledge (Einstein).

Such explanations of the external world (Nature) constitute scientific knowledge. However, present scientific understanding of the world is fairly recent (just a few hundred years old), and prior to that people used other 'non-scientific' explanations or 'beliefs'. Note that scientific theories and beliefs are inductive as they both explain repeatable events.

The Philosophy of Science is concerned with the relationship between the objective world (Nature) and human ideas (mental models) describing this world. The main questions in Philosophy of Science (PS) are:

(PS1)   Is human knowledge formed mainly by experience (or sense-perceptions) or by pure thought (or reasoning)?

(PS2)   What is the process of knowledge formation (knowledge discovery) from empirical data (or sense-perceptions) and mental constructs?

(PS3)   The *problem of induction* or inductive inference: Is it possible to obtain assured knowledge from empirical data or observations?

(PS4)   What are general conditions (principles) for distinguishing between scientific theories, and non-scientific explanations or beliefs? Such a criterion is known as the *'demarcation principle'* in philosophy. So when a new theory is proposed, this demarcation principle would help to classify it as true scientific theory, or just a belief.

These questions have been posed and discussed by philosophers over many centuries. However, only recently with the advent of computer technology, have these issues become increasingly relevant for engineers, biologists and scientists estimating predictive models from data. The demarcation principle has become very relevant in modern life, where beliefs and opinions supported by observed correlations in the data are often presented as scientific findings.

The growing use of digital technology in modern society is changing the nature of human knowledge. Knowledge can be broadly defined as a relationship between facts and ideas. Classical science aims at describing many facts (observations) using just a few fundamental principles. Typically, this *first-principle* knowledge is in the form of deterministic relationships between a few basic concepts. Such knowledge has several characteristic properties:

- it describes well simple deterministic systems. Here 'simple' refers to conceptual simplicity, rather than 'technical' system complexity. For instance, a mechanical system may have many moving objects interacting with each other; however each object is described by simple equations that involve just a few variables.
- The number of facts (data samples, experimental observations) initially used to derive such knowledge is small.
- The cost of collecting or generating these observations is high.

In the modern world the classical balance between facts and ideas has totally shifted. Today, we are flooded with data and are expected to act upon it. With the advent of the Internet, the cost of acquiring, generating and transmitting information has become negligible (practically zero). Nowadays, most data comes from digital devices and sensors, rather than human sense-perceptions as in classical philosophy.

According to philosophical view of naïve realism, abundance of data should generate unprecedented growth of knowledge. This is reflected in a popular view:

The data deluge makes the scientific method obsolete. We can stop looking for (scientific) models. We can analyze the data without hypotheses about what it might show. We can throw the numbers into the biggest computing clusters the world ever seen and let statistical algorithms find patterns where science cannot [1].

The reality is more sober, as usual. Fast accrual of new information (facts) has not translated into any significant growth of knowledge. Even worse, many 'discovered patterns' have questionable scientific value. Classical statistical estimation approaches proved to be inadequate for describing complex data-rich systems. Estimation of useful dependencies in such systems requires new methodologies, where the goal of modeling is to act (or predict) well rather than accurate estimation of probabilistic data models.

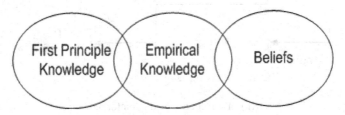

Fig. 1. Three types of knowledge

We may loosely define *empirical knowledge* as useful dependencies estimated from data or derived from experience. As argued later in this section, useful dependencies usually have predictive (or generalization) capability. In contrast to first-principle knowledge, empirical knowledge typically:

- Describes certain properties of complex systems that lack credible first-principle models. Here 'complexity' usually refers to a large number of observed parameters (variables);
- is statistical in nature, i.e., allows to make non-deterministic predictions, at best;
- has a quantifiable practical utility for a given application.

We emphasize that our definition requires such knowledge to be *useful* in the context of a given application. This is consistent with general notion of learning, which implies accomplishing a specific task, i.e. learning to drive, or learning to play piano. Empirical knowledge has been used by humans in medicine for centuries (i.e., herbal folk medicine), however its role has dramatically increased in our digital age.

It is important to differentiate between three types of knowledge: first-principle, empirical and beliefs, as shown in Fig. 1. Here *beliefs* refer to mental models that are neither first-principle nor instrumental empirical knowledge. The distinction between first-principle knowledge and beliefs is usually easy to make. Examples of true scientific theories versus pseudo-scientific beliefs are: chemistry vs. alchemy and astronomy vs. astrology. The distinction between empirical knowledge and beliefs is not so clear, as both are usually supported by statistical correlations in observed data. Yet this distinction is of great practical importance, because rational humans prefer to act upon knowledge rather than beliefs.

Modern science and engineering are based on the *first-principle* models for physical, biological, and social systems. Such an approach starts with a basic scientific model (e.g., Newton's laws of mechanics or Maxwell's laws) and then builds upon

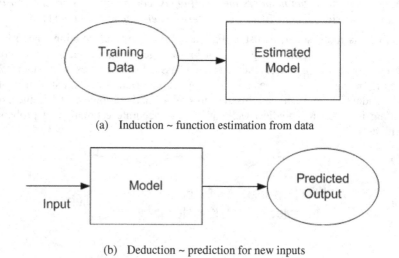

(a)   Induction ~ function estimation from data

(b)   Deduction ~ prediction for new inputs

Fig. 2. Learning or function estimation interpreted as induction-deduction process

it various applications (e.g., mechanical engineering or electrical engineering). In classical science, the experimental data (measurements) are used to test (or verify) the underlying first-principle models and, sometimes, to estimate certain model parameters that are difficult to measure directly. However, in many applications the underlying first principles are unknown or the systems under study are too complex to be mathematically described. With the growing use of computers and low-cost sensors, there is a great amount of data being generated by such systems. In the absence of first-principle models, such readily available data can be used to derive models by estimating useful relationships between observed system variables (i.e., inputs and outputs). Thus there is currently a paradigm shift from the classical modeling based on first principles to developing empirical (data-driven) models.

All data-analytic models usually pursue two goals: explanation of available (training) data and prediction of future data. For most applications, the main practical goal is prediction. Estimation of predictive models with finite data is a challenging problem. It can be immediately related to the problem of induction and to inductive-deductive reasoning in classical philosophy [2], [3]. See Figure 2.

According to popular interpretation of data-analytic modeling, an estimated predictive model (in Fig. 2a) represents new knowledge 'discovered from data'. This view is rather simplistic, because there are thousands of learning algorithms that can be applied to the same data, resulting in millions of potentially plausible models. Of course, the pragmatic solution to this dilemma advocated by practitioners is to use many learning algorithms and then select the best predictive model, typically using resampling for evaluating prediction (generalization) performance. Such a 'solution' is not very satisfactory because:

- It is highly dependent on the experience of human modelers who are very experienced in tuning their favorite modeling technique. This explains a strange phenomenon when researchers propose a new learning algorithm and present empirical comparisons to demonstrate its superiority over other methods. Typically, there are at least 5-10 such new superior learning algorithms routinely introduced at every neural network or machine learning conference.
- modeling results may be quite sensitive to particular implementation of resampling, especially for high-dimensional data.
- there may be several good predictive models that have completely different parameterization (see examples discussed later in Section 3).

Statistical data-analytic approach to knowledge discovery raises many important issues:

1. What is the role/relative importance of the data vs. prior knowledge?
2. What constitutes the prior knowledge and how it can be combined with data?
3. Is it possible, in principle, to estimate models that generalize well, from finite number of samples?
4. How to differentiate between several 'good' empirical models that explain well the same past data?

These issues are clearly similar to the main questions in the philosophy of science (PS1)-(PS4).

This paper describes methodological aspects of predictive learning and relates predictive data-analytic modeling to important philosophical ideas. Section 2 describes several philosophical concepts important for knowledge discovery. Section 3 presents methodological framework of predictive learning and its philosophical interpretation. In particular, we emphasize methodological differences between classical statistical and predictive (VC-theoretical) approaches to data-analytic modeling. These differences are further illustrated in Section 4 describing understanding / interpretation of black-box predictive models. Section 5 presents summary and conclusions.

2   Classical Philosophy of Science

Western philosophy of science has been shaped by monumental advances in natural sciences (physics, chemistry, biology) and their practical applications that have totally transformed the human society. These advances are based on the first-principles scientific knowledge, such as Newton's laws or Maxwell's equations that completely define the state of a physical system. This knowledge is universal, causal and deterministic, and it represents a perfect example of 'assured knowledge'. For example, Newton's laws should apply everywhere in the Universe. Newton's law of gravity can *explain and predict* the movement of planets and asteroids. Moreover, Newton's laws are interpretable, as they show deterministic relationships between just a few fundamental concepts. These first-principle laws in classical science are akin to axioms in mathematics. That is, classical science can be developed by applying logical reasoning to first-principle laws, in a manner similar to Euclidean geometry.

Philosophy of science deals with the relationship between the external world, and the human mind. Connections between natural sciences and philosophy are very deep and profound. Many great scientists contemplated about philosophical implications of their discoveries. Sir Isaac Newton, the father of theoretical physics, considered his work as natural philosophy. Later scientific discoveries, such as Einstein's relativity and the quantum mechanics, also had a profound effect on philosophy.

The philosophical school known as *'realism'* (or *materialism*) presumes an existence of objective physical reality, which is perceived via sensory inputs (facts, observations). This sensory data is processed in a human mind, to form some mental constructs (scientific models, beliefs, etc.). The primary role of physical reality is reflected in the 'information flow' shown in Fig. 3a. Note that according to realists, the physical world exists objectively and independently of human observers. Realism is essential to common sense, but it *cannot be proven* by logical arguments [4, 5].

An opposite view is taken by the school known as *idealism*, where the primary role belongs to ideas or mental constructs, and the physical reality is viewed just a by-product of the mind. Idealism asserts that (scientific) knowledge can be attained by mere reflection and reasoning. In its extreme form, idealism dismisses the role of empirical observations and therefore is inconsistent with natural science.

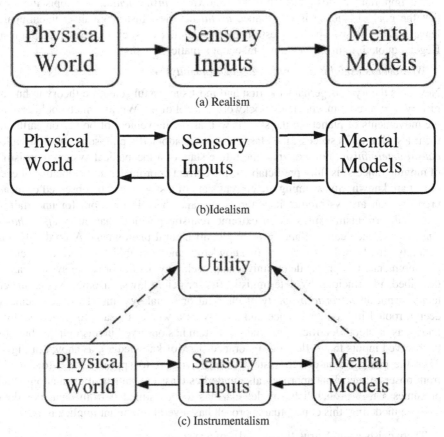

(a) Realism

(b)Idealism

(c) Instrumentalism

Fig. 3. Three main philosophical schools: (a) Realism, (b) Idealism, (c) Instrumentalism

Even though Realism and Idealism are presented as mutually exclusive in Fig. 3a and Fig. 3b, there are also many intermediate philosophical views. For example, Hegel (1770-1831) believed that *reality* and *mind* are parts of a complex system, and thus cannot be considered separately. According to Hegel:

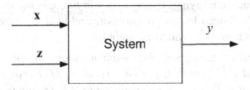

Fig. 4. Unknown system with observed inputs **x** and unobserved inputs **z**

Whatever exists (is real) is rational, and whatever is rational is real.

In this quotation, 'rational' refers to a mental construct of human mind, and 'real' refers to physical reality.

An important philosophical position known as *instrumentalism* adopts the view that the goal of science is to produce *useful* theories. Instrumentalists intentionally leave out of discussion the question of *truthfulness* of such theories. Paraphrasing Hegel's quotation, instrumentalists take a pragmatic view that:

Whatever is useful is also rational and (hopefully) real.

Newton's theory was perhaps the first and most successful scientific theory in human history. It offered a mechanical 'clockwork' model of the World, which could explain the movements of planets in the sky, as well as the movement of bodies on earth and the tides. Due to its success, the classical science adopted a philosophical view called *causal determinism*. For example, the future state of a mechanical system (consisting of moving objects) is fully predictable if the current coordinates and velocities of each object are known. So assuming the current state of a system can be observed or measured, we can always predict its future state, and there is no room for uncertainty. Causal determinism expresses an extreme realistic position that *every effect has a cause*, so science can explain, in principle, all natural phenomena. Accordingly, uncertainty simply reflects our lack of knowledge and/or inability to perform accurate measurements. Under this deterministic approach, any natural or social system can be described, in principle, by deterministic first-principle laws. In other words, uncertainty is not an *intrinsic property* of physical or social systems. This view remains deeply rooted in modern science and society as a whole. It leads to a methodology known as 'system identification' where a system has observed inputs/outputs, but also unobserved inputs that reflect our ignorance (lack of knowledge), as shown in Fig. 4. Then the goal of data-driven system identification is to approximate a true model from observations. This approach also underlies classical statistical estimation which presumes a true probabilistic model that has to be estimated from data. For data-analytic modeling, this conceptual approach has several important implications:

1. There exists a single true (best) model of a system;
2. this model can be accurately estimated from observed data;
3. more data leads to better knowledge, i.e. closer approximation to the truth.

The main issue in the philosophy of science was to explain the process of knowledge discovery, i.e. discovery of the first-principle laws. Newton himself believed that the basic concepts and laws of his system could be derived from experience (or observations), as evident from him saying: *Hypotheses non fingo*. However, this realistic view was immediately challenged by the philosophers and later, by other physicists in 19th and 20th century. As argued by Albert Einstein [6]:

Experience may suggest the appropriate mathematical concepts, but they most certainly cannot be deduced from it. Experience remains, of course, the sole criterion of the physical utility of a theory... In a certain sense, I hold true that pure thought can grasp reality, as the ancients dreamed.

Moreover, later physical theories such as quantum mechanics explicitly contradict Newton's realistic position that physical reality can be modeled independent of a human observer. According to Werner Heisenberg:

What we observe is not nature itself, but nature exposed to our method of questioning.

Similarly, many philosophers believe that scientific theories represent interpretation of observations in the light of theories invented by human mind. As Immanuel Kant put it:

Our intellect does not draw its laws from nature, but tries – with varying degrees of success – to impose upon nature laws which it freely invents.

These quotes suggest that first-principle scientific knowledge is a creation of human mind. However, this knowledge also reflects certain objective properties of the real world, so that it can be tested by experiments. This philosophical position is more akin to Instrumentalism (in Fig. 3c) where the 'utility' of our knowledge can be empirically verified. Note that the instrumentalist view is quite different from system identification, as it:

- rejects the notion of a single best (true) theory or model;
- emphasizes the importance of a method of questioning, i.e., the idealistic component of knowledge discovery;
- allows for multiple good scientific models describing the same phenomenon.

All of these properties also hold true for data-analytic knowledge discovery, as shown later in Sections 3 and 4.

The growing importance of science and technology has also influenced philosophical understanding of inference. The main advantage of the Scientific Method pioneered by Galileo is its *objectivity*, so that scientific theories hold irrespective of the personal bias of an observer (scientist), and that scientific experiments can be *repeated* by other scientists. Science does not study any observable events, but only recurrent (repeatable) phenomena. This immediately brings up the traditional philosophical *problem of induction* or *inductive inference*:

— How/why we can justify the belief (or scientific theory) that the future experiments (or observations) will be similar to the past?

Ill-posed nature of empirical inference has been a subject of many lively discussions in philosophy over many centuries, dating back to Hume and Kant. David Hume (1711-1776) was interested in the question of whether our beliefs or knowledge can be justified by 'sufficient reasons'. He noted that many important abstract concepts, such as the notion of causality, *cannot* be gained from empirical observations or sense-perceptions. So if sense-perceptions are the only source of human knowledge, as asserted by Newton, then our 'knowledge' simply represents beliefs and expectations, rather than 'assured knowledge' like Euclidean geometry. Further, Hume made an important distinction between a logical and a psychological problem of induction. *The logical problem* relates to logical justification of generalization (or inference) from repeated observed instances (events, experiments, facts) to other future instances. According to Hume, the logical induction cannot be justified, even when the number of past observations is very large.

Yet people often form generalizations based on past observations of repeatable events. So Hume's *problem of psychological induction* is: Why do most reasonable people believe in generalization based on frequent repetition of events? – His explanation is that people are conditioned by repetitions (of observed events) and believe in such inference due to custom or habit. This is similar to inference mechanism in biological systems known as 'learning by association' or Pavlov's conditional reflex.

Hume's ideas dominate modern philosophical views on empirical inference as nicely summarized by Ludwig Wittgenstein (1889-1951):

The process of induction is the process of assuming the simplest law that can be made to harmonize with our experience. This process, however, has no logical foundation, but only a psychological one. It is clear that there are no grounds for believing that the simplest course of events will really happen.

Philosophical treatment of the psychological induction does not differentiate between the instrumental empirical knowledge and beliefs (as shown in Fig.1). This is because the philosophy of science is concerned with classical first-principle knowledge, whereas the empirical knowledge has become important only recently, in the past 10-15 years. This empirical knowledge is formed by statistical dependencies which have useful predictive value. Estimation of such dependencies is provided by the methodology of predictive learning discussed in the next section.

3   Predictive Learning and Knowledge Discovery

Knowledge discovery always involves *inference*, i.e. the process of deriving a conclusion based on existing knowledge and/ or observations (or already known facts). Epistemological interpretation of inference clearly depends on the general philosophical framework (i.e., idealism vs. realism) and also on the philosophical interpretation of uncertainty. Classical philosophy adopts a deterministic view of 'assured knowledge' or 'true Laws of Nature'. So it is mainly concerned with understanding and discovery of the first-principle knowledge. In contrast, most modern data-driven applications deal with estimation of predictive statistical models, or *statistical inference*. In spite of the multitude of existing machine learning and statistical algorithms for data-driven modeling, there is no credible philosophical treatment of these methods. This section describes application of philosophical concepts to predictive statistical inference. Our objective is to describe general conceptual framework for machine learning and statistical methodologies for data-analytic predictive modeling. Our approach is largely based on the ideas from Vapnik-Chervonenkis (VC) theory [2, 7].

Philosophical ideas develop in response to scientific and technological advances. Data-analytic modeling has been influenced by two theoretical developments:

- Classical statistics developed by R. Fisher in the first half of the 20th century [8].
- VC-theory developed in the 1970's by Vapnik and Chervonenkis [2]. The VC-theory provides the mathematical analysis and conditions for estimating predictive models from data.

The two related technological advances are:

- The field of applied statistics based on Fisher's probabilistic modeling;
- Various applied disciplines (machine learning, data mining, artificial neural networks) concerned with predictive data-analytic modeling. All these fields have been originally introduced as ad hoc learning algorithms. However, later it became clear that they fall under the predictive learning framework and hence can be conceptually described using VC-theoretical methodology [9].

Next we discuss two main philosophical aspects of predictive data modeling:

- What is a proper philosophical interpretation of predictive modeling? Is it different from classical statistical modeling?
- What is the philosophical interpretation of mathematical conditions for statistical inference developed in VC-theory?

The first question is a variation of the philosophical problem (PS1). What is primary in data-driven knowledge discovery: observation (data) or hypothesis (method of questioning)? Also, a more subtle related question: What is the meaning of 'hypothesis' or 'method of questioning'? According to Karl Popper [4,5], knowledge discovery does not start from observations but always from problems. An observation is always preceded by a particular interest, a question, or something theoretical. This view is consistent with classical statistics where a parametric model (or 'hypothesis') is given a priori, and then available data is used to estimate model parameters.

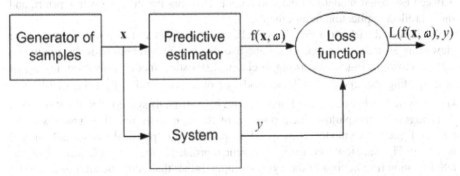

Fig. 5. Predictive Estimator uses (input, output) observations of the unknown System to approximate (or imitate) its output

Philosophical interpretation of model-free estimation methods in machine learning is less straightforward, because these methods specify (infinitely) many parameterizations. In this case, 'hypothesis' refers to the existence of some unspecified predictive model that can be estimated from available data. Using mathematical notation, model-free methods adopt flexible model parameterization, i.e., a set of admissible models or functions $f(\mathbf{x}, \omega)$ where $\omega \in \Omega$ denotes an abstract set of parameters. An estimated model is a function $f : \mathbf{x} \rightarrow y$ from this set, which predicts an output for future (test) inputs. Examples of flexible parameterizations include MLP networks, RBF networks, decision trees and SVM methods [3, 10]. An important VC-theoretical result states that prediction (generalization) with finite data depends on the complexity of a set of admissible models $f(\mathbf{x}, \omega)$, rather than on actual parameterization. Therefore, two different parameterizations, say MLP and RBF networks, or SVM classifiers using different types of kernels, may produce similar prediction accuracy, as long as their complexity is well controlled [3]. So, in contrast to classical statistics, actual model parameterization is not very important, relative to complexity control.

An important aspect of predictive data modeling is the notion of the learning problem setting, which can be regarded as the 'method of questioning'. The learning problem setting is an important methodological concept that reflects general assumptions underlying available data and the goals of modeling (or learning). For example, the *inductive learning setting* is shown in Fig. 5. Under this setting, unknown system that has observed inputs (shown as vector $\mathbf{x}$) and an output y. Then past (or training) data is in the form $(\mathbf{x}_i, y_i), i = 1, 2, ..., n$, where n is the number of training samples. There may also exist unobserved inputs $\mathbf{z}$, responsible for the random nature of unknown system, so that it may produce different output values y, for the same values of inputs $\mathbf{x}$. The goal of learning is to estimate unknown dependency between the input $(\mathbf{x})$ and output (y) variables, from a set of past observations of $(\mathbf{x}, y)$ values. The estimated model, or function, $f(\mathbf{x})$ is later used for predicting the output for new (test) inputs. The quality of prediction is evaluated as expected test error (for future test data). The test error is specified via a loss function that measures the discrepancy between estimated model and the true output. This loss function is given a priori, and should reflect application requirements.

As discussed earlier (in Fig. 2), model estimation from known training data can be viewed as an inductive step. There are two distinct interpretations of the predictive system shown in Fig.5. According to classical statistics, model estimation has a goal of estimating the true probabilistic model of observed data $(\mathbf{x}, y)$ generated by unknown system. That is, the goal of modeling is (statistical) system identification. This is analogous to the philosophical position of Realism, assuming that 'unknown system' and 'predictive model' correspond, respectively, to 'physical world' and 'mental model' in Fig. 3a. However, for most practical problems, the goal of learning is accurate imitation (prediction) of the System output, rather than identification of unknown system generating the data. This leads to another interpretation of predictive system in Fig. 5 where the goal of learning is to estimate certain properties of unknown data distribution which yield good generalization (or small test error). This system imitation approach is adopted in VC-theory, and it corresponds to the philosophical view of Instrumentalism [2]. That is, Physical World, Mental Model and Utility in Fig. 3c correspond, respectively, to the System, Predictive Estimator and Loss Function in Fig. 5. For finite sample estimation problems, it can be shown, both theoretically and empirically, that the goal of system imitation often yields good predictive models, even when estimation of 'true' underlying probabilistic models is intrinsically impossible. Specification of an appropriate learning problem setting is critical for most real-life applications [3, 11].

Note that the two approaches, probabilistic *system identification* and predictive *system imitation*, are fundamentally different, even though both estimate models from data. This distinction is clearly stated in VC-theory, but it has not been widely understood in statistics and machine learning at large. The difficulty stems from the fact that researchers in these fields focus on the development of constructive learning algorithms (and omit the problem setting), and that such learning algorithms can be usually motivated under either approach [11]. For example, support vector machines (SVM) have been originally introduced under predictive (VC-theoretical) setting, but later presented and 'explained' by statisticians under probabilistic (Bayesian) setting. This lack of clarity often leads to methodological and conceptual confusion among practitioners applying predictive data-analytic tools. Simply stated, one cannot effectively use these tools without understanding the 'method of questioning', i.e. the problem setting. A leading modern statistician Leo Breiman [12] refers to this distinction as the two cultures in data modeling:

There are two cultures in the use of statistical modeling to reach conclusions from data. One assumes that the data are generated by a given stochastic data model. The other uses algorithmic models and treats the data mechanism as unknown. The statistical community has been committed to the almost exclusive use of data models. This commitment has led to irrelevant theory, questionable conclusions and has kept statisticians from working on a large range of interesting current problems.

As convincingly argued by Breiman [12], 'algorithmic black-box' models yield better prediction accuracy than classical statistical techniques. However, adoption of these new modeling methods requires understanding of a new data modeling culture by statisticians. Breiman's term 'new culture' essentially refers to the instrumentalist philosophy of predictive learning. This new modeling philosophy raises new issues, such as the multiplicity of good predictive models and the resulting difficulties in interpreting black-box models. These issues are discussed later in Section 4.

Next we present important assumptions underlying standard inductive learning setting. This setting is used in most learning (or model estimation) algorithms developed in statistics, machine learning, artificial neural networks, signal processing etc. Inductive learning system shown in Fig. 5 follows three steps:

1. Observe a System, in order to collect training data samples.
2. Estimate predictive model from the training data (inductive step).
3. Apply the estimated model to make predictions (deductive step).

Such an inductive inference process is very general and hides many important details. For one thing, a modeler chooses what to observe in Step 1. So we usually assume meaningful (informative) specification of the input and output variables. The second issue is the fundamental problem of inductive inference: is it possible to estimate a good predictive model from observed data? Note, that a model estimated during inductive step tries to 'explain' the training data via some data fitting procedure; however this can also result in fitting the noise. In order to address the main question (is generalization possible?) we need to introduce certain modeling assumptions and theoretical concepts. These include:

- statistical assumptions about past (training) and future (test) data in the learning system shown in Fig. 5;
- quantifiable (mathematical) definition of generalization.

VC-theory provides such generic assumptions that make predictive learning possible. Specifically, standard inductive learning assumes that:

- Training and future data samples $(\mathbf{x}, y)$ are sampled independently from the same distribution $P(\mathbf{x}, y)$.
- This distribution $P(\mathbf{x}, y)$ is unknown but fixed or 'stationary'.

The requirement that training and test data originate from the same (unknown) distribution reflects a common-sense notion that the future is statistically similar to the past. An assumption about independent and identically distributed (i.i.d.) sampling guarantees that each new training sample yields maximum information.

The goal of learning is to estimate a function $f : \mathbf{x} \rightarrow y$ which predicts an output for future (test) inputs. This function (or estimated model) is selected from the set of admissible functions $f(\mathbf{x}, \omega)$ given a priori. The quality of estimated model is measured via non-negative loss function $L(y, f(\mathbf{x}, \omega))$, so that the best model is the one minimizing the 'prediction risk' functional:

$$R(\omega) = \int L\big(y, f(\mathbf{x}, \omega)\big) dP(\mathbf{x}, y) \qquad (1)$$

Commonly used loss functions include: squared loss (for regression problems with real-valued output y) and 0/1 loss for binary classification problems (where y is a binary class label). Prediction risk (1) can be evaluated using large test set.

Note that under standard inductive learning setting:

- Test samples are not known/used during model estimation.
- The quality of learning (generalization) is evaluated using some loss function that should be specified a priori, *before* a predictive model is evaluated from data. This loss function has to be driven by application domain requirements.
- Predictive black-box models generally do not assume causality [3,10]. Also, interpretation/explanation of estimated models is outside the framework of predictive learning. In particular, there may be several good predictive models estimated from the same training data.

The difficulty of learning is due to the fact that the distribution $P(\mathbf{x}, y)$ is unknown, so the prediction risk (1) cannot be directly minimized. All that is known is finite training sample $(\mathbf{x}_i, y_i)$ $i = 1, 2, \ldots n$. So it reasonable to use model estimation procedures that choose a model that has the smallest fitting error on the training set. The VC-theory provides general conditions on a set of admissible models $f(\mathbf{x}, \omega)$, under which such data fitting approach yields good generalization. The main idea is that we should favor 'simple' models describing training data. This is, of course, consistent with old philosophical ideas such as Occam's razor. However, the VC-theory provides a new measure of 'complexity' called the VC-dimension, which is different from statistical indices such as the number of free parameters or degrees-of-freedom.

VC-theoretical inductive learning setting corresponds to a new type of statistical inference appropriate for many applications. This VC-inference is clearly different from the logical and psychological induction in classical philosophy. It is also different from the probabilistic inference in classical statistics developed for large-sample (asymptotic) settings. VC-inference provides precise characterization of empirical knowledge that can be defined as a statistical predictive model estimated from data. This empirical knowledge is different from deterministic first-principle knowledge and from beliefs (see Fig. 1). In particular, 'beliefs' can be regarded as data-analytic models that describe well the past data, but cannot predict well.

VC-theory also provides general mathematical conditions under which generalization (prediction) with finite samples is possible. The main practical result of VC-theory can be stated as follows:

If a model explains well past (training) data, and it is simple, then it is likely to have good generalization for future (test) data.

Or, in more technical terms:

Good generalization (prediction) is possible if the training error is small and the VC-dimension (of a set of admissible models) is small.

Here model simplicity corresponds to low VC-dimension. This condition for generalization has an interesting philosophical interpretation. Vapnik [2] interprets finite VC-dimension, which is the main condition for model-free learning, using Popper's falsifiability. That is, available training data corresponds to "facts" or assertions known to be true. A set of admissible models $f(\mathbf{x}, \omega)$ corresponds to all possible generalizations. Each function from this set is a model or hypothesis about unknown dependency. Generalization (from these known facts) amounts to selecting a particular model from the set of all admissible models. According to Popper's notion of falsifiability, a true inductive theory cannot explain all possible facts. In data modeling, known facts are training samples, and their 'explanation' corresponds to data fitting. So the ability to explain (or fit) a given number of training samples is equivalent to the notion of *shattering* for binary classification [3, 7]. The finiteness of the VC-dimension implies that a set of functions $f(\mathbf{x}, \omega)$ cannot explain an arbitrary large data set, so it can be interpreted as Popper's falsifiability [7].

This leads to another interpretation of VC-theoretical conditions for generalization:

If a model explains well the training data and can be easily falsified, then it has good generalization (for future data).

This interpretation is sometimes stated as the principle of VC falsifiability for estimating predictive models from finite samples [2]:

Select the model that explains available data and is easy to falsify.

Note that VC falsifiability is a well-defined quantity (measured via VC-dimension). Popper has also defined a complexity index (that he called a *characteristic number* of a theory) and, using this index, related the model simplicity (~ number of parameters) to falsifiability [13]. However, the VC-dimension is different from the number of free parameters, so Popper's complexity index does not really ensure good generalization, except for a trivial case of linear estimators [2, 3].

Philosophical interpretation of predictive modeling is important for understanding the difference between classical first-principle knowledge and empirical knowledge in Fig. 1. Classical first-principle knowledge is consistent with the causal deterministic view of a simple world developed in 18-th and 19-th centuries. Such simple deterministic models may not be adequate for describing complex physical and social systems. However, it may be still possible to model certain properties of complex systems in a modern data-rich world. This leads to a new type of empirical knowledge, in the form of statistical models estimated from data. Practical utility of such models is usually related to their prediction capability, defined in the context of a particular application. VC-theory provides mathematical and methodological basis for estimating predictive models from finite samples. Now one can see that classical knowledge and empirical knowledge are two different concepts. Classical knowledge is universal and its growth is understood in terms of accumulation. Empirical knowledge is more conditional as it describes an empirical relationship derived from observations of a complex system. This knowledge is instrumental, and it is useful a limited application domain. In many cases, empirical knowledge is transient, because an underlying system itself changes with time (i.e., as in financial markets). Hence, understanding empirical knowledge requires understanding of the predictive learning methodology developed in VC-theory.

4   Practical Aspects of Predictive Data-Analytic Modeling

Philosophical and methodological aspects of predictive learning are very important for understanding and using data-analytic models. The main issue is that empirical data-driven knowledge is qualitatively different from the classical first-principle knowledge and from beliefs, as shown in Fig.1. Yet, this distinction is often overlooked by practitioners and users of data-analytic tools. Two prominent application areas where this commonly happens are economics (financial engineering) and life sciences. Economists often design mathematical models for social systems (such as financial markets). Even though such models employ sophisticated math, they fail to reflect the complexity of social systems, and usually have questionable predictive value. In many cases, these models can be regarded as beliefs. The main philosophical dilemma is a popular view (among economists) that in economic systems all uncertainty about the future is quantifiable. This assumption clearly contradicts the methodology of predictive learning.

In life sciences, a similar fallacy occurs when researchers apply classical statistical methodology for estimating data-analytic models, and interpreted them as causal first-principle models. This is evident in a glut of reported studies in medicine, bioinformatics and cognitive science, where scientists 'discover' knowledge via uncritical application of classical statistical techniques to real-life data. Often, these studies arrive to conclusions using goodness-of-fit tests and residual analysis. Unfortunately, just because a study can claim the statistical significance does not mean it can be reproduced by others, i.e. it has little predictive value [14].This occurs because classical statistical assumptions do not hold true for real-life applications [3,12]. These assumptions include unbiasedness, large sample size (relative to model complexity) etc. In classical statistics, these 'favorable' assumptions are typically ensured by a good experimental design, which is not possible with high-dimensional observational data.

This paper makes clear distinction between different types of knowledge:

- *first-principle knowledge* describes universal causal relationships. These relationships (laws) are deterministic and they have both the predictive and explanatory value, as they involve just a few concepts (variables). In philosophy, knowledge discovery is related to the problem of logical induction.
- *empirical knowledge* describes statistical dependencies (derived from observed data) that have predictive properties. According to VC-theoretical methodology, explanation of such empirical models is outside the scope of predictive modeling.
- *beliefs* are statistical dependencies (derived from observed data) that have explanatory (or descriptive) value but little predictive value. In philosophical terms, beliefs can be regarded as a form of psychological induction.

Classical science has both explanatory and predictive value. However, in natural sciences it is well-understood that prediction (rather than interpretability) is the main property of scientific theory. This point is stated by a famous physicist R. Feynman:

It is whether or not the theory gives predictions that agree with experiment. It is not a question of whether a theory is philosophically delightful, or easy to understand, or perfectly reasonable from the point of view of common sense.

In contrast, in social and life sciences statistical models explaining observations (or empirical data) are often regarded as scientific knowledge, even when such models are not used for prediction. These explanatory models are *simple*, and they usually refer to statistical correlations discovered from data. For example, many studies report discovery of a single gene associated with a particular type of cancer. Such studies naively presume the intelligibility (simplicity) of complex biological phenomena.

Classical parametric statistics is methodologically biased towards simple interpretable data-analytic models. In real-life applications, prediction accuracy and simplicity (interpretability) are usually in conflict [12]. Traditional approach to reducing model complexity is reducing the number of prediction variables (via feature selection) leading to the final low-dimensional model. Modern machine learning approaches (SVMs, random forests [15] etc.) using all input variables, achieve better prediction, but are not easily interpretable. This poses a real challenge for many applications. For example, physicians are accustomed to simple logistic regression, because it produces a linear model with weights that give an indication of the variable importance. However, they cannot interpret a black-box predictive model, even if it achieves superior prediction. In fact, doctors and practitioners in many other fields often do not even understand the assumptions underlying classical statistical and modern predictive data modeling approaches. So, interpretation of predictive data-analytic models should be always preceded with understanding modeling assumptions (e.g., standard inductive learning setting).

Next we present two examples illustrating potential difficulties in interpreting predictive models. The first example is concerned with interpretation of high-dimensional SVM models. The second example shows the multiplicity of predictive models, which is possible even with low-dimensional data. These examples reinforce the point that interpretability is closely related to the methodological/philosophical aspects of data-analytic modeling, rather than its technical aspects.

Both examples assume standard binary classification problem under inductive learning setting described in Section 3. The first example uses MNIST digit recognition data set, where the goal is to classify handwritten digits '5' vs. '8', where each digit is represented as 784 (28x28) grey-scale pixels encoded as 784-dimensional input. Nonlinear SVM classifier is estimated using 1,000 training samples (500 per class). The nonlinear SVM uses RBF kernel of the form $K(\mathbf{x}, \mathbf{x}') = \exp(-\gamma \|\mathbf{x} - \mathbf{x}'\|^2)$. The SVM approach is known to yield very good generalization (prediction) performance for this data set. The problem, however, is to provide simple interpretation of a high-dimensional nonlinear SVM model. Most existing SVM model interpretation methods [16] represent nonlinear SVM decision boundary via some interpretable parameterization, such as a set of *if … then …else* rules. Each rule is defined over a small number of input features (e.g., pixels, in this example). Actual methods adopt different approaches for such rule extraction techniques. For example, see [16, 17] for a survey of SVM rule extraction techniques.

There are three critical issues for such SVM interpretation approaches:

- Large number of input variables adversely affects interpretation. Complex SVM models with thousands of input variables can be only approximated by a large number of rules. So interpretability often requires some form of feature selection or dimensionality reduction, at the expense of prediction accuracy [12].
- Multiplicity of good predictive models estimated from the same data. This makes model interpretation questionable.
- Interpretable approximations (of black-box SVM models) often provide significantly lower generalization performance. This becomes especially critical for modeling High-Dimensional Low Sample Size (HDLSS) data.

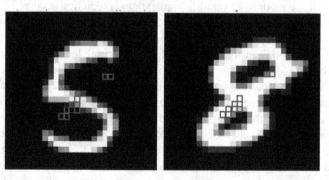

Fig. 6. Top 10 most informative features (pixels) selected using Fisher's criterion

Table 1. Test/Training errors averaged over 10 random realizations of the data (std. deviation shown in parenthesis)

Method	Test Error (%)	Training Error (%)
SVM	1.08(0.23)	0(0)
FISHER+SVM	7.28(0.85)	4.93(1.35)
PCA+SVM	6.22(1.45)	6.18(1.97)

Next we elaborate on the trade-off between feature selection and the prediction accuracy. The goal is to compare three SVM-based modelling approaches:

- *RBF SVM* – where the model is estimated using all 784 input features.
- *FISHER+RBF SVM* - for this method, as a part of pre-processing, we select the top 10 features using the Fisher criterion [18] on the training data, and discard the others. Then the RBF SVM model is estimated in this 10-dimensional reduced feature space.
- *PCA+RBF SVM* – for this method the preprocessing step involves selecting the first 3 principal components from (unlabeled) training data [10]. Then the RBF SVM model is estimated in this 3-dimensional reduced feature space.

Clearly, SVM models using small number of features are easier to interpret. For example, ten most informative features (pixels) selected using the Fisher index are shown in Fig.6. Then it may be possible to construct a simple set of rules that 'explain' the SVM model in terms of these 10 pixel values. However, the generalization performance of these three methods will be quite different. Table 1 shows the test error rate evaluated on independent test set (1,866 test samples). The SVM model parameters (for all methods) are tuned using an independent validation data set (1,000 samples). The experimental comparisons are performed 10 times using different realizations of the training and validation data, and average training/test error rates are shown in Table 1. As expected, SVM model using all 784 input features yields much better prediction accuracy than low-dimensional (interpretable) models.

Cherkassky and Dhar [17, 19, 20] argue that interpretation of high-dimensional SVM models cannot be separated from understanding SVM-related concepts responsible for generalization, such as margin. So they advocate presentation of a complex SVM model, in a simple graphical form using the univariate histogram of projections technique [17, 19, 20]. This technique provides simple graphical representation of the training data and the estimated SVM decision boundary for binary classification:

Univariate Histogram of Projections *is the histogram of the projection values of the data samples onto the normal direction of the trained SVM decision boundary.*

For the MNIST data set used in this example, the 784-dimensional nonlinear SVM model represented as a univariate histogram of projections shown in Fig. 7a. This figure displays projections of the training data for the RBF SVM with optimal parameters (e.g., C and RBF kernel parameter) tuned using an independent validation set of 1,000 samples. As evident from Fig. 7a, the training samples are well separable in the optimally chosen RBF kernel space. Also, the histogram of projections clearly illustrates the clustering of data samples at the margin borders. This effect, called data piling, is typical for high-dimensional data [3, 21].

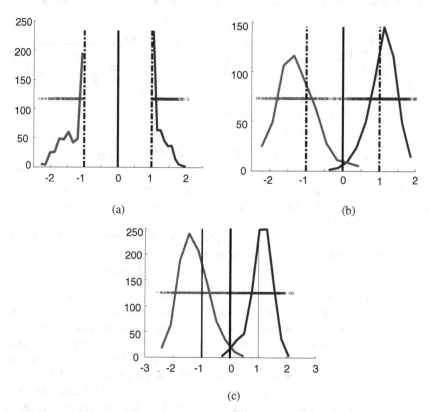

Fig. 7. Univariate histogram of projections for MNIST data set (a) training data; (b) validation data (validation error 1.7%); (c) test data (test error 1.23%). Note that SVM decision boundary is marked as 0, and SVM margin borders are labelled as -1 and +1 on the histogram axis.

However, this separability of the high-dimensional training data does not imply the separability of validation or test data. This can be seen from the projections of validation and test samples in Fig.7b and Fig. 7c. The SVM optimization algorithm tries to achieve high separability of the training data by penalizing the samples that are inside the soft margin. Hence the histogram in Fig. 7a where many training samples are outside the soft margin is typical. However, during model selection we are only concerned with validation samples that are correctly classified by the model. Hence we may select a model that allows the validation/test samples to be within the soft-margin as long as it provides small validation error. This explains the overlapping histogram regions for validation and test data, as shown in Fig. 7b and Fig. 7c. Hence, the histogram of projections technique enables better understanding of the SVM decision making procedure and its model selection strategy. Such an understanding, however, does not require traditional model parameterization in the form of *if ... then ...else* rules using individual input features (pixels).

Another potential advantage is that the histogram of projections technique helps a general user to quantify the confidence in SVM predictions, according to the distance from the margin border. For instance, referring to Fig. 7c, if a test input is projected *inside* the margin borders, the confidence is low. Alternatively, if a test input is *outside* the margin borders, the confidence in SVM predictions is high.

Second example illustrates the multiplicity of predictive black-box models. This problem arises due to predictive setting where the goal is to estimate a model yielding good generalization performance, rather than to estimate the 'true model' of observed data [12, 17]. This is illustrated next using a real-life financial application called timing (or daily trading) of mutual funds [22, 23]. The practice of timing mutual funds tries to profit from daily price changes, under the assumption that the next-day price change can be statistically predicted from today's market indices. For international mutual funds (investing in foreign stocks) this assumption turns out to be true, due to (a) different closing time for US, European and Asian markets, and (b) the fact that foreign markets tend to follow US markets. In early 2000's, this practice of timing international mutual funds, has resulted in scandals in the mutual fund industry [22].

Empirical validation of market timing is presented next for an international mutual fund called American Century International Fund (symbol TWIEX), using predictive data-analytic modeling [24]. A trading strategy generates a BUY or SELL signal at the end of each trading day, i.e. right before US market close at 4 p.m Eastern Standard Time. Effectively, a BUY order *today* is a bet that the price of this mutual price will go up *tomorrow*. Such a trading strategy can be formalized as a binary classifier. The two input indicators used for prediction are the daily percentage price changes of:

- SP 500 stock index (symbol ^GSPC);
- Euro-to-dollar exchange rate (symbol EURUSD).

There are practical reasons for choosing these predictor variables for trading international mutual funds. Namely, the first input reflects an assumption that foreign markets closely follow US market. The second input is chosen because international mutual funds are priced in US dollars. The predictive model uses today's values of these two inputs to predict tomorrow's price change (UP or DOWN) of this international fund. In this study, a predictive model was estimated using Year 2004 data, and then tested on Year 2005 data.

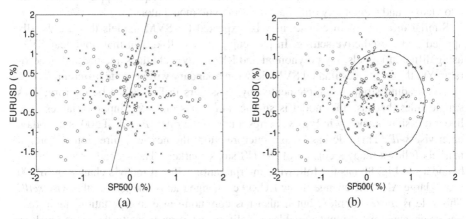

(a)   (b)

Fig. 8. Linear and quadratic decision boundaries estimated from training data (Year 2004). (a) Fisher Linear Discriminant Analysis. (b) Quadratic Discriminant Analysis.

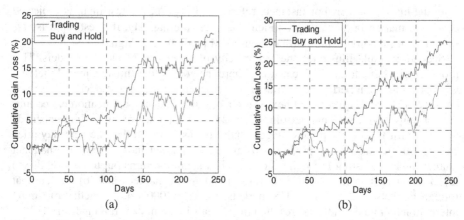

Fig. 9. Performance of the linear and quadratic decision models for test period (Year 2005). (a) Fisher Linear Discriminant Analysis. (b) Quadratic Discriminant Analysis.

Two parametric models used in this study [24] include Fisher's Linear Discriminant Analysis (LDA) and a Quadratic Discriminant Analysis (QDA). Estimated decision boundaries are shown, along with Year 2004 training data, in Fig. 8. Clearly, these two models look very different. Yet, they provide very similar performance for the test period (Year 2005), as shown in Fig. 9. The performance of trading strategies is compared to the benchmark 'Buy-and-Hold' scenario, when an account is 100% invested in TWIEX *all the time*. This figure shows that *both models* yield very good performance, i.e. consistently outperform the Buy-and-Hold strategy. This superior performance is achieved at lower risk, since the trading account is out of the market (in cash) about 40% of the time. So the usual question (posed by classical statisticians) which model most accurately describes the training data is difficult to resolve [12]. In fact, these models reflect two *different* successful trading strategies. Both strategies can be explained and probably make sense to financial experts and traders. However, understanding and explanation of these models requires application domain knowledge, and cannot rely only on a data-analytic model alone.

Similar difficulties, of course, may be expected for SVM models that are usually applied under predictive setting. In particular, it is well-known that nonlinear SVM using different kernels, say polynomial and RBF, may yield the same prediction accuracy. Yet the form of estimated SVM decision boundary will be quite different.

Two trading strategies implemented by classifiers in Fig. 8 can be interpreted as simple rules, because the model is simple *and* it has only two input variables. The linear decision boundary in Fig. 8a corresponds to the rule: '*Buy* if SP500 is up today, otherwise *sell*'. This rule has simple interpretation: the next-day direction of foreign markets follows today's change of the US stock market. The second-order decision boundary in Fig. 8b has the following interpretation: '*Buy* if today's change in SP500 is not_large AND the change in the EURO exchange rate is not_large, otherwise *sell*'. This rule is more complex, but it also has common-sense interpretation: large (extreme) changes of the input variables usually occur in response to the news (such as earnings reports, economic statistics), that are released in the morning when European markets are still open. Hence, such information should be already reflected in the closing prices of European equities. Further, during training and test periods (2004-2005), the two input variables have low correlation and can be regarded as independent. So the first (linear) model can have *causal* interpretation, i.e. foreign markets *follow* the direction of US stock market. Note that all these interpretations are based on understanding of the problem at hand, i.e. knowledge of financial markets, opening/closing time of the US and European markets etc. This knowledge cannot be derived from a black-box predictive model alone [17, 24].

Traditional model interpretation techniques originate from the classical probabilistic setting that has an ultimate goal of quantifying the effect (or importance) of individual input variables of the output. In many studies, the effect of individual inputs on the output is presented under causal setting, as the final objective of data modeling. This often leads to considerable confusion, since black-box models may not be causal [3, 12]. We argue that interpretation of predictive models should reflect understanding of an application domain *and* a good grasp of predictive modeling methodology, rather than just the technical analysis of the functional form of estimated models. Under predictive setting, the 'importance' of each input variable should be measured in terms of its effect on model's generalization performance.

5   Summary and Conclusions

We presented the philosophical interpretation of predictive data-analytic modeling. This interpretation is closely related to the methodological aspects of machine learning. Hence, it may be useful for practitioners who apply learning algorithms to data and interpret the resulting models. This paper makes an important distinction between classical first-principle knowledge, empirical knowledge and (empirical) beliefs. In particular, we emphasize the predictive aspects of empirical knowledge and the importance of predictive methodology. Predictive (VC-theoretical) framework also has a clear philosophical (instrumentalist) interpretation which is different from the causal deterministic position adopted in classical science. This distinction leads to major methodological differences for users of data-analytic tools. For most practical applications, this paper advocates the predictive learning (system imitation) view of data-analytic modeling.

The analysis presented in this paper contradicts the popular view that 'the data deluge makes the scientific method obsolete' [1]. However, the data deluge does, in fact, lead to a new type of (empirical) knowledge which is different from classical first-principle knowledge. Therefore, it is increasingly important to differentiate between first-principle knowledge, empirical knowledge and beliefs. This distinction is commonly missing in application studies that interpret data-analytic models as first-principle knowledge.

Let us recall an earlier quotation from Section 1:

Man has an intense desire for assured knowledge (Einstein).

Here 'assured knowledge' refers to the first-principle scientific knowledge. In our digital age, the human desire for 'assured knowledge' does not change. However, the meaning of what constitutes 'assured knowledge' has changed dramatically. That is, the main property of empirical data-analytic knowledge is its prediction (generalization) capability. In many cases, this data-analytic knowledge lacks simple interpretability, due to large number of variables and/or multiplicity of good predictive models.

An important aspect of 'assured' data-driven knowledge is a set of general mathematical assumptions about available data and the goals of learning/modeling. These assumptions are collectively known as 'the learning problem setting'. It corresponds to the 'method of questioning' in the philosophy of science. In particular, classical statistics and predictive learning use different underlying assumptions and pursue different goals of data-analytic modeling [2, 3, 15].

Selecting or specifying an appropriate learning problem setting (for a given application) is a creative part of predictive modeling which cannot be formalized. This paper describes the standard inductive learning setting which is used by most machine learning and statistical algorithms. However, there also exist several powerful nonstandard learning settings, e.g. transduction, semi-supervised learning, multi-task learning, Universum learning, and learning using hidden information [2, 3, 7, 11].

Acknowledgement. This work was supported, in part, by NSF grant ECCS-0802056. Parts of this paper are adapted from [11]. Empirical results presented in this paper are based on a joint work with Sauptik Dhar from the University of Minnesota.

References

1. The End of Science. Wired Magazine 16 (2007),
   http://www.wired.com/wired/issue/16-07
2. Vapnik, V.N.: Estimation of Dependencies Based on Empirical Data. In: Empirical Inference Science: Afterword of 2006. Springer, New York (2006)
3. Cherkassky, V., Mulier, F.: Learning from Data: Concepts, Theory and Methods. Wiley, NY (2007)
4. Popper, K.: Objective Knowledge. An Evolutionary Approach. Oxford University Press (1979)
5. Popper, K.: Conjectures and Refutations: The Growth of Scientific Knowledge. Routledge Press, London (2000)
6. Einstein, A.: Ideas and Opinions. Bonanza Books, New York (1988)
7. Vapnik, V.N.: Statistical Learning Theory. Wiley, New York (1998)
8. Fisher, R.A.: Contributions to Mathematical Statistics. Wiley, New York (1952)
9. Cherkassky, V., Muller, F.: Learning from Data: Concepts, Theory, and Methods. Wiley, NY (1998)
10. Hastie, T.J., Tibshirani, R.J., Friedman, J.: The Elements of Statistical Learning. Springer, New York (2001)
11. Cherkassky, V.: Introduction to Predictive Learning (to appear, 2012)
12. Breiman, L.: Statistical Modeling: the Two Cultures. Statistical Science 16(3), 199–231 (2001)
13. Popper, K.: The Logic of Scientific Discovery, 2nd edn. Harper Torch Books, New York (1968)

14. Ioannidis, J.P.A.: Contradicted and initially stronger effects in highly cited clinical research. JAMA 294(2), 218–228 (2005)
15. Breiman, L.: Random Forests. Machine Learning 45, 5–32 (2001)
16. Diederich, J.: Rule Extraction from Support Vector Machines. Springer (2008)
17. Dhar, S., Cherkassky, V.: Understanding Black Box Data-Analytic Models. Neural Networks (2011) (submitted)
18. Fisher, R.: The use of multiple measurements in Taxonomic problems. Annals of Eugenics 7, 179–188 (1936)
19. Cherkassky, V., Dhar, S.: Simple Method for Interpretation of High-Dimensional Nonlinear SVM Classification Models. In: The 6th International Conference on Data Mining (July 2010)
20. Cherkassky, V., Dhar, S., Dai, W.: Practical Conditions for Effectiveness of the Universum Learning. IEEE Trans. on Neural Networks 22, 1241–1255 (2011)
21. Ahn, J., Marron, J.S.: The Maximal Data Piling Direction for Discrimination. Biometrika 97, 254–259 (2010)
22. Zitzewitz, E.: Who cares about shareholders? Arbitrage proofing mutual funds. Journal of Law, Economics and Organization 19(4), 245–280 (2003)
23. Frankel, T., Cunningham, L.A.: The mysterious ways of mutual funds: market timing. Annual Review of Banking and Financial Law 25(1) (2006)
24. Cherkassky, V., Dhar, S.: Market Timing of International Mutual Funds: A Decade after the Scandal. In: Proc. CIFEr (2012)

Evolving Spiking Neural Networks and Neurogenetic Systems for Spatio- and Spectro-Temporal Data Modelling and Pattern Recognition

Nikola Kasabov

Knowledge Engineering and Discovery Research Institute - KEDRI,
Auckland University of Technology,
and Institute for Neuroinformatics, INI, ETH and University of Zurich
nkasabov@aut.ac.nz, www.kedri.info,
ncs.ethz.ch/projects/evospike

Abstract. Spatio- and spectro-temporal data (SSTD) are the most common types of data collected in many domain areas, including engineering, bioinformatics, neuroinformatics, ecology, environment, medicine, economics, etc. However, there is lack of methods for the efficient analysis of such data and for spatio-temporal pattern recognition (STPR). The brain functions as a spatio-temporal information processing machine and deals extremely well with spatio-temporal data. Its organisation and functions have been the inspiration for the development of new methods for SSTD analysis and STPR. The brain-inspired spiking neural networks (SNN) are considered the third generation of neural networks and are a promising paradigm for the creation of new intelligent ICT for SSTD. This new generation of computational models and systems are potentially capable of modelling complex information processes due to their ability to represent and integrate different information dimensions, such as time, space, frequency, and phase, and to deal with large volumes of data in an adaptive and self-organising manner. The paper reviews methods and systems of SNN for SSTD analysis and STPR, including single neuronal models, evolving spiking neural networks (eSNN) and computational neuro-genetic models (CNGM). Software and hardware implementations and some pilot applications for audio-visual pattern recognition, EEG data analysis, cognitive robotic systems, BCI, neurodegenerative diseases, and others are discussed.

Keywords: spatio-temporal data, spectro-temporal data, pattern recognition, spiking neural networks, gene regulatory networks, computational neuro-genetic modeling, probabilistic modeling, personalized modeling, EEG data.

1 Spatio- and Spectro-Temporal Data Modelling and Pattern Recognition

Most problems in nature require spatio- or/and spectro-temporal data (SSTD) that include measuring spatial or/and spectral variables over time. SSTD is described by a

J. Liu et al. (Eds.): WCCI 2012 Plenary/Invited Lectures, LNCS 7311, pp. 234–260, 2012.

triplet (X,Y,F), where: X is a set of independent variables measured over consecutive discrete time moments t; Y is the set of dependent output variables, and F is the association function between whole segments ('chunks') of the input data, each sampled in a time window d_t and the output variables belonging to Y:

$$F: \quad X(d_t) \to Y, \text{ where: } X(t) = \quad (x_1(t), x_2(t), \dots, x_n(t)), \ t=1,2,\dots; \quad (1)$$

It is important for a computational model to capture and learn *whole* spatio- and spectro-temporal patterns from data streams in order to predict most accurately future events for new input data. Examples of problems involving SSTD are: brain cognitive state evaluation based on spatially distributed EEG electrodes [70, 26, 51, 21, 99, 100] (fig.1a); fMRI data [102] (fig.1b); moving object recognition from video data [23, 60, 25] (fig.15); spoken word recognition based on spectro-temporal audio data [93, 107]; evaluating risk of disease, e.g. heart attack [20]; evaluating response of a disease to treatment based on clinical and environmental variables, e.g. stroke [6]; prognosis of outcome of cancer [62]; modelling the progression of a neuro-degenerative disease, such as Alzheimer's Disease [94, 64]; modelling and prognosis of the establishment of invasive species in ecology [19, 97]. The prediction of events in geology, astronomy, economics and many other areas also depend on accurate SSTD modelling.

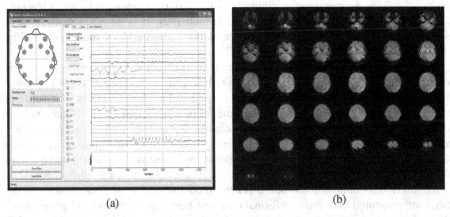

(a)   (b)

Fig. 1. (a) EEG SSTD recorded with the use of Emotive EEG equipment (from McFarland, Anderson, Müller, Schlögl, Krusienski, 2006); (b) fMRI data (from http://www.fmrib.ox.ac.uk)

The commonly used models for dealing with temporal information based on Hidden Markov Models (HMM) [88] and traditional artificial neural networks (ANN) [57] have limited capacity to achieve the integration of complex and long temporal spatial/spectral components because they usually either ignore the temporal dimension or over-simplify its representation. A new trend in machine learning is currently emerging and is known as *deep machine learning* [9, 2-4, 112]. Most of the proposed models still learn SSTD by entering single time point frames rather than learning whole SSTD patterns. They are also limited in addressing adequately the interaction between temporal and spatial components in SSTD.

The human brain has the amazing capacity to learn and recall patterns from SSTD at different time scales, ranging from milliseconds to years and possibly to millions of years (e.g. genetic information, accumulated through evolution). Thus the brain is the ultimate inspiration for the development of new machine learning techniques for SSTD modelling. Indeed, brain-inspired Spiking Neural Networks (SNN) [32, 33, 68] have the potential to learn SSTD by using trains of spikes (binary temporal events) transmitted among spatially located synapses and neurons. Both spatial and temporal information can be encoded in an SNN as locations of synapses and neurons and time of their spiking activity respectively. Spiking neurons send spikes via connections that have a complex dynamic behaviour, collectively forming an SSTD memory. Some SNN employ specific learning rules such as Spike-Time-Dependent-Plasticity (STDP) [103] or Spike Driven Synaptic Plasticity (SDSP) [30]. According to the STDP a connection weight between two neurons increases when the pre-synaptic neuron spikes before the postsynaptic one. Otherwise, the weight decreases.

Models of single neurons as well as computational SNN models, along with their respective applications, have been already developed [33, 68, 73, 7, 8, 12], including evolving connectionist systems and evolving spiking neural networks (eSNN) in particular, where an SNN learns data incrementally by one-pass propagation of the data via creating and merging spiking neurons [61, 115]. In [115] an eSNN is designed to capture features and to aggregate them into audio and visual perceptions for the purpose of person authentification. It is based on four levels of feed-forward connected layers of spiking neuronal maps, similarly to the way the *cortex* works when learning and recognising images or complex input stimuli [92]. It is an SNN realization of some computational models of vision, such as the 5-level HMAX model inspired by the information processes in the cortex [92].

However, these models are designed for (static) object recognition (e.g. *a picture of a cat*), but not for moving object recognition (e.g. *a cat jumping to catch a mouse*). If these models are to be used for SSTD, they will still process SSTD as a sequence of static feature vectors extracted in single time frames. Although an eSNN accumulates incoming information carried in each consecutive frame from a pronounced word or a video, through the increase of the membrane potential of output spike neurons, they do not learn complex spatio/spectro-temporal associations from the data. Most of these models are deterministic and do not allow to model complex stochastic SSTD.

In [63, 10] a computational neuro-genetic model (CNGM) of a single neuron and SNN are presented that utilize information about how some proteins and genes affect the spiking activities of a neuron, such as fast excitation, fast inhibition, slow excitation, and slow inhibition. An important part of a CNGM is a dynamic gene regulatory network (GRN) model of genes/proteins and their interaction over time that affect the spiking activity of the neurons in the SNN. Depending on the task, the genes in a GRN can represent either biological genes and proteins (for biological applications) or some system parameters including probability parameters (for engineering applications). Recently some new techniques have been developed that allow the creation of new types of computational models, e.g.: probabilistic spiking neuron models [66, 71]; probabilistic optimization of features and parameters of eSNN [97, 96]; reservoir computing [73, 108]; personalized modelling frameworks [58, 59]. This paper reviews methods and systems for SSTD that utilize the above and some other contemporary SNN techniques along with their applications.

2   Single Spiking Neuron Models

2.1   A Biological Neuron

A single biological neuron and the associated synapses is a complex information processing machine, that involves short term information processing, long term information storage, and evolutionary information stored as genes in the nucleus of the neuron (fig.2).

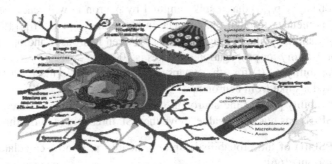

Fig. 2. A single biological neuron with the associated synapses is a complex information processing machine (from Wikipedia)

2.2   Single Neuron Models

Some of the-state-of-the-art models of a spiking neuron include: early models by Hodgkin and Huxley [41] 1952; more recent models by Maas, Gerstner, Kistler, Izhikevich and others, e.g.: Spike Response Models (SRM) [33, 68]; Integrate-and-Fire Model (IFM) [33, 68]; Izhikevich models [52-55], adaptive IFM, and others.

The most popular for both biological modelling and engineering applications is the IFM. The IFM has been realised on software-hardware platforms for the exploration of patterns of activities in large scale SNN under different conditions and for different applications. Several large scale architectures of SNN using IFM have been developed for modelling brain cognitive functions and engineering applications. Fig. 3a and b illustrate the structure and the functionality of the Leaky IFM (LIFM) respectively. The neuronal post synaptic potential (PSP), also called membrane potential $u(t)$, increases with every input spike at a time t multiplied to the synaptic efficacy (strength) until it reaches a threshold. After that, an output spike is emitted and the membrane potential is reset to an initial state (e.g. 0). Between spikes, the membrane potential leaks, which is defined by a parameter.

An important part of a model of a neuron is the model of the synapses. Most of the neuronal models assume scalar synaptic efficacy parameters that are subject to learning, either on-line or off-line (batch mode). There are models of dynamics synapses (e.g. [67, 71, 72]), where the synaptic efficacy depends on synaptic parameters that change over time, representing both long term memory (the final efficacy after learning) and short term memory – the changes of the   synaptic efficacy over a shorter time period not only during learning, but during recall as well.

One generalization of the LIFM and the dynamic synaptic models is the probabilistic model of a neuron [66] as shown in fig.4a, which is also a biologically plausible model [45, 68, 71]. The state of a spiking neuron n_i is described by the sum $PSP_i(t)$ of the inputs received from all m synapses. When the $PSP_i(t)$ reaches a firing threshold $\vartheta_i(t)$, neuron n_i fires, i.e. it emits a spike. Connection weights ($w_{j,i}$, j=1,2,...,m) associated with the synapses are determined during the learning phase using a learning rule. In addition to the connection weights $w_{j,i}(t)$, the probabilistic spiking neuron model has the following three probabilistic parameters:

- A probability $p_{cj,i}(t)$ that a spike emitted by neuron n_j will reach neuron n_i at a time moment t through the connection between n_j and n_i. If $p_{cj,i}(t)=0$, no connection and no spike propagation exist between neurons n_j and n_i. If $p_{cj,i}(t) = 1$ the probability for propagation of spikes is 100%.
- A probability $p_{sj,i}(t)$ for the synapse $s_{j,i}$ to contribute to the $PSP_i(t)$ after it has received a spike from neuron n_j.
- A probability $p_i(t)$ for the neuron n_i to emit an output spike at time t once the total $PSP_i(t)$ has reached a value above the PSP threshold (a noisy threshold).

The total $PSP_i(t)$ of the probabilistic spiking neuron n_i is now calculated using the following formula [66]:

$$PSP_i(t) = \sum_{p=t_0,,,t} \quad (\sum_{j=1,..,m} e_j f_1(p_{cj,i}(t-p)) f_2(p_{sj,i}(t-p)) w_{j,i}(t) + \eta(t-t_0)) \qquad (2)$$

where: e_j is 1, if a spike has been emitted from neuron n_j, and 0 otherwise; $f_1(p_{cj,i}(t))$ is 1 with a probability $p_{cji}(t)$, and 0 otherwise; $f_2(p_{sj,i}(t))$ is 1 with a probability $p_{sj,i}(t)$, and 0 otherwise; t_0 is the time of the last spike emitted by n_i; $\eta(t-t_0)$ is an additional term representing decay in the PSP_i. As a special case, when all or some of the probability parameters are fixed to "1", the above probabilistic model will be simplified and will resemble the well known IFM. A similar formula will be used when a leaky IFM is used as a fundamental model, where a time decay parameter is introduced.

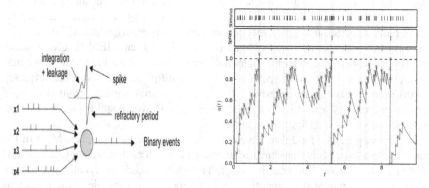

Fig. 3. (a) Example of a LIFM. (b) Input spikes, output spikes and PSP dynamics of a LIFM.

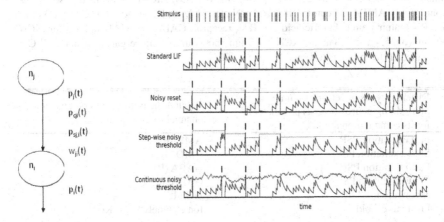

Fig. 4. (a) A simple probabilistic spiking neuron model (from [66]); (b) Different types of noisy thresholds have different effects on the output spikes (from [99, 98])

It has been demonstrated that SNN that utilise the probabilistic neuronal model can learn better SSTD than traditional SNN with simple IFM, especially in a nosy environment [98, 83]. The effect of each of the above three probabilistic parameters on the ability of a SNN to process noisy and stochastic information was studied in [98]. Fig. 4b presents the effect of different types of nosy thresholds on the neuronal spiking activity.

2.3   A Neurogenetic Model of a Neuron

A neurogenetic model of a neuron is proposed in [63] and studied in [10]. It utilises information about how some proteins and genes affect the spiking activities of a neuron such as *fast excitation, fast inhibition, slow excitation, and slow inhibition*. Table 1 shows some of the proteins in a neuron and their relation to different spiking activities. For a real case application, apart from the GABAB receptor some other metabotropic and other receptors could be also included. This information is used to calculate the contribution of each of the different synapses, connected to a neuron n_i, to its post synaptic potential PSPi(t):

$$\varepsilon_{ij}^{synapse}(s) = A^{synapse}\left(\exp\left(-\frac{s}{\tau_{decay}^{synapse}}\right) - \exp\left(-\frac{s}{\tau_{rise}^{synapse}}\right)\right) \qquad (3)$$

where: $\tau_{decay/rise}^{synapse}$ are time constants representing the rise and fall of an individual synaptic PSP; A is the PSP's amplitude; $\varepsilon_{ij}^{synapse}$ represents the type of activity of the synapse between neuron j and neuron i that can be measured and modelled separately for a fast excitation, fast inhibition, slow excitation, and slow inhibition (it is affected by different genes/proteins). External inputs can also be added to model background noise, background oscillations or environmental information.

Table 1. Neuronal action potential parameters and related proteins and ion channels in the computational neuro-genetic model of a spiking neuron: AMPAR - (amino- methylisoxazole-propionic acid) AMPA receptor; NMDR - (N-methyl-D-aspartate acid) NMDA receptor; $GABA_AR$ - (gamma-aminobutyric acid) $GABA_A$ receptor, $GABA_BR$ - $GABA_B$ receptor; SCN - sodium voltage-gated channel, KCN - kalium (potassium) voltage-gated channel; CLC - chloride channel (from Benuskova and Kasabov, 2007)

Different types of action potential of a spiking neuron used as parameters for its computational model	Related neurotransmitters and ion channels
Fast excitation PSP	AMPAR
Slow excitation PSP	NMDAR
Fast inhibition PSP	$GABA_AR$
Slow inhibition PSP	$GABA_BR$
Modulation of PSP	mGluR
Firing threshold	Ion channels SCN, KCN, CLC

An important part of the model is a dynamic gene/protein regulatory network (GRN) model of the dynamic interactions between genes/proteins over time that affect the spiking activity of the neuron. Although biologically plausible, a GRN model is only a highly simplified general model that does not necessarily take into account the exact chemical and molecular interactions. A GRN model is defined by:

(a) A set of genes/proteins, G= $(g_1, g_2, \ldots, g_k)$;
(b) An initial state of the level of expression of the genes/proteins G(t=0);
(c) An initial state of a connection matrix L = $(L_{11}, \ldots, L_{kk})$, where each element L_{ij} defines the known level of interaction (if any) between genes/proteins g_j and g_i;
(d) Activation functions f_i for each gene/protein g_i from G. This function defines the gene/protein expression value at time (t+1) depending on the current values G(t), L(t) and some external information E(t):

$$g_i(t+1)= f_i (G(t), L(t), E(t)) \qquad (4)$$

3   Learning and Memory in a Spiking Neuron

3.1   General Classification

A learning process has an effect on the synaptic efficacy of the synapses connected to a spiking neuron and on the information that is memorized. Memory can be:

- Short-term, represented as a changing PSP and temporarily changing synaptic efficacy;
- Long-term, represented as a stable establishment of the synaptic efficacy;
- Genetic (evolutionary), represented as a change in the genetic code and the gene/protein expression level as a result of the above short-term and long term memory changes and evolutionary processes.

Learning in SNN can be:

- Unsupervised - there is no desired output signal provided;
- Supervised – a desired output signal is provided;
- Semi-supervised.

Different tasks can be learned by a neuron, e.g:

- Classification;
- Input-output spike pattern association.

Several biologically plausible learning rules have been introduced so far, depending on the type of the information presentation:

- Rate-order learning, that is based on the average spiking activity of a neuron over time [18, 34, 43];
- Temporal learning, that is based on precise spike times [44, 104, 106, 13, 42];
- Rank-order learning, that takes into account the order of spikes across all synapses connected to a neuron [105, 106].

Rate-order information representation is typical for cognitive information processing [18]. Temporal spike learning is observed in the auditory [93], the visual [11] and the motor control information processing of the brain [13, 90]. Its use in neuro-prosthetics is essential, along with applications for a fast, real-time recognition and control of sequence of related processes [14]. Temporal coding accounts for the precise time of spikes and has been utilised in several learning rules, most popular being Spike-Time Dependent Plasticity (STDP) [103, 69] and SDSP [30, 14]. Temporal coding of information in SNN makes use of the exact time of spikes (e.g. in milliseconds). Every spike matters and its time matters too.

3.2   The STDP Learning Rule

The STDP learning rule uses Hebbian plasticity [39] in the form of long-term potentiation (LTP) and depression (LTD) [103, 69]. Efficacy of synapses is strengthened or weakened based on the timing of post-synaptic action potential in relation to the pre-synaptic spike (example is given in fig.5a). If the difference in the spike time between the pre-synaptic and post-synaptic neurons is negative (pre-synaptic neuron spikes first) than the connection weight between the two neurons increases, otherwise it decreases. Through STDP, connected neurons learn consecutive temporal associations from data. Pre-synaptic activity that precedes post-synaptic firing can induce long-term potentiation (LTP), reversing this temporal order causes long-term depression (LTD).

3.3   Spike Driven Synaptic Plasticity (SDSP)

The SDSP is an unsupervised learning method [30, 14], a modification of the STDP, that directs the change of the synaptic plasticity V_{w0} of a synapse w_0 depending on the time of spiking of the pre-synaptic neuron and the post-synaptic neuron. V_{w0} increases or decreases, depending on the relative timing of the pre- and post-synaptic spikes.

If a pre-synaptic spike arrives at the synaptic terminal before a postsynaptic spike within a critical time window, the synaptic efficacy is increased (potentiation). If the post-synaptic spike is emitted just before the pre-synaptic spike, synaptic efficacy is decreased (depression). This change in synaptic efficacy can be expressed as:

$$\Delta V_{w0} = \frac{I_{pot}(t_{post})}{C_p} \Delta t_{spk} \qquad \text{if } t_{pre} < t_{post} \tag{5}$$

$$\Delta V_{w0} = -\frac{I_{dep}(t_{post})}{C_d} \Delta t_{spk} \qquad \text{if } t_{post} < t_{pre} \tag{6}$$

where: Δt_{spk} is the pre- and post-synaptic spike time window.

The SDSP rule can be used to implement a supervised learning algorithm, when a teacher signal, that copies the desired output spiking sequence, is entered along with the training spike pattern, but without any change of the weights of the teacher input.

The SDSP model is implemented as an VLSI analogue chip [49]. The silicon synapses comprise bistability circuits for driving a synaptic weight to one of two possible analogue values (either potentiated or depressed). These circuits drive the synaptic-weight voltage with a current that is superimposed on that generated by the STDP and which can be either positive or negative. If, on short time scales, the synaptic weight is increased above a set threshold by the network activity via the STDP learning mechanism, the bi-stability circuits generate a constant weak positive current. In the absence of activity (and hence learning) this current will drive the weight toward its potentiated state. If the STDP decreases the synaptic weight below the threshold, the bi-stability circuits will generate a negative current that, in the absence of spiking activity, will actively drive the weight toward the analogue value, encoding its depressed state. The STDP and bi-stability circuits facilitate the implementation of both long-term and short term memory.

3.4   Rank-Order Learning

The rank-order learning rule uses important information from the input spike trains – the rank of the first incoming spike on each synapse (fig.5b). It establishes a priority of inputs (synapses) based on the order of the spike arrival on these synapses for a particular pattern, which is a phenomenon observed in biological systems as well as an important information processing concept for some STPR problems, such as computer vision and control [105, 106].

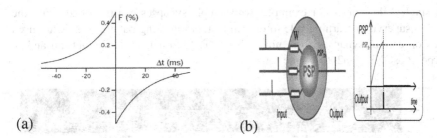

Fig. 5. a,b: (a) An example of synaptic change in a STDP learning neuron [103]; (b) Rank-order LIF neuron

This learning makes use of the extra information of spike (event) order. It has several advantages when used in SNN, mainly: fast learning (as it uses the extra information of the order of the incoming spikes) and asynchronous data entry (synaptic inputs are accumulated into the neuronal membrane potential in an asynchronous way). The learning is most appropriate for AER input data streams [23] as the events and their addresses are entered into the SNN 'one by one', in the order of their happening.

The postsynaptic potential of a neuron i at a time t is calculated as:

$$PSP(i,t) = \sum mod^{order(j)} w_{j,i} \tag{7}$$

where: *mod* is a modulation factor; j is the index for the incoming spike at synapse j,i and wj,i is the corresponding synaptic weight; order(j) represents the order (the rank) of the spike at the synapse j,i among all spikes arriving from all m synapses to the neuron i. The *order(j)* has a value 0 for the first spike and increases according to the input spike order. An output spike is generated by neuron i if the PSP (i,t) becomes higher than a threshold PSPTh (i).

During the training process, for each training input pattern (sample, example) the connection weights are calculated based on the order of the incoming spikes [105]:

$$\Delta w_{j,i}(t) = mod^{\text{order }(j,i\,(t))} \tag{8}$$

3.5 Combined Rank-Order and Temporal Learning

In [25] a method for a combined rank-order and temporal (e.g. SDSP) learning is proposed and tested on benchmark data. The initial value of a synaptic weight is set according to the rank-order learning based on the first incoming spike on this synapse. The weight is further modified to accommodate following spikes on this synapse with the use of a temporal learning rule – SDSP.

4   STPR in a Single Neuron

In contrast to the distributed representation theory and to the widely popular view that a single neuron cannot do much, some recent results showed that a single neuronal model can be used for complex STPR.

A single LIF neuron, for example, with simple synapses can be trained with the STDP unsupervised learning rule to discriminate a repeating pattern of synchronised spikes on certain synapses from noise (from: T. Masquelier, R. Guyonneau and S. Thorpe, PlosONE, Jan2008) – see fig. 6.

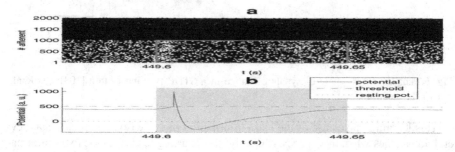

Fig. 6. A single LIF neuron with simple synapses can be trained with the STDP unsupervised learning rule to discriminate a repeating pattern of synchronised spikes on certain synapses from noise (from: T. Masquelier, R. Guyonneau and S. Thorpe, PlosONE, Jan2008))

Single neuron models have been introduced for STPR, such as: Temportron [38]; Chronotron [28]; ReSuMe [87]; SPAN [76, 77]. Each of them can learn to emit a spike or a spike pattern (spike sequence) when a certain STP is recognised. Some of them can be used to recognise multiple STP per class and multiple classes [87, 77, 76]. Fig.7c,d shows the use of a single SPAN neuron for the classification of 5 STP belonging to 5 different classes [77]. The accuracy of classification is rightly lower for the class 1 (the neuron emits a spike at the very beginning of the input pattern) as there is no sufficient input data - fig.7d) [77].

(a)

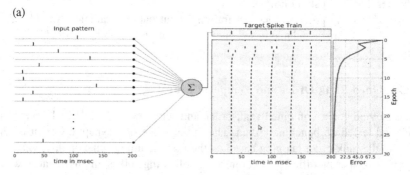

A single output neuron is trained to respond with a temporally precise output spike train to a specific spatio-temporal input.

Fig. 7. (a) The SPAN model [77]

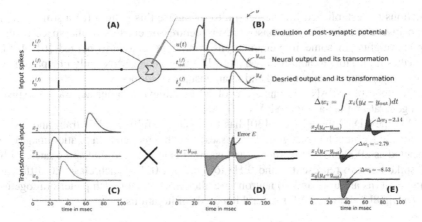

Illustration of the proposed training algorithm.

(b)

Fig. 7. (b) The Widrow-Hoff Delta learning rule applied to learn to associate an output spike sequence to an input STP [77, 30]

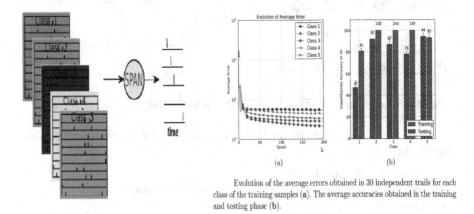

Evolution of the average errors obtained in 30 independent trails for each class of the training samples (**a**). The average accuracies obtained in the training and testing phase (**b**).

Fig. 7. c,d. The use of a single SPAN neuron for the classification of 5 STP belonging to 5 different classes [77]. The accuracy of classification is rightly lower for the class 1 – spike at the very beginning of the input pattern as there is no sufficient input data - fig.7d).

5 Evolving Spiking Neural Networks

Despite the ability of a single neuron to conduct STPR, a single neuron has a limited power and complex STPR tasks will require multiple spiking neurons.

One approach is proposed in the evolving spiking neural networks (eSNN) framework [61, 111]. eSNN evolve their structure and functionality in an on-line manner, from incoming information. For every new input pattern, a new neuron is dynamically allocated and connected to the input neurons (feature neurons). The neurons connections are established for the neuron to recognise this pattern (or a similar one) as a positive example. The neurons represent centres of clusters in the space of the synaptic weights. In some implementations similar neurons are merged [61, 115]. That makes it possible to achieve a very fast learning in an eSNN (only one pass may be necessary), both in a supervised and in an unsupervised mode.

In [76] multiple SPAN neurons are evolved to achieve a better accuracy of spike pattern generation than a single SPAN – fig.8a.

In [14] the SDSP model from [30] has been successfully used to train and test a SNN for 293 character recognition (classes). Each character (a static image) is represented as 2000 bit feature vector, and each bit is transferred into spike rates, with 50Hz spike burst to represent 1 and 0 Hz to represent 0. For each class, 20 different training patterns are used and 20 neurons are allocated, one for each pattern (altogether 5860) (fig.8b) and trained for several hundreds of iterations.

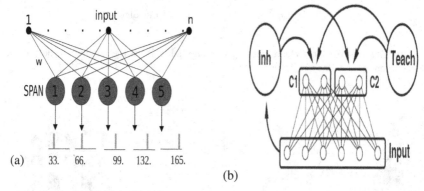

Fig. 8. a,b: (a) Multiple SPAN neurons [76]; (b) Multiple SDSP trained neurons [14]

A general framework of eSNN for STPR is shown in fig.9. It consists of the following blocks:

- Input data encoding block;
- Machine learning block (consisting of several sub-blocks);
- Output block.

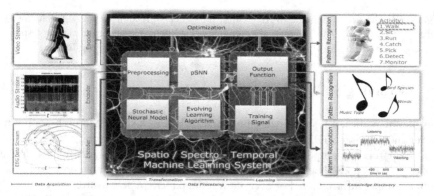

Fig. 9. The eSNN framework for STPR (from: http://ncs.ethz.ch/projects/evospike)

In the input block continuous value input variables are transformed into spikes. Different approaches can be used:

- population rank coding [13] – fig.10a;
- thresholding the input value, so that a spike is generated if the input value (e.g. pixel intensity) is above a threshold;
- Address Event Representation (AER) - thresholding the difference between two consecutive values of the same variable over time as it is in the artificial cochlea [107] and artificial retina devices [23] – fig.10b.

The input information is entered either on-line (for on-line, real time applications) or as a batch data. The *time* of the input data is in principal different from the internal SNN *time* of information processing.

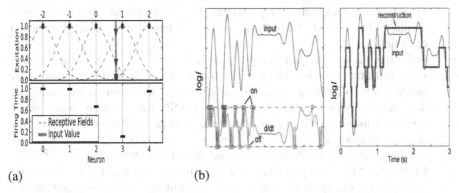

(a)   (b)

Fig. 10. a,b: (a) Population rank order coding of input information; (b) Address Event Representations (AER) of the input information [23]

Long and complex SSTD cannot be learned in simple one-layer neuronal structures as the examples in fig.8a,b. They require neuronal 'buffers' as shown in fig.11a. In [82] a 3D buffer was used to store spatio-temporal 'chunks' of input data before the data is classified. In this case the size of the chunk (both in space and time) is fixed by the size of the reservoir. There are no connections between the layers in the buffer. Still, the system outperforms traditional classification techniques as it is demonstrated on sign language recognition, where eSNN classifier was applied [61, 115]. Reservoir computing [73, 108] has already become a popular approach for SSTD modelling and pattern recognition. In the classical view a 'reservoir' is a homogeneous, passive 3D structure of probabilistically connected and fixed neurons that in principle has no learning and memory, neither it has an interpretable structure – fig.11b. A reservoir, such as a Liquid State Machine (LSM) [73, 37], usually uses *small world recurrent connections* that do not facilitate capturing explicit spatial and temporal components from the SSTD in their relationship, which is the main goal of learning SSTD. Despite difficulties with the LSM reservoirs, it was shown on several SSTD problems that they produce better results than using a simple classifier [95, 73, 99, 60]. Some publications demonstrated that probabilistic neurons are suitable for reservoir computing especially in a noisy environment [98, 83]. In [81] an improved accuracy of

LSM reservoir structure on pattern classification of hypothetical tasks is achieved when STDP learning was introduced into the reservoir. The learning is based on comparing the liquid states for different classes and adjusting the connection weights so that same class inputs have closer connection weights. The method is illustrated on the phone recognition task of the TIMIT data base phonemes – spectro-temporal problem. 13 MSCC are turned into trains of spikes. The metric of separation between liquid states representing different classes is similar to the Fisher's t-test [27].

After a presentation of input data example (or a 'chunk' of data) the state of the SNN reservoir S(t) is evaluated in an output module and used for classification purposes (both during training and recall phase). Different methods can be applied to capture this state:

-   Spike rate activity of *all* neurons at a certain time window: The state of the reservoir is represented as a vector of n elements (n is the number of neurons in the reservoir), each element representing the spiking probability of the neuron within a time window. Consecutive vectors are passed to train/recall an output classifier.
-   Spike rate activity of spatio-temporal clusters C_1, C_2, ... C_k of close (both in space and time) neurons: The state $S_{Ci}(t)$ of each cluster Ci is represented by a single number, reflecting on the spiking activity of the neurons in the cluster in a defined time window (this is the internal SNN time, usually measured in '*msec*'). This is interpreted as the current spiking probability of the cluster. The states of all clusters define the current reservoir state S(t). In the output function, the cluster states $S_{Ci}(t)$ are used differently for different tasks.
-   Continuous function representation of spike trains: In contrast to the above two methods that use spike rates to evaluate the spiking activity of a neuron or a neuronal cluster, here the train of spikes from each neuron within a time window, or a neuronal cluster, is transferred into a continuous value temporal function using a kernel (e.g. α-kernel). These functions can be compared and a continuous value error measured.

In [95] a comparative analysis of the three methods above is presented on a case study of Brazilian sign language gesture recognition (see fig.18) using a LSM as a reservoir.

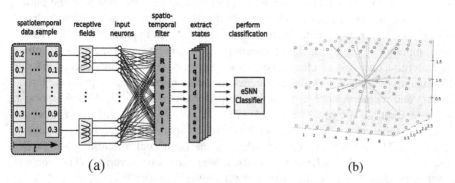

(a)   (b)

Fig. 11. a,b. (a) An eSNN architecture for STPR using a reservoir; (b) The structure and connectivity of a reservoir

Different adaptive classifiers can be explored for the classification of the reservoir state into one of the output classes, including: statistical techniques, e.g. regression techniques; MLP; eSNN; nearest-neighbour techniques; incremental LDA [85]. State vector transformation can be done with the use of adaptive incremental transformation functions, such as incremental PCA [84].

6   Computational Neurogenetic Models (CNGM)

Here, the neurogenetic model of a neuron [63, 10] is utilized. A CNGM framework is shown in fig.12 [64].

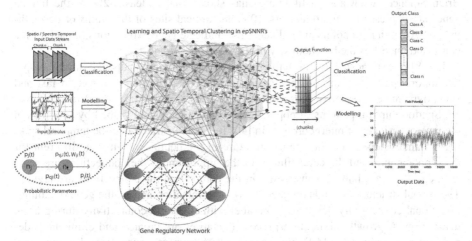

Fig. 12. A schematic diagram of a CNGM framework, consisting of: input encoding module; output function for SNN state evaluation; output classifier; GRN (optional module). The framework can be used to create concrete models for STPR or data modelling (from [64]).

The CNGM framework comprises a set of methods and algorithms that support the development of computational models, each of them characterized by:

- Two-tire, consisting of an eSNN at the higher level and a gene regulatory network (GRN) at the lower level, each functioning at a different time-scale and continuously interacting between each other;
- Optional use of probabilistic spiking neurons, thus forming an epSNN;
- Parameters in the epSNN model are defined by genes/proteins from the GRN;
- Can capture in its internal representation both spatial and temporal characteristics from SSTD streams;
- The structure and the functionality of the model evolve in time from incoming data;
- Both unsupervised and supervised learning algorithms can be applied in an on-line or in a batch mode.
- A concrete model would have a specific structure and a set of algorithms depending on the problem and the application conditions, e.g.: classification of SSTD; modelling of brain data.

The framework from fig.12 supports the creation of a multi-modular integrated system, where different modules, consisting of different neuronal types and genetic parameters, represent different functions (e.g.: vision; sensory information processing; sound recognition; motor-control) and the whole system works in an integrated mode.

The neurogenetic model from fig.12 uses as a main principle the analogy with biological facts about the relationship between spiking activity and gene/protein dynamics in order to control the learning and spiking parameters in a SNN when SSTD is learned. Biological support of this can be found in numerous publications (e.g. [10, 40, 117, 118]).

The Allen Human Brain Atlas (www.brain-map.org) of the Allen Institute for Brain Science (www.alleninstitute.org) has shown that at least 82% of the human genes are expressed in the brain. For 1000 anatomical sites of the brains of two individuals 100 mln data points are collected that indicate gene expressions of each of the genes and underlies the biochemistry of the sites.

In [18] it is suggested that both the firing rate (rate coding) and spike timing as spatiotemporal patterns (rank order and spatial pattern coding) play a role in fast and slow, dynamic and adaptive sensorimotor responses, controlled by the cerebellar nuclei. Spatio-temporal patterns of population of Purkinji cells are shaped by activities in the molecular layer of interneurons. In [40] it is demonstrated that the temporal spiking dynamics depend on the spatial structure of the neural system (e.g. different for the hippocampus and the cerebellum). In the hippocampus the connections are scale free, e.g. there are hub neurons, while in the cerebellum the connections are regular. The spatial structure depends on genetic pre-determination and on the gene dynamics. Functional connectivity develops in parallel with structural connectivity during brain maturation. A growth-elimination process (synapses are created and eliminated) depend on gene expression [40], e.g. glutamatergic neurons issued from the same progenitors tend to wire together and form ensembles, also for the cortical GABAergic interneuron population. Connections between early developed neurons (mature networks) are more stable and reliable when transferring spikes than the connections between newly created neurons (thus the probability of spike transfer). Postsynaptic AMPA-type glutamate receptors (AMPARs) mediate most fast excitatory synaptic transmissions and are crucial for many aspects of brain function, including learning, memory and cognition [10, 31].

[65] shows the dramatic effect of a change of single gene, that regulates the τ parameter of the neurons, on the spiking activity of the whole SNN of 1000 neurons – see fig.13.

The spiking activity of a neuron may affect as a feedback the expressions of genes [5]. As pointed in [118] on a longer time scales of minutes and hours the function of neurons may cause the changes of the expression of hundreds of genes transcribed into mRNAs and also in microRNAs, which makes the short-term, the long-term and the genetic memories of a neuron linked together in a global memory of the neuron and further - of the whole neural system.

A major problem with the CNGM from fig.12 is how to optimize the numerous parameters of the model. One solution could be using evolutionary computation, such as PSO [75, 83] and the recently proposed quantum inspired evolutionary computation techniques [22, 97, 96]. The latter can deal with a very large dimensional space as each quantum-bit chromosome represents the whole space, each point to certain probability. Such algorithms are faster and lead to a close solution to the global optimum in a very short time. In one approach it may be reasonable to use same parameter values (same GRN) for all neurons in the SNN or for each of different types of neurons (cells) that will results in a significant reduction of the parameters to be optimized. This can be interpreted as 'average' parameter value for the neurons of the same type. This approach corresponds to the biological notion to use one value (average) of a gene/protein expression for millions of cells in bioinformatics.

Another approach to define the parameters of the probabilistic spiking neurons, especially when used in biological studies, is to use prior knowledge about the association of spiking parameters with relevant genes/proteins (neuro-transmitter, neuro-receptor, ion channel, neuro-modulator) as described in [64]. Combination of the two approaches above is also possible.

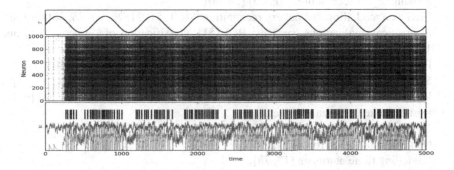

Fig. 13. A GRN interacting with a SNN reservoir of 1000 neurons. The GRN controls a single parameter, i.e. the τ parameter of all 1000 LIF neurons, over a period of five seconds. The top diagram shows the evolution of   τ. The response of the SNN is shown as a raster plot of spike activity.   A black point in this diagram indicates a spike of a specific neuron at a specific time in the simulation. The bottom diagram presents the evolution of the membrane potential of a single neuron from the network (green curve) along with its firing threshold ϑ (red curve). Output spikes of the neuron are indicated as black vertical lines in the same diagram (from [65]).

7   SNN   Software and Hardware Implementations to Support STPR

Software and hardware realisations of SNN are already available to support various applications of SNN for STPR. Among the most popular software/hardware systems are [24, 16, 29]:

- jAER (http://jaer.wiki.sourceforge.net) [23];
- Software simulators, such as Brian [16], NEST, NeMo [79],etc;
- Silicon retina camera [23];

- Silicon cochlea [107];
- SNN hardware realisation of LIFM and SDSP [47-50];
- The SpiNNaker hardware/software environment [89, 116];
- FPGA implementations of SNN [56];
- The IBM LIF SNN chip, recently announced.

Fig.14 shows a hypothetical engineering system using some of the above tools (from [47, 25]).

8   Current and Future Applications of eSNN and CNGM for STPR

Numerous are the applications of eSNN for STPR. Here only few of them are listed:

- Moving object recognition (fig. 15) [23, 60];
- EEG data modelling and pattern recognition [70, 1, 51, 21, 26, 99, 35, 36] directed to practical applications, such as: BCI [51], classification of epilepsy [35, 36, 109] - (fig.16);
- Robot control through EEG signals [86] (fig.17) and robot navigation [80];
- Sign language gesture recognition (e.g. the Brazilian sign language – fig.18) [95];
- Risk of event evaluation, e.g. prognosis of establishment of invasive species [97] – fig.19; stroke occurrence [6], etc.
- Cognitive and emotional robotics [8, 64];
- Neuro-rehabilitation robots [110];
- Modelling finite automata [17, 78];
- Knowledge discovery from SSTD [101];
- Neuro-genetic robotics [74];
- Modelling the progression or the response to treatment of neurodegenerative diseases, such as Alzheimer's Disease [94, 64] – fig.20. The analysis of the obtained GRN model in this case could enable the discovery of unknown interactions between genes/proteins related to a brain disease progression and how these interactions can be modified to achieve a desirable effect.
- Modelling financial and economic problems (neuro-economics) where at a 'lower' level the GRN represents the dynamic interaction between time series variables (e.g. stock index values, exchange rates, unemployment, GDP, prize of oil), while the 'higher' level epSNN states represents the state of the economy or the system under study. The states can be further classified into pre-define classes (e.g. buy, hold, sell, invest, likely bankruptcy) [113];
- Personalized modelling, which is concerned with the creation of a single model for an individual input data [58, 59, 62]. Here as an individual data a whole SSTD pattern is taken rather than a single vector.

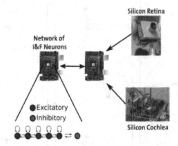

Fig. 14. A hypothetical neuromorphic SNN application system (from http://ncs.ethz.ch)

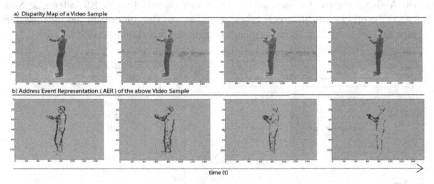

Fig. 15. Moving object recognition with the use of AER [23]

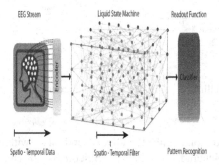

Fig. 16. EEG recognition system

Fig. 17. Robot control and navigation through EEG signals

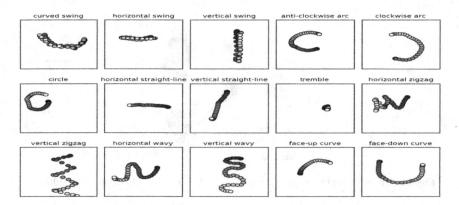

Fig. 18. A single sample for each of the 15 classes of the LIngua BRAsileira de Sinais (LIBRAS) - the official Brazilian sign language is shown. The colour indicates the time frame of a given data point (black/white corresponds to earlier/later time points) [95].

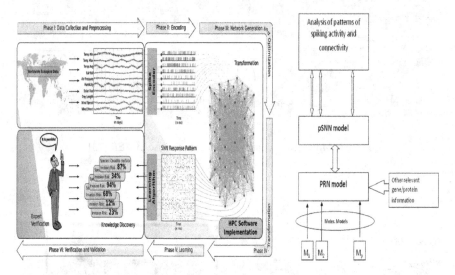

Fig. 19. A prognostic system for ecological modelling [97]

Fig. 20. Fig.20.Hierarchical CNGM [64]

Acknowledgment. I acknowledge the discussions with G.Indivery and also with: A.Mohemmed, T.Delbruck, S-C.Liu, N.Nuntalid, K.Dhoble, S.Schliebs, R.Hu, R.Schliebs, H.Kojima, F.Stefanini. The work on this paper is sponsored by the Knowledge Engineering and Discovery Research Institute, KEDRI (www.kedri.info) and the EU FP7 Marie Curie International Incoming Fellowship project PIIF-GA-2010-272006 *EvoSpike,* hosted by the Institute for Neuroinformatics – the Neuromorphic Cognitive Systems Group, at the University of Zurich and ETH Zurich (http://ncs.ethz.ch/projects/evospike). Diana Kassabova helped with the proofreading. A preliminary publication of this paper appeared in "Natural Intelligence: the INNS Magazine", vol.1, issue 2, 23-37, 2012.

References

1. Acharya, R., Chua, E.C.P., Chua, K.C., Min, L.C., Tamura, T.: Analysis and Automatic Identification of Sleep Stages using Higher Order Spectra. Int. Journal of Neural Systems 20(6), 509–521 (2010)
2. Arel, I., Rose, D.C., Karnowski, T.P.: Deep Machine Learning: A New Frontier in Artificial Intelligence Research. IEEE Comput. Intelligence Magazine 5(4), 13–18 (2010)
3. Arel, I., Rose, D., Coop, B.: DeSTIN: A deep learning architecture with application to high-dimensional robust pattern recognition. In: Proc. 2008 AAAI Workshop Biologically Inspired Cognitive Architectures, BICA (2008)
4. Arel, I., Rose, D., Karnovski, T.: Deep Machine Learning – A New Frontier in Artificial Intelligence Research. IEEE CI Magazine, 13–18 (November 2010)
5. Barbado, M., Fablet, K., Ronjat, M., De Waard, M.: Gene regulation by voltage-dependent calcium channels. Biochimica et Biophysica Acta 1793, 1096–1104 (2009)
6. Barker-Collo, S., Feigin, V.L., Parag, V., Lawes, C.M.M., Senior, H.: Auckland Stroke Outcomes Study. Neurology 75(18), 1608–1616 (2010)
7. Belatreche, A., Maguire, L.P., McGinnity, M.: Advances in Design and Application of Spiking Neural Networks. Soft Comput. 11(3), 239–248 (2006)
8. Bellas, F., Duro, R.J., Faiña, A., Souto, D.: MDB: Artificial Evolution in a Cognitive Architecture for Real Robots. IEEE Transactions on Autonomous Mental Development 2, 340–354 (2010)
9. Bengio, Y.: Learning Deep Architectures for AI. Found. Trends. Mach. Learning 2(1), 1–127 (2009)
10. Benuskova, L., Kasabov, N.: Computational neuro-genetic modelling, 290 pages. Springer, New York (2007)
11. Berry, M.J., Warland, D.K., Meister, M.: The structure and precision of retinal spike-trains. PNAS 94(10), 5411–5416 (1997)
12. Bohte, S., Kok, J., LaPoutre, J.: Applications of spiking neural networks. Information Processing Letters 95(6), 519–520 (2005)
13. Bohte, S.M.: The evidence for neural information processing with precise spike-times: A survey. Natural Computing 3 (2004)
14. Brader, J., Senn, W., Fusi, S.: Learning real-world stimuli in a neural network with spike-driven synaptic dynamics. Neural Computation 19(11), 2881–2912 (2007)
15. Brader, J.M., Senn, W., Fusi, S.: Learning Real-World Stimuli in a Neural Net-work with Spike-Driven Synaptic Dynamics. Neural Comput. 19(11), 2881–2912 (2007)
16. Brette, R., Rudolph, M., Carnevale, T., Hines, M., Beeman, D., Bower, J.M., Diesmann, M., Morrison, A., Goodman, P.H., Harris, F.C., Zirpe, M., Natschläger, T., Pecevski, D., Ermentrout, B., Djurfeldt, M., Lansner, A., Rochel, O., Vieville, T., Muller, E., Davison, A.P., Boustani, S.E., Destexhe, A.: Simulation of networks of spiking neurons: a review of tools and strategies. J. Comput. Neurosci. 23, 349–398 (2007)
17. Buonomano, D., Maass, W.: State-dependent computations: Spatio-temporal processing in cortical networks. Nature Reviews, Neuroscience 10, 113–125 (2009)
18. De Zeeuw, C.I., Hoebeek, F.E., Bosman, L.W.J., Schonewille, M.: Spati-otemporal firing patterns in the cerebellum. Nature Reviews Neurosc. 12, 327–344 (2011)
19. Shortall, C.R., Moore, A., Smith, E., Hall, M.J., Woiwod, I.P., Harrington, R.: Long-term changes in the abundance of flying insects. Insect Conservation and Diversity 2(4), 251–260 (2009)
20. Cowburn, P.J., Cleland, J.G.F., Coats, A.J.S., Komajda, M.: Risk stratifica-tion in chronic heart failure. Eur. Heart J. 19, 696–710 (1996)

21. Craig, D.A., Nguyen, H.T.: Adaptive EEG Thought Pattern Classifier for Advanced Wheelchair Control, Engin. In: Medicine and Biology Society, EMBS 2007, pp. 2544–2547 (2007)
22. Defoin-Platel, M., Schliebs, S., Kasabov, N.: Quantum-inspired Evolutionary Algorithm: A multi-model EDA. IEEE Trans. Evolutionary Computation 13(6), 1218–1232 (2009)
23. Delbruck, T.: jAER open source project (2007), http://jaer.wiki.sourceforge.net
24. Douglas, R., Mahowald, M.: Silicon Neurons. In: Arbib, M. (ed.) The Handbook of Brain Theory and Neural Networks, pp. 282–289. MIT Press (1995)
25. Dhoble, K., Nuntalid, N., Indivery, G., Kasabov, N.: Online Spatio-Temporal Pattern Recognition with Evolving Spiking Neural Networks utilising Address Event Representation, Rank Order, and Temporal Spike Learning. In: Proc. IJCNN 2012, Brisbane. IEEE (June 2012)
26. Ferreira, A., Almeida, C., Georgieva, P., Tomé, A., Silva, F.: Advances in EEG-based Biometry. In: Campilho, A., Kamel, M. (eds.) ICIAR 2010, Part II. LNCS, vol. 6112, pp. 287–295. Springer, Heidelberg (2010)
27. Fisher, R.A.: The use of multiple measurements in taxonomic problems. Annals of Eugenics 7, 179–188 (1936)
28. Florian, R.V.: The chronotron: a neuron that learns to fire temporally-precise spike patterns (2010)
29. Furber, S., Temple, S.: Neural systems engineering, Interface. J.of the Royal Society 4, 193–206 (2007)
30. Fusi, S., Annunziato, M., Badoni, D., Salamon, A., Amit, D.: Spike-driven synaptic plasticity: theory, simulation, VLSI implementation. Neural Computation 12(10), 2227–2258 (2000)
31. Gene and Disease (2005), NCBI, http://www.ncbi.nlm.nih.gov
32. Gerstner, W.: Time structure of the activity of neural network models. Phys. Rev 51, 738–758 (1995)
33. Gerstner, W.: What's different with spiking neurons? In: Mastebroek, H., Vos, H. (eds.) Plausible Neural Networks for Biological Modelling, pp. 23–48. Kluwer Academic Publs. (2001)
34. Gerstner, W., Kreiter, A.K., Markram, H., Herz, A.V.M.: Neural codes: firing rates and beyond. Proc. Natl. Acad. Sci. USA 94(24), 12740–12741 (1997)
35. Ghosh-Dastidar, S., Adeli, H.: A New Supervised Learning Algorithm for Multiple Spiking Neural Networks with Application in Epilepsy and Seizure Detection. Neural Networks 22(10), 1419–1431 (2009)
36. Ghosh-Dastidar, S., Adeli, H.: Improved Spiking Neural Networks for EEG Classification and Epilepsy and Seizure Detection. Integrated Computer-Aided Engineering 14(3), 187–212 (2007)
37. Goodman, E., Ventura, D.: Spatiotemporal pattern recognition via liquid state ma-chines. In: International Joint Conference on Neural Networks, IJCNN 2006, Vancouver, BC, pp. 3848–3853 (2006)
38. Gutig, R., Sompolinsky, H.: The tempotron: a neuron that learns spike timing-based decisions. Nat. Neurosci. 9(3), 420–428 (2006)
39. Hebb, D.: The Organization of Behavior. John Wiley and Sons, New York (1949)
40. Henley, J.M., Barker, E.A., Glebov, O.O.: Routes, destinations and delays: recent advances in AMPA receptor trafficking. Trends in Neurosc. 34(5), 258–268 (2011)
41. Hodgkin, A.L., Huxley, A.F.: A quantitative description of membrane current and its application to conduction and excitation in nerve. Journal of Physiology 117, 500–544 (1952)

42. Hopfield, J.: Pattern recognition computation using action potential timing for stimulus representation. Nature 376, 33–36 (1995)
43. Hopfield, J.J.: Neural networks and physical systems with emergent collective computational abilities. PNAS USA 79, 2554–2558 (1982)
44. Hugo, G.E., Ines, S.: Time and category information in pattern-based codes. Frontiers in Computational Neuroscience 4(0) (2010)
45. Huguenard, J.R.: Reliability of axonal propagation: The spike doesn't stop here. PNAS 97(17), 9349–9350 (2000)
46. Iglesias, J., Villa, A.E.P.: Emergence of Preferred Firing Sequences in Large Spiking Neural Networks During Simulated Neuronal Development. Int. Journal of Neural Systems 18(4), 267–277 (2008)
47. Indiveri, G., Linares-Barranco, B., Hamilton, T., Van Schaik, A., Etienne-Cummings, R., Delbruck, T., Liu, S., Dudek, P., Häfliger, P., Renaud, S., et al.: Neuromorphic silicon neuron circuits. Frontiers in Neuroscience 5 (2011)
48. Indiveri, G., Chicca, E., Douglas, R.J.: Artificial cognitive systems: From VLSI networks of spiking neurons to neuromorphic cognition. Cognitive Computation 1(2), 119–127 (2009)
49. Indiveri, G., Stefanini, F., Chicca, E.: Spike-based learning with a generalized inte-grate and fire silicon neuron. In: 2010 IEEE Int. Symp. Circuits and Syst (ISCAS 2010), Paris, May 30-June 02, pp. 1951–1954 (2010)
50. Indiviery, G., Horiuchi, T.: Frontiers in Neuromorphic Engineering. Frontiers in Neuroscience 5, 118 (2011)
51. Isa, T., Fetz, E.E., Muller, K.: Recent advances in brain-machine interfaces. Neural Networks, Brain-Machine Interface 22(9), 1201–1202 (2009)
52. Izhikevich, E.: Simple model of spiking neurons. IEEE Trans. on Neural Networks 14(6), 1569–1572 (2003)
53. Izhikevich, E.M.: Which model to use for cortical spiking neurons? IEEE TNN 15(5), 1063–1070 (2004)
54. Izhikevich, E.M., Edelman, G.M.: Large-Scale Model of Mammalian Thalamocortical Systems. PNAS 105, 3593–3598 (2008)
55. Izhikevich, E.: Polychronization: Computation with Spikes. Neural Computation 18, 245–282 (2006)
56. Johnston, S.P., Prasad, G., Maguire, L., McGinnity, T.M.: FPGA Hard-ware/software co-design methodology - towards evolvable spiking networks for robotics application. Int. J. Neural Systems 20(6), 447–461 (2010)
57. Kasabov, N.: Foundations of Neural Networks. In: Fuzzy Systems and Knowledge Engineering, p. 550. MIT Press, Cambridge (1996)
58. Kasabov, N., Hu, Y.: Integrated optimisation method for personalised modelling and case study applications. Int. Journal of Functional Informatics and Personalised Medicine 3(3), 236–256 (2010)
59. Kasabov, N.: Data Analysis and Predictive Systems and Related Methodologies – Person-alised Trait Modelling System. PCT/NZ2009/000222, NZ Patent
60. Kasabov, N., Dhoble, K., Nuntalid, N., Mohemmed, A.: Evolving Probabilistic Spiking Neural Networks for Spatio-temporal Pattern Recognition: A Preliminary Study on Moving Object Recognition. In: Lu, B.-L., Zhang, L., Kwok, J. (eds.) ICONIP 2011, Part III. LNCS, vol. 7064, pp. 230–239. Springer, Heidelberg (2011)
61. Kasabov, N., Evolving connectionist systems: The knowledge engineering approach. Springer (2003, 2007)

62. Kasabov, N.: Global, local and personalised modelling and profile discovery in Bioinformatics: An integrated approach. Pattern Recogn. Letters 28(6), 673–685 (2007)
63. Kasabov, N., Benuskova, L., Wysoski, S.: A Computational Neurogenetic Model of a Spiking Neuron. In: Proc. IJCNN 2005 Conf., vol. 1, pp. 446–451. IEEE Press (2005)
64. Kasabov, N., Schliebs, R., Kojima, H.: Probabilistic Computational Neurogenetic Framework: From Modelling Cognitive Systems to Alzheimer's Disease. IEEE Trans. Autonomous Mental Development 3(4), 1–12 (2011)
65. Kasabov, N., Schliebs, S., Mohemmed, A.: Modelling the Effect of Genes on the Dynamics of Probabilistic Spiking Neural Networks for Computational Neurogenetic Modelling. In: Proc. 6th meeting on Computational Intelligence for Bioinformatics and Biostatistics, CIBB 2011, Gargangio, Italy, June 30-July 2. LNCS (LNBI). Springer (2011)
66. Kasabov, N.: To spike or not to spike: A probabilistic spiking neuron model. Neural Netw. 23(1), 16–19 (2010)
67. Kilpatrick, Z.P., Bresloff, P.C.: Effect of synaptic depression and adaptation on spatiotemporal dynamics of an excitatory neural networks. Physica D 239, 547–560 (2010)
68. Kistler, G., Gerstner, W.: Spiking Neuron Models - Single Neurons, Populations, Plasticity. Cambridge Univ. Press (2002)
69. Legenstein, R., Naeger, C., Maass, W.: What Can a Neuron Learn with Spike-Timing-Dependent Plasticity? Neural Computation 17(11), 2337–2382 (2005)
70. Lotte, F., Congedo, M., Lécuyer, A., Lamarche, F., Arnaldi, B.: A review of classification algorithms for EEG-based brain–computer interfaces. J. Neural Eng. 4(2), R1–R15 (2007)
71. Maass, W., Markram, H.: Synapses as dynamic memory buffers. Neural Network 15(2), 155–161 (2002)
72. Maass, W., Zador, A.M.: Computing and learning with dynamic synapses. In: Pulsed Neural Networks, pp. 321–336. MIT Press (1999)
73. Maass, W., Natschlaeger, T., Markram, H.: Real-time computing without stable states: A new framework for neural computation based on perturbations. Neural Computation 14(11), 2531–2560 (2002)
74. Meng, Y., Y. Jin, J. Yin, and M. Conforth (2010) Human activity detection using spiking neural networks regulated by a gene regulatory network. Proc. Int. Joint Conf. on Neural Net-works (IJCNN), IEEE Press, pp.2232-2237, Barcelona, July 2010.
75. Mohemmed, A., Matsuda, S., Schliebs, S., Dhoble, K., Kasabov, N.: Optimization of Spiking Neural Networks with Dynamic Synapses for Spike Sequence Generation using PSO. In: Proc. Int. Joint Conf. Neural Networks, California, USA, pp. 2969–2974. IEEE Press (2011)
76. Mohemmed, A., Schliebs, S., Matsuda, S., Kasabov, N.: Evolving Spike Pattern Association Neurons and Neural Networks. Neurocomputing (in print)
77. Mohemmed, A., Schliebs, S., Matsuda, S., Kasabov, N.: SPAN: Spike Pattern Association Neuron for Learning Spatio-Temporal Sequences. International Journal of Neural Systems (in print, 2012)
78. Natschläger, T., Maass, W.: Spiking neurons and the induction of finite state machines. Theoretical Computer Science - Natural Computing 287(1), 251–265 (2002)
79. NeMo spiking neural network simulator, http://www.doc.ic.ac.uk/~akf/nemo/index.html
80. Nichols, E., McDaid, L.J., Siddique, N.H.: Case Study on Self-organizing Spiking Neural Networks for Robot Navigation. International Journal of Neural Systems 20(6), 501–508 (2010)
81. Norton, D., Ventura, D.: Improving liquid state machines through iterative refinement of the reservoir. Neurocomputing 73, 2893–2904 (2010)

82. Nuzlu, H., Kasabov, N., Shamsuddin, S., Widiputra, H., Dhoble: An Extended Evolving Spiking Neural Network Model for Spatio-Temporal Pattern Classification. In: Proc. IJCNN, California, USA, pp. 2653–2656. IEEE Press (2011)

83. Nuzly, H., Kasabov, N., Shamsuddin, S.: Probabilistic Evolving Spiking Neural Network Optimization Using Dynamic Quantum Inspired Particle Swarm Optimization. In: ICONIP 2010, Part I. LNCS, vol. 6443 (2010)

84. Ozawa, S., Pang, S., Kasabov, N.: Incremental Learning of Chunk Data for On-line Pattern Classification Systems. IEEE Trans. Neural Networks 19(6), 1061–1074 (2008)

85. Pang, S., Ozawa, S., Kasabov, N.: Incremental Linear Discriminant Analysis for Classification of Data Streams. IEEE Trans. SMC-B 35(5), 905–914 (2005)

86. Pfurtscheller, G., Leeb, R., Keinrath, C., Friedman, D., Neuper, C., Guger, C., Slater, M.: Walking from thought. Brain Research 1071(1), 145–152 (2006)

87. Ponulak, F., Kasinski, A.: Supervised learning in spiking neural networks with ReSuMe: sequence learning, classification, and spike shifting. Neural Computation 22(2), 467–510 (2010)

88. Rabiner, L.R.: A tutorial on hidden Markov models and selected applications in speech recognition. Proc. IEEE 77(2), 257–285 (1989)

89. Rast, A.D., Jin, X., Galluppi, F., Plana, L.A., Patterson, C., Furber, S.: Scalable Event-Driven Native Parallel Processing: The SpiNNaker Neuromimetic System. In: Proc. of the ACM International Conference on Computing Frontiers, Bertinoro, Italy, May 17-19, pp. 21–29 (2010) ISBN 978-1-4503-0044-5

90. Reinagel, P., Reid, R.C.: Precise firing events are conserved across neurons. Journal of Neuroscience 22(16), 6837–6841 (2002)

91. Reinagel, R., Reid, R.C.: Temporal coding of visual information in the thalamus. Journal of Neuroscience 20(14), 5392–5400 (2000)

92. Riesenhuber, M., Poggio, T.: Hierarchical Model of Object Recognition in Cortex. Nature Neuroscience 2, 1019–1025 (1999)

93. Rokem, A., Watzl, S., Gollisch, T., Stemmler, M., Herz, A.V., Samengo, I.: Spike-timing precision underlies the coding efficiency of auditory receptor neurons. J. Neurophysiol. (2005)

94. Schliebs, R.: Basal forebrain cholinergic dysfunction in Alzheimer´s disease – interrelationship with β-amyloid, inflammation and neurotrophin signaling. Neurochemical Research 30, 895–908 (2005)

95. Schliebs, S., Hamed, H.N.A., Kasabov, N.: Reservoir-Based Evolving Spiking Neural Network for Spatio-temporal Pattern Recognition. In: Lu, B.-L., Zhang, L., Kwok, J. (eds.) ICONIP 2011, Part II. LNCS, vol. 7063, pp. 160–168. Springer, Heidelberg (2011)

96. Schliebs, S., Kasabov, N., Defoin-Platel, M.: On the Probabilistic Optimization of Spiking Neural Networks. International Journal of Neural Systems 20(6), 481–500 (2010)

97. Schliebs, S., Defoin-Platel, M., Worner, S., Kasabov, N.: Integrated Feature and Parameter Optimization for Evolving Spiking Neural Netw.: Exploring Heterogeneous Probabilistic Model. Neural Netw. 22, 623–632 (2009)

98. Schliebs, S., Mohemmed, A., Kasabov, N.: Are Probabilistic Spiking Neural Networks Suitable for Reservoir Computing? In: Int. Joint Conf. Neural Networks, IJCNN, San Jose, pp. 3156–3163. IEEE Press (2011)

99. Schliebs, S., Nuntalid, N., Kasabov, N.: Towards Spatio-Temporal Pattern Recognition Using Evolving Spiking Neural Networks. In: Wong, K.W., Mendis, B.S.U., Bouzerdoum, A. (eds.) ICONIP 2010, Part I. LNCS, vol. 6443, pp. 163–170. Springer, Heidelberg (2010)

100. Schrauwen, B., Van Campenhout, J.: BSA, a fast and accurate spike train encoding scheme. In: Proceedings of the International Joint Conference on Neural Networks, vol. 4, pp. 2825–2830. IEEE (2003)

101. Soltic, S., Kasabov, N.: Knowledge extraction from evolving spiking neural networks with rank order population coding. International Journal of Neural Systems 20(6), 437–445 (2010)

102. Sona, D., Veeramachaneni, H., Olivetti, E., Avesani, P.: Inferring cognition from fMRI brain images. In: Proc. of IJCNN. IEEE Press (2011)

103. Song, S., Miller, K., Abbott, L., et al.: Competitive hebbian learning through spike-timing-dependent synaptic plasticity. Nature Neuroscience 3, 919–926 (2000)

104. Theunissen, F., Miller, J.P.: Temporal encoding in nervous systems: a rigorous definition. Journal of Computational Neuroscience 2(2), 149–162 (1995)

105. Thorpe, S., Gautrais, J.: Rank order coding. Computational Neuroscience: Trends in Research 13, 113–119 (1998)

106. Thorpe, S., Delorme, A., et al.: Spike-based strategies for rapid processing. Neural Netw. 14(6-7), 715–725 (2001)

107. van Schaik, A., Shih-Chii Liu, L.: AER EAR: a matched silicon cochlea pair with address event representation interface. In: Proc. of ISCAS - IEEE Int. Symp. Circuits and Systems, May 23-26, vol. 5, pp. 4213–4216 (2005)

108. Verstraeten, D., Schrauwen, B., D'Haene, M., Stroobandt, D.: An experimental unification of reservoir computing methods. Neural Networks 20(3), 391–403 (2007)

109. Villa, A.E.P., et al.: Cross-channel coupling of neuronal activity in parvalbumin-deficient mice susceptible to epileptic seizures. Epilepsia 46(suppl. 6), 359 (2005)

110. Wang, X., Hou, Z.G., Zou, A., Tan, M., Cheng, L.: A behavior controller for mobile robot based on spiking neural networks. Neurocomputing 71(4-6), 655–666 (2008)

111. Watts, M.: A Decade of Kasabov's Evolving Connectionist Systems: A Review. IEEE Trans. Systems, Man and Cybernetics- Part C: Appl. and Reviews 39(3), 253–269 (2009)

112. Weston, I., Ratle, F., Collobert, R.: Deep learning via semi-supervised embedding. In: Proc. 25th Int. Conf. Machine Learning, pp. 1168–1175 (2008)

113. Widiputra, H., Pears, R., Kasabov, N.: Multiple Time-Series Prediction through Multiple Time-Series Relationships Profiling and Clustered Recurring Trends. In: Huang, J.Z., Cao, L., Srivastava, J. (eds.) PAKDD 2011, Part II. LNCS, vol. 6635, pp. 161–172. Springer, Heidelberg (2011)

114. Widrow, B., Lehr, M.: 30 years of adaptive neural networks: perceptron, madaline, and backpropagation. Proceedings of the IEEE 78(9), 1415–1442 (1990)

115. Wysoski, S., Benuskova, L., Kasabov, N.: Evolving spiking neural networks for audiovisual information processing. Neural Networks 23(7), 819–835 (2010)

116. Jin, X., Lujan, M., Plana, L.A., Davies, S., Temple, S., Furber, S.: Modelling Spiking Neural Networks on SpiNNaker. Computing in Science & Engineering 12(5), 91–97 (2010) ISSN 1521-961

117. Yu, Y.C., et al.: Specific synapses develop preferentially among sister excitatory neurons in the neocortex. Nature 458, 501–504 (2009)

118. Zhdanov, V.P.: Kinetic models of gene expression including non-coding RNAs. Phys. Reports 500, 1–42 (2011)

Uncovering the Neural Code Using a Rat Model during a Learning Control Task

Chenhui Yang, Hongwei Mao, Yuan Yuan, Bing Cheng, and Jennie Si

Arizona State University, Tempe AZ 85287, USA

Abstract. How neuronal firing activities encode meaningful behavior is an ultimate challenge to neuroscientists. To make the problem tractable, we use a rat model to elucidate how an ensemble of single neuron firing events leads to conscious, goal-directed movement and control. This study discusses findings based on single unit, multi-channel simultaneous recordings from rats frontal areas while they learned to perform a decision and control task. To study neural firing activities, first and foremost we needed to identify single unit firing action potentials, or perform spike sorting prior to any analysis on the ensemble of neural activities. After that, we studied cortical neural firing rates to characterize their changes as rats learned a directional paddle control task. Single units from the rat's frontal areas were inspected for their possible encoding mechanism of directional and sequential movement parameters. Our results entail both high level statistical snapshots of the neural data and more detailed neuronal roles in relation to rat's learning control behavior.

1 Introduction

The neural events leading to a voluntary movement, or an intentional purposeful movement, may be characterized by three complex processes: target identification, plan of action, and execution. Several distinct regions of the cerebral cortex are believed to be involved in governing these processes, including the posterior parietal cortex, the premotor areas (PM) of the frontal cortex, and the primary motor cortex (M1) [Kandel et al. (2000)]. Adaptation represented in neural firing events has been observed in motor cortical areas which correlate with improved behavioral parameters [Kargo et al. (2004)]. Premotor and parietal areas appear to participate in a fundamental event necessary to purposeful movement: the translation of sensory inputs into motor coordinates needed to specify precise movements [Andersen et al. (2004)]. On the other hand, there has been growing evidence of M1's involvement in sequential tasks using a monkey model [Ben-Shaul et al. (2004); Carpenter et al. (2004); Kakei et al. (1999); Li et al. (1999); Lu et al. (1999); Shima et al. (2000)]. The study in [Shima et al. (2000)] showed that supplementary (SMA) played roles in task execution but presupplementary (pre-SMA) area was responsible for learning new aspects of a task in memorized tasks. In a rat model, existing studies were based on short-duration behavioral tasks (e.g. 20ms). However, there is little anatomical or functional evidence that rats have a well delineated pre-SMA.

Even though the frontal areas of a rat is not as elaborate as a primate, it is however well observed, including our own data, that rats do have the capability to derive abstract

J. Liu et al. (Eds.): WCCI 2012 Plenary/Invited Lectures, LNCS 7311, pp. 261–279, 2012.
© Springer-Verlag Berlin Heidelberg 2012

control strategy via associative learning. Therefore, in this study, we aim to perform a functional study to examine neural coding in the primary motor cortex (M1) and the premotor cortex (PM) during a rat's natural movement in response to a cognitive control task that requires multiple presses at a control paddle. This study centers on investigating the following three aspects. First, neural adaptation may be reflected in the mean firing rate of a motor cortical neural ensemble during learning of a cognitive control task. Second, a larger percentage of PM neurons may be involved in interpreting sensory stimuli and motor planning than M1 neurons. Third, M1 neuronal responses vary according to the movement context in a multiple press task.

In the following, we first introduce an automated action potential detection algorithm which is an important first step to perform any analysis on single unit based analysis of neuronal firing activities. Spike rate based analyses will then be carried out using the sorted single unit spikes. The aims of the analyses are to characterize the rat's behavioral learning parameters by providing a neural substrate based on simultaneously recorded multiple neurons in the rat's motor cortical areas.

2   Single Unit Recording from Behaving Rats

In this section we provide details of our experimental set up, from behavioral training to simultaneous electrophysiolocal recording of single unit neural activities from an ensemble of neurons in the rat's motor cortical areas.

2.1   Animal Handling and Training Procedures

All procedures involving animals were conducted according to the National Guidelines on Animal Experiments and were approved by the Arizona State University Institutional Animal Care and Use Committee.

Recording electrode implant surgeries were performed when the rats reached a weight of 390-500 grams and they were proficient (with an accuracy of 90% or higher) at pressing the control paddle inside the Skinner training chamber for at least 3 consecutive days. Once recording began, rats were food restricted to a daily diet of 12-15 grams of food pellets including the amount of reward collected during the recording session. Food restricted rats were monitored for their weights to be above 80% of the average weight at their respective age.

2.2   Surgical Procedures

Figure 1(a) illustrates the craniotomy for the electrode array to be placed. Additionally, 3 anchor holes were drilled between bregma and lambda: 2 in the right hemisphere and 1 in the left hemisphere, for mounting bone screws which serve as signal ground and also provide fixation to secure the head cap. A 16 channel microwire array (Omnetics or ZIF-Clip, TDT Corporate, Florida) was then lowered slowly into the craniotomy while neural signals were monitored in real time. The target depth was about 1.8-2.3 mm from dura aiming for layer 5 pyramidal neurons. The final depth was determined by optimal spiking activities on majority of the recording channels.

2.3   Electrophysiology

Neural waveforms were recorded through 16-channel microwire arrays connected with an omnetics headstage or a Zif-clip headstage by TDT (Medusa Connector LP16CH or ZIF-Clip ZC16). Analog waveforms passed through a unity gain preamplifier (Medusa PreAmp RA16PA, TDT Corporate), which also provides a band-pass filter (2.2 Hz to 7.5 kHz). The waveforms were digitally sampled at 24414 Hz and then sent over a fiber optic link to a DSP device, where they were filtered (band-pass 300 Hz to 3 kHz), and processed (cross channel denoising) in real-time (RX5/RX7, TDT Corporate). The stored waveforms were spike sorted offline into single unit action potentials using a multi-scale correlation of wavelet coefficients (MCWC) spike detection algorithm [Yang et al. (2011)] followed by a template matching sorting procedure. Events in the behavioral task such as cue on, paddle release, paddle press and food reward were registered simultaneously and time stamped by the TDT system.

2.4   Behavioral Task

Rats were freely moving inside a Skinner box when not performing the designed task. The task is self-paced, which is for the rat to associate light cues with control paddles. The chamber is dark with a 0.5 watt infrared light illumination for video recording. Figure 1(b) is a top view of the recording chamber. When working on the task, the rat faced the front panel of the chamber where 5 red LED lights were placed. At most one cue light was lit at any given time. The 3 control paddles were to be used by the rat to

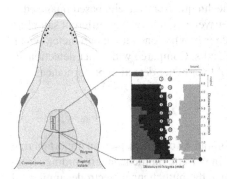
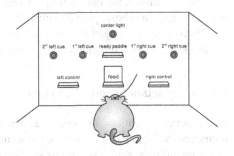

(a) Implant site: a 16-channel array was placed in the frontal area of the rat with the center at 3 mm anterior and 2 mm lateral from bregma. Numbered circles indicate electrode positions. Black area is the primary cortex, and white area is the secondary motor cortex, according to the rat brain atlas.

(b) The recording chamber. The rat pressed the center paddle to signal the start of a new trial. As one of the cue light appeared, the rat had to make a decision of pressing either the left or right control paddle to control the movement of the light. Each left/right press of the control paddle moved the light in the right/left direction by one step.

Fig. 1. Experimental setup

complete the association task. A center control paddle was placed for the rat to press as a signal of a new trial start. The two control paddles on each side of the central paddle were for controlling the movement of the light positions. A food pellet dispenser was located in the center paddle for rewarding the rat. The goal of the task was to move the light position to the center by pressing the control paddles and remain there for 1 second. The rat was not able to start a new trial until a lapse of 8 seconds for successful trials and 15 seconds for failed or timed-out trials, respectively. Upon pressing of the center paddle by the rat to start a new trial, one of the 5 cue lights was lit. Two seconds later, both control paddles were released. Right paddle moved the light to the left by one light position and similarly for the right side. The rats were naive initially and they learned the task by trial and error. If they managed to keep the light remain in the center position for 1 second, they would be rewarded with food pellets. If the rats did not respond by pressing any paddle within the time allowance of 1 second or if the light moved out of range, the trial was deemed a failure.

3   Spike Detection Based on Wavelet Transform

Extracellular chronic recordings have been used as important evidence in neuroscientific studies to unveil the fundamental neural network mechanisms in the brain. Spike detection is the very first step in the analysis of the recorded neural waveforms to decipher useful information and to provide useful signals for brain machine interface applications. This multiscale correlation of wavelet coefficients (MCWC) is an automated spike detection algorithm, which leverages a technique from wavelet based image edge detection. It utilizes the correlation between wavelet coefficients at different sampling scales to create a robust spike detector. The algorithm has one tuning parameter, which potentially reduces subjectivity of detection results. Compared with other detection algorithms, the proposed method has a comparable or better detection performance.

3.1   Introduction to Spike Detection

Neural action potentials, also known as nerve impulses or spikes, play an important role in understanding the central nervous system. In chronic multichannel recordings from behaving animals, action potentials are obtained by multichannel electrodes implanted in brain areas of interest. As such, noise from brain tissues, muscle movement, and other biological and instrumental interferences are inevitable [Musial et al. (2002)]. As the first step of neuroscientific studies and engineering applications such as brain machine interfaces, identifying real neural spikes from noisy recordings is essential.

The wavelet transform is a technique for representing a time domain signal by a set of functions that are scaled and time-translated from a mother wavelet. Its characteristics make it a natural candidate for transient signal representation and thus spike detection applications. Several spike detection algorithms based on wavelet transforms have been proposed [Yang et al. (1988),Kim et al. (2003),Hulat et al. (2000),Hulat et al. (2002),Quiroga et al. (2004),Oweiss et al. (2002a), Oweiss et al. (2002b),Nenadic et al. (2005), Benitez et al. (2008)].

It is noticed that simple threshold based detection methods are intuitive in principle and easy to implement. This is echoed by its popularity including commercial realizations of the algorithms. However as pointed out in [Wood et al. (2004)], the detection results are variable and subjective to users, in addition to high false alarm rates. Other than these direct thresholding of the recorded neural waveforms as a function of time, the idea of thresholding was also an important part of several other approaches, such as the threshold applied to wavelet coefficients in [Yang et al. (1988)] and [Kim et al. (2003)], a higher than usual threshold to gather spikes as the ground truth in [Song et al. (2006)], the threshold applied to the output of minimum average correlation energy (MACE) filter in [Dedual et al. (2007)], the threshold for selecting potential neural spikes in [Hulat et al. (2002)] and [Hulat et al. (2000)], two thresholds used in multi-resolution generalized likelihood ratio test (MRGLRT) in [Oweiss et al. (2002a)] and [Oweiss et al. (2002b)], and the threshold used for separating neural spikes from noise in [Nenadic et al. (2005)]. It is worth pointing out that several algorithms rely on a Gaussian noise assumption to make an optimal detection statement. On one hand, it gives users some assurance of optimality, but unfortunately, noise profile is rarely Gaussian in recorded neural waveforms.

The MCWC aims at providing robust detection performance with high detection rate and low false alarm. The goal is to alleviate subjectivity and variability in detection results. In doing so, we made use of the observation that a sharp rise of neural waveform signifying the onset of a neural spike in a 1-D neural signal is similar in characteristic to an edge in a 2-D image. Therefore, the MCWC algorithm is a wavelet based approach, inspired by image edge detection. In [Xu et al. (1994)], an edge detection algorithm makes use of a property in wavelet transform coefficients that the wavelet transform coefficients of image edges usually have higher magnitudes than the coefficients from noise. As shown later in this study, the wavelet coefficient magnitudes of neural recordings preserve similar properties with a properly selected wavelet function: coefficients of neural spikes have higher magnitudes than those coefficients of noise.

The MCWC utilizes continuous wavelet transform as that in wavelet detection method (WDM) [Nenadic et al. (2005)] and [Benitez et al. (2008)], however with different wavelet functions in the respective implementations. Another major difference between the two algorithms is that while WDM performs detection at individual wavelet scales prior to fusing the results from multiple levels for a final spike detection, our approach fuses wavelet transforms from multiple scales first at each scale level and then perform a single detection by hypothesis testing. We only introduce one free parameter, which in turn helps reduce the subjectivity of the algorithm.

3.2   Working Principle of the Multiscale Correlation of Wavelet Coefficients (MCWC)

We chose wavelet function "coiflets" based on the following considerations. When the time support of the wavelet function matches the duration of a neural waveform, the corresponding wavelet transform coefficients become high. But the waveforms of noise usually do not resemble the wavelet function. Therefore the coefficients from noise have small or close to zero magnitudes. By inspecting waveforms corresponding to high wavelet transform coefficients, we can detect neural spikes.

The MCWC spike detection algorithm is based on continuous wavelet transform. It takes a multiscale approach by first calculating the wavelet coefficients at each scale, and correlates (by multiplication) wavelet coefficients from multiple scales, and then perform a hypothesis test for spike detection. It is motivated by a robust image edge detection algorithm [Xu et al. (1994)] where in this case, a sharp spike is considered a 1-D edge. What follows is a step by step development of the MCWC algorithm.

3.2.1 Computing Normalized Correlation of Wavelet Coefficients.
Consider a neural waveform $x(t)$. Let J be the width of the observation window of the waveform under consideration which is used as the integration interval in the calculation of wavelet coefficients. And let N be the number of samples in the observation window J. With scale factor $\{a_i\} = \{0.5, 0.6, \cdots, 1.5ms\}$, and time translation $\{b_j\} = \{0, 1, 2, \cdots, N - 1\}$, we obtain

$$Tx(a_i, b_j) = \int_J x(t) \frac{1}{\sqrt{a_i}} \psi\left(\frac{t - b_j}{a_i}\right) dt. \tag{1}$$

$$r_S(a_i, b_j) = \prod_{k=0}^{S-1} Tx(a_{i+k}, b_j). \tag{2}$$

$$P_{r_S}(a_i) = \sum_{j \in J} r_S(a_i, b_j)^2, \tag{3}$$

$$P_{Tx}(a_i) = \sum_{j \in J} Tx(a_i, b_j)^2, \tag{4}$$

$$r'_S(a_i, b_j) = r_S(a_i, b_j) \times \sqrt{\frac{P_{Tx}(a_i)}{P_{r_S}(a_i)}}. \tag{5}$$

Where $\psi(t)$ is wavelet function, $Tx(a, b)$ denotes the wavelet transform of $x(t)$, S is the number of sampling scales in a continuous wavelet transform.

3.2.2 Spike Detection Using Hypothesis Testing.
H_0: $x(t)$ contains no spikes in the small window $[t_0, t_1]$ belonging to J under consideration (Fig. 2),

H_1: $x(t)$ contains a spike at b_j in the small window $[t_0, t_1]$ belonging to J under consideration (Fig. 2).

Specifically, H_0 holds, or no spike is detected if

$$\left| \frac{r'_S(a_i, b_j)}{Tx(a_i, b_j)} \right| \leq 1 \tag{6}$$

and H_1 holds, or a spike is detected if (7) is satisfied,

$$\left| \frac{r'_S(a_i, b_j)}{Tx(a_i, b_j)} \right| > 1 \tag{7}$$

Fig. 3 illustrates the principle of spike detection proposed in this study.

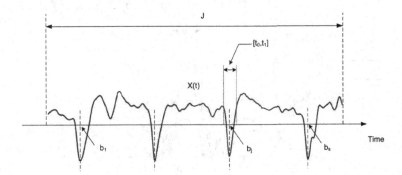

Fig. 2. The integration window for computing wavelet coefficients and illustration of finding spike instants: b_j, $j = 1, 2, \cdots, s$, represent the instants of neural spike peaks

Let H_1 hold inside the small interval $[t_0, t_1]$ at specific points of b_j where there can possibly be more than one b_j's. Let t_d be the instant of a spike within $[t_0, t_1]$ (refer to Fig. 3). Then a spike is detected at t_d within $[t_0, t_1]$ from the following

$$t_d = \arg \max_{\substack{b_j \in [t_0, t_1] \\ a_i \in \{0.5, \cdots, 1.5\}}} |Tx(a_i, b_j)|. \tag{8}$$

The width of the spike detected at t_d, τ, is estimated by

$$\tau = \arg \max_{a_i \in \{0.5, \cdots, 1.5\}} |Tx(a_i, t_d)|. \tag{9}$$

This effectively implies that no other spikes exist within a distance of τ from t_d.

3.2.3 Detection Principle: Adaptive Thresholding.

We are now ready to demonstrate that the MCWC detection algorithm actually is an adaptive thresholding method. The threshold level changes as the signal-to-noise ratio or the noise covariance varies.

First, consider the case of $S = 2$. Re-write (7) into the following by assuming that $Tx(a_i, b_j)$ is non-zero, which is commonly true.

$$\left| Tx(a_i, b_j) Tx(a_{i+1}, b_j) \sqrt{\frac{\sum\limits_{j \in J} Tx(a_i, b_j)^2}{\sum\limits_{j \in J} Tx(a_i, b_j)^2 Tx(a_{i+1}, b_j)^2}} \right| > |Tx(a_i, b_j)|. \tag{10}$$

Define $\overline{T}x(a_i, S)|_{S=2}$ as in (11),

$$\overline{T}x(a_i, S)|_{S=2} \triangleq \frac{\sum\limits_{j \in J} Tx(a_i, b_j)^2 Tx(a_{i+1}, b_j)^2}{\sum\limits_{j \in J} Tx(a_i, b_j)^2}. \tag{11}$$

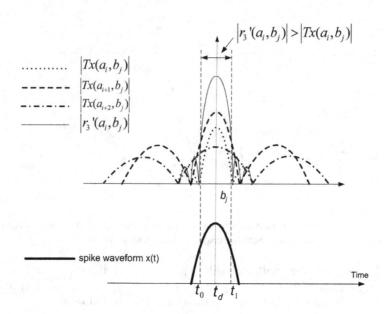

Fig. 3. Demonstration of MCWC detection principle: The multiplication of multi-scale wavelet coefficients enhances the detection of a neural spike. Inequality $|r'_S(a_i, b_j)| > |Tx(a_i, b_j)|_{S=3}$ for $t \in [t_0, t_1]$ indicates that hypothesis H_1 passes the test in this interval. The detection of a neural spike at time instant t_d is declared.

Re-arranging (10) by substituting the newly defined term $\overline{T}x(a_i, S)|_{S=2}$, we obtain the following new form of spike detection criterion,

$$Tx(a_{i+1}, b_j)^2 > \overline{T}x(a_i, S)|_{S=2}. \tag{12}$$

Under a similar assumption to that in [Nenadic et al. (2005)] at the ith scale level, $\{Tx(a_i, b_j)\}$ are independent Gaussian random variables and comply with the following distributions,

$Tx(a_i, b_j) \sim N(0, \sigma^2)$ given H_0 holds, which implies that given H_0, $Tx(a_i, b_j)$ complies with a Gaussian distribution with zero mean and σ^2 as its variance.

$Tx(a_i, b_j) \sim N(\mu, \sigma^2)$ given H_1 holds, which implies that given H_1, $Tx(a_i, b_j)$ complies with a Gaussian distribution with μ as its mean and σ^2 as its variance.

Define a weighting coefficient w_i as shown below,

$$w_i \triangleq \frac{Tx(a_i, b_j)^2}{\sum\limits_{j \in J} Tx(a_i, b_j)^2} = \frac{Tx(a_i, b_j)^2/\sigma^2}{\sum\limits_{j \in J} [Tx(a_i, b_j)^2/\sigma^2]}. \tag{13}$$

Let $P(H_0)$ be the prior probability associated with hypothesis H_0 and $P(H_1)$ be that with hypothesis H_1. Then for real neural recordings, it is reasonable to assume that $P(H_0) \gg P(H_1)$ since majority of the time course of a neural recording corresponds with noise [Nenadic et al. (2005)]. Given that H_0 holds, then $Tx(a_i, b_j)$ complies with

$N(0, \sigma^2)$. Consequently the new variable $Tx(a_i, b_j)^2/\sigma^2$ complies with the chi-square distribution with 1 degree of freedom, i.e., $Tx(a_i, b_j)^2/\sigma^2 \sim \chi_1^2(1)$. Therefore $E\left[Tx(a_i, b_j)^2/\sigma^2\right] = 1$ remains valid most of the time. It also is approximately true if H_1 holds but the mean μ in $Tx(a_i, b_j)$ is relatively low. Since $E\left[Tx(a_i, b_j)^2/\sigma^2\right] \approx 1$, then $\sum_{j \in J}\left[Tx(a_i, b_j)^2/\sigma^2\right] \approx M$, where M is the cardinalities of J. Thus the weight $w_i \approx 1/M$.

Since $Tx(a_i, b_j)$ for a given scale may be viewed as an independent Gaussian variable with zero mean most of the time especially when it corresponds with noise, the maximum likelihood estimate of the variance of the noise sequence $\{Tx(a_{i+1}, b_j)\}$ is $\frac{1}{M}\sum_J Tx(a_{i+1}, b_j)^2$. To see that, refer to (13) and that $w_i \approx 1/M$. Therefore the threshold $\overline{T}x(a_i, S)|_{S=2}$ defined in (12) can be viewed as an approximation of the maximum likelihood estimation of the noise variance since $P(H_0) \gg P(H_1)$. When $Tx(a_{i+1}, b_j)^2 > \overline{T}x(a_i, S)|_{S=2}$, it implies that the correlation between the neural waveform and the wavelet is greater than the noise variance, and therefore, a neural spike is likely to be present, and that H_1 is true. For $S \geq 3$ case, readers are referred to [Yang et al. (2011)].

Fig. 4(a) and Fig. 4(b) illustrate how the adaptive threshold values vary as a function of the SNR or the noise co-variance and the scale level S.

3.3 Detection Performance Evaluation

In this section, we provide detailed performance evaluation on the multiscale correlation of wavelet coefficient (MCWC) algorithm. While comparisons are conducted for a few algorithms including direct thresholding, our focus is on comparing MCWC and WDM

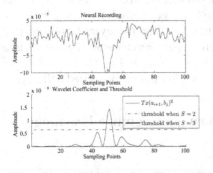
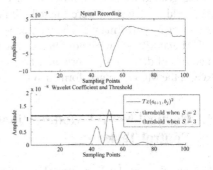

(a) Detection thresholds at two S levels when neural signal has a low SNR or high noise covariance

(b) Detection thresholds at two S levels when neural signal has a high SNR or low noise covariance

Fig. 4. Illustration of adaptive threshold in MCWC, which varies with SNRs or noise covariances and S values. From (a) and (b), at a given SNR, when S is high, the threshold levels are high and vice versa. The SNR in (a) is lower than that in (b), therefore at a given S, the threshold level is higher when SNR is higher.

since both are wavelet based, and WDM has been shown outperforming several other approaches [Nenadic et al. (2005)] and [Santaniello et al. (2008)].

18 artificially generated neural waveforms span 50 seconds, and they are sampled at 20KHz. Half of the 18 artificial data sets (A1-1 to A1-9) were obtained with 1dB signal-to-noise ratio (SNR), while the other half (A2-1 to A2-9) with 10dB SNR.

Each artificial neural data set was generated the same way as in [Smith (2006)].

When applying the WDM algorithm, one is required to select a parameter L that determines the cost ratio between the false alarms and missed detections.

3.3.1   Comparison of Detection Performance among Thresholding, MCWC, and WDM Using Artificial Neural Data Sets

In this section, we compare detection performances among MCWC, WDM, and thresholding provided in Plexon's Offline Sorter. Eighteen artificial data sets, A1-1 to A1-9, and A2-1 to A2-9 are used. The thresholding method used was the "Signal Energy" in Offline Sorter as described below,

$$energy(i) = \frac{1}{W} \sum_{j=i-W/2}^{i+W/2} v^2(j), \tag{14}$$

where $v(j)$ is the raw neural recording at time j. W is the window width used in averaging. In this study, $W = 7$ is used.

To make results comparable, we manually selected the threshold value in Offline Sorter such that the total number of detected neural spikes was close to that of the ground truth. The L and S parameters in WDM and MCWC, respectively, were chosen similarly such that the total number of detected spikes by each algorithm was close to the ground truth. To remove low frequency noise, the artificial data sets used in thresholding detection were filtered with a band-pass butterworth filter, which usually enhances its performance. The pass band is [100, 6000]Hz. However, the data used in WDM and MCWC were not filtered. The receiver operating characteristics (ROC) as a measure of detection performance are shown in Fig. 5(a) and Fig. 5(b). Based on the ROCs, the MCWC outperformed WDM and thresholding at the two tested SNR levels.

4   Cortical Neural Modifications during a Cognitive Learning Control Task

The neural events leading to a voluntary movement, or an intentional purposeful movement, may be characterized by three complex processes: target identification, plan of action, and execution. Several distinct regions of the cerebral cortex are believed to be involved in governing these processes, including the posterior parietal cortex, the premotor areas (PM) of the frontal cortex, and the primary motor cortex (M1) [Kandel et al. (2000)]. Premotor and parietal areas appear to participate in a fundamental event necessary to purposeful movement - the translation of sensory inputs into motor coordinates needed to specify precise movements [Andersen et al. (2004)]. And adaptation represented in neural firing events has been observed in motor cortical areas which correlates

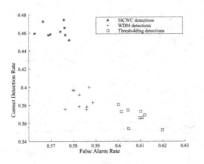 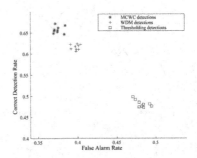

(a) Detection performance with data sets A1-1 to A1-9 (SNR=1dB)

(b) Detection performance with data sets A2-1 to A2-9 (SNR=10dB)

Fig. 5. ROC performance for MCWC, WDM and Plexon thresholding: MCWC has a better performance than WDM and Plexon thresholding in low and high SNR scenarios.

with improved behavioral parameters [Kargo et al. (2004)]. To investigate the neural mechanism of cognitive control, here we studied various aspects of cortical neural firing rates to characterize their changes as rats learned a directional paddle control task. Five rats learned to press one of two paddles (left or right) which extend at 2 seconds after the onset of a directional light cue. By trial and error, most subjects improved their behavioral accuracy to 85% or above in 5 weeks. Both primary motor (MI) and premotor (PM) cortical neurons were recorded from rat's left brain (Fig. 1(a)) during the entire course of learning.

4.1 Characterizing Cortical Neural Modifications Using Firing Rates

The rat's behavioral learning control process in our experiments was divided into 3 stages: Naive, Improving, and Stable according to the behavioral accuracy (Fig. 6(a), rat W09 for example). As it usually takes the rat several weeks to improve the accuracy to a high and stable level, the reaction time, in response to both the center ready paddle (to start a trial) and the control paddle (to control the lights), became stabilized in only a few days (Fig. 6(b)). This may indicate that it is the cognitive aspect of the behavioral task, rather than the motor skill, that the rat learned in this experiment.

Correct trials were grouped as L and R trials, representing either left or right movement directions. A task trial was sliced to form 4 epochs, which are cue on (CO), movement onset (MO), getting ready (RE) (0-400ms, 400-800ms and 1400-1800ms, respectively, after cue onset), and preparing to press (PP, 400-0ms before first press).

For MI neurons, significant mean firing rate changes ($p < 0.001$, ANOVA) through the three learning stages were observed in right (contralateral to the recording sites) but not left trials (Fig. 7). For example, the mean firing rates over the 3 stages in the CO epoch of R trials are 33.5, 36.2 and 38.5 Hz, and the mean rates of L trials in the same epoch are 37.7, 36.9 and 37.4 Hz. In PM neurons, there was statistically significant firing rate change in all epochs. The rate changes were more pronounced in the first

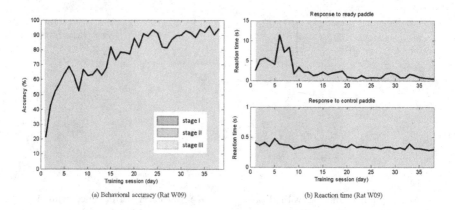

(a) Behavioral accuracy (Rat W09)   (b) Reaction time (Rat W09)

Fig. 6. Behavioral results: (a) accuracy and (b) reaction time

three epochs (e.g. PM neurons R trials, CO: 41.9 ±25.1, 39.9 ±23.9, 49.0 ±22.8 Hz) than in PP (PM neurons R trials: 43.0 ±27.4, 41.0 ±25.5, 43.3 ±22.9 Hz). The PP epoch is immediately before control paddle press, which had been a familiar motor skill before the rat was recorded. But the first three epochs, during which the rat observed the cue light and made decision of movement direction, were believed to involve more cognitive effort. So the changes in mean firing rates found in the same period might be associated with the cognitive learning process.

The mean Fano factor (FF, variance over mean of trial based firing rates) value of both MI and PM neuron ensembles increased with learning (Fig. 8). Meanwhile, decrement of the standard deviation of the FF values was observed in some cases (e.g. 2.05, 1.71 and 1.53, MI neurons in L trials and RE epoch). A potential explanation for this is that single neurons increased their firing variability as a means of characterizing plasticity during the acquisition of a new task.

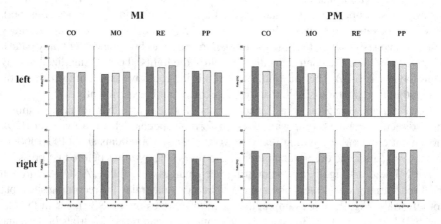

Fig. 7. Mean firing rates of the three learning stages, Naive (red), Improving (green), and Stable (blue), in four trial epochs for MI and PM neurons

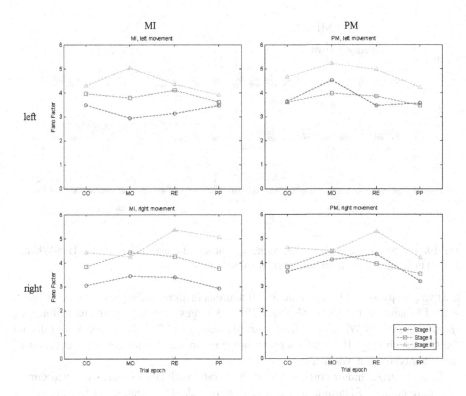

Fig. 8. Median Fano Factor value of the three learning stages, in four trial epochs for MI and PM neurons

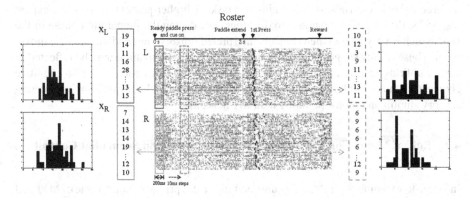

Fig. 9. An illustration of measuring statistical firing rate differences between L and R trials

The difference in firing rates between L and R trials was also measured. As shown in Fig. 9, a 200ms window was moved at 10ms steps through the trial time. The number of spikes within the window in each trial was counted and grouped for L and R trials. Mann-Whitney U-test was then performed to evaluate the statistical difference between

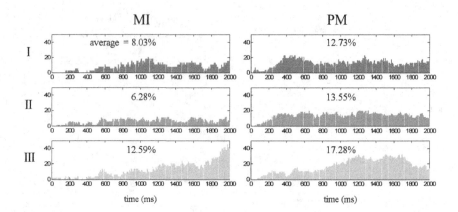

Fig. 10. The percentage of neurons showing significant difference ($P < 0.01$, Mann-Whitney U-test) in firing rates between left and right movement.

the two groups. And this was done for all neurons in all recording sessions. As a result, more PM neurons (7.5%, 8.1% and 5.7% in 3 stages) showed significant differences ($p < 0.01$, Mann-Whitney U-test) than MI neurons (1.2%, 2.5% and 1.3%) during the CO epoch (Fig. 10), when sensory information was processed and a decision of movement direction was made.

To summarize, motor cortical neural firing rate modulations were observed during the entire course of learning a cognitive control task. The mean firing rate change of motor cortical neural ensemble may be an indication of cognitive development. The increment of neuronal Fano factor value indicated that individual neurons fired more differently between trials after learning the task. A higher percentage of PM neurons responded differently in response to different cues during the preparation phase of the task, which is very likely when subjects interpreted sensory cues and planned the movement. These results suggested the followings, (1) neural adaptation may be reflected in the mean firing rate of a motor cortical neural ensemble during learning of a cognitive control task, and (2) a larger percentage of PM neurons may be involved in interpreting sensory stimuli and motor planning than M1 neurons.

4.2 Role of Motor Cortical Neurons in a Directional and Sequential Control Task

In order to examine the roles of neural coding in the primary motor cortex (M1) and the premotor cortex (PM) [Paxinos et al. (2007)] of rats during natural movement, for example cognitive control by multiple paddle presses in our experiment, we hypothesize that M1 neuronal responses change as a function of the movement context.

There is growing evidence of M1's involvement in sequential tasks using a monkey model [Ben-Shaul et al. (2004); Carpenter et al. (2004); Kakei et al. (1999); Li et al. (1999); Lu et al. (1999); Shima et al. (2000)] in memorized tasks and learning new aspects of a task [Shima et al. (2000)]. However, in a rat model, existing studies were only based on short-duration behavioral tasks (e.g. 20ms) and therefore unable to make similar statements. Furthermore, little anatomical or functional evidence shows that rats have a pre-SMA and it is still inconclusive that rats having a mixture of ventral premotor (PMv) and SMA, and thus leaving the question open if rat's motor cortical neurons are present and encode sequential information.

In our experiment, the experiment apparatus provides five light cue positions at center (C), single left (L), and double left (LL); similarly for the right side. A rat was to press one of the two directional paddles to move the cue light to the center. As a trial started, one of the five cues was lit for the rat to respond in 2s. In response to the cue light position, the rat made a single or double presses on the paddle with each press move the light to the respective direction once. The rat worked for food rewards when he moved the light cue to the center position and kept there for 1s. Neural waveforms were recorded in the primary motor (M1) and the premotor (PM) areas while rats learned to perform this control task from a naive state to finally mastering it in about 30 sessions (days) on average.

The motor cortical neurons displayed some unique patterns. Using three-way ANOVA (direction (left or right) vs. task sequence length (single or double) vs. order (first or second)) analysis on the neural firing rates (20ms bin). 26% of M1 neurons and 26% of PM neurons were found to represent directional selectivity ($p < 0.01$ for left and right) as shown in Fig. 11, and an example raster is shown in Fig. 12(a). However, only 5% neurons showed significant difference between the L and LL trials ($p < 0.01$ for single versus double presses) when the rat responded to the cue and made his first press. Similar for R and RR trials. Among all recorded neurons, 28% of M1 neurons and 29% of PM neurons showed order selectivity ($p < 0.01$ for the L, LL 1st press versus the 2nd press in LL), as shown in Fig. 12(b). Figure 12(c) is an example of another kind of neurons, which has a tendency of both directional and sequential selectivity.

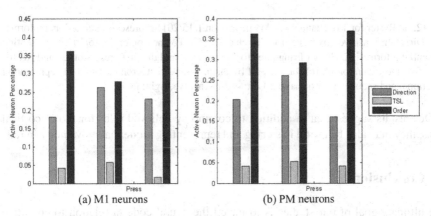

(a) M1 neurons   (b) PM neurons

Fig. 11. 3-way ANOVA of M1 and PM neurons on 3 adjacent windows (length: 20ms) around press on direction (left or right), task sequence length (single or double), and order (first or second). (a) 26% of M1 neurons were found to represent directional selectivity, and 5% and 28% were task sequence length and order selective, respectively. (b) 26% of PM neurons were found to represent directional selectivity, and 6% and 29% were task sequence length and sequence selective, respectively.

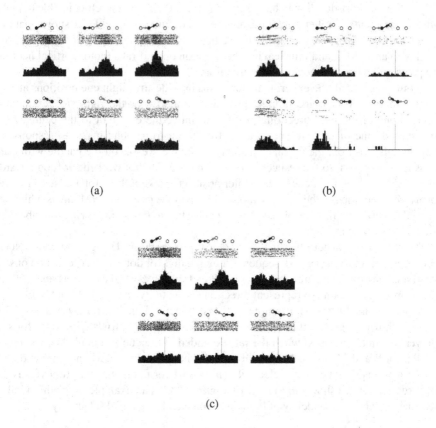

Fig. 12. (a) Raster for T10 channel 4, a M1 neuron on 1/15/2011 representing directional informa-tion. Direction sensitive. (b) Raster for T10 channel 15, a PM neuron on 12/15/2010 representing sequential information. Higher firing rates for single press and the first presses of LL and RR tri-als before movement onset. (c) Raster for T10 channel 4, a M1 neuron on 1/6/2011 representing directional and sequential information. Left: order sensitive; right: none.

Our results suggest that in addition to commonly believed roles for motor cortical areas, they may also be useful in storing and representing sequential movements in rats.

5   Conclusion

The ultimate goal of our studies is to unveil the neural code in relation to cognitive control behaviors. Toward this end, we made use of a rat model and recorded the rats' motor cortical areas using multiple electrode to obtain single unit recordings while the rats performed a directional control task through associative learning. We began our study by introducing a new, automated spike detection algorithm, MCWC, which makes use of correlation and comparison among continuous wavelet transform coefficients at multiple scales. This algorithm provided us the freedom to conduct analysis on large volume of neural recordings with objective measures for spike detection and sorting. Based on this result, we were able to conduct several further analysis using spike rates to analyze spike firing patterns in association with the behavioral learning process and the many aspects of the control task, such as directional control, sequential control, and so on. We found that the role for M1 in a multiple press task is beyond that of controlling movements; the context of a movement contributes to shaping M1 representations, and that the PM is more actively involved in the learning aspect of the cognitive control task.

References

Andersen, R., Musallam, S., Pesaran, B.: Selecting the signals for a brain-machine interface. Current Opinion in Neurobiology 14, 720–726 (2004)

Bao, P., Zhang, L.: Noise reduction for magnetic resonance images via adaptive multiscale products thresholding. IEEE Transactions on Medical Imaging 22(9), 1089–1099 (2003)

Benitez, R., Nenadic, Z.: Robust Unsupervised Detection of Action Potentials With Probabilistic Models. IEEE Transactions on Biomedical Engineering 55(4), 1344–1354 (2008)

Ben-Shaul, Y., Drori, R., Asher, I., Stark, E., Nadasdy, Z., Abeles, M.: Neuronal activity in motor cortical areas reflects the sequential context of movement. J. Neurophysiol. 91, 1726–1748 (2004)

Carpenter, A., Georgopoulos, A., Pellizzer, G.: Motor cortical encoding of serial order in a context-recall task. Science 283, 1752–1757 (1999)

Dedual, N., Ozturk, M., Sanchez, J., Principe, J.: An associative memory readout in ESN for neural action potential detection. In: International Joint Conference on Neural Networks, IJCNN 2007, p. 2295 (2007)

Fee, M.S., Mitra, P.P., Kleinfeld, D.: Variability of extracellular spike waveforms of cortical neurons. Journal of Neurophysiology 76(6), 3823–3833 (1996)

Hulata, E., Segev, R., Ben-Jacob, E.: A method for spike sorting and detection based on wavelet packets and shannon's mutual information. Journal of Neuroscience Methods 117(1), 1–12 (2002)

Hulata, E., Segev, R., Shapira, Y., Benveniste, M., Ben-Jacob, E.: Detection and sorting of neural spikes using wavelet packets. Phys. Rev. Lett. 85(21), 4637–4640 (2000)

Humphrey, D., Schmidt, E.: Extracellular Single-Unit Recording Methods. Neurophysiological Techniques: Applications to Neural Systems 15, 1–64 (1991)

Kakei, S., Hoffman, D., Strick, P.: Muscle and movement representations in the primary motor cortex. Science 285, 2136–2139 (1999)

Kandel, E., Schwartz, J., Jessell, T.: Principles of Neural Science (2000)

Kargo, W., Nitz, D.: Improvements in the Signal-to-Noise Ratio of Motor Cortex Cells Distinguish Early versus Late Phases of Motor Skill Learning. J. Neurosci. 24, 5560–5569 (2004)

Kim, K.H., Kim, S.J.: A wavelet-based method for action potential detection from extracellular neural signal recording with low signal-to-noise ratio. IEEE Transactions on Biomedical Engineering 50(8), 999–1011 (2003)

Kreiter, A.K., Aertsen, A.M., Gerstein, G.L.: A low-cost single-board solution for real-time, unsupervised waveform classification of multineuron recordings. Journal of Neuroscience Methods 30(1), 59–69 (1989)

Li, C., Padoa-Schioppa, C., Bizzi, E.: Neuronal correlates of motor performance and motor learning in the primary motor cortex of monkeys adapting to an external force field. Neuron 30, 593–607 (2001)

Lu, X., Ashe, J.: Anticipatory activity in primary motor cortex codes memorized movement sequences. Neuron 45, 967–973 (2005)

Mallat, S.: A wavelet tour of signal processing. Academic Press, San Diego (1989)

Matsuzaka, Y., Picard, N., Strick, P.: Skill Representation in the Primary Motor Cortex After Long-Term Practice. J. Neurophysiol. 97, 1819–1832 (2007)

Musial, P., Baker, S., Gerstein, G., King, E., Keating, J.: Signal-to-noise ratio improvement in multiple electrode recording. Journal of Neuroscience Methods 115, 29–43 (2002)

Naundorf, B., Wolf, F., Volgushev, M.: Unique features of action potential initiation in cortical neurons. Nature 440(7087), 1060–1063 (2006)

Nenadic, Z.: Spike detection with the continuous wavelet transform, matlab software. University of California, Irvine, Center for BioMedical Signal Processing and Computation (2005), http://cbmspc.eng.uci.edu

Nenadic, Z., Burdick, J.: Spike detection using the continuous wavelet transform. IEEE Transactions on Biomedical Engineering 52(1), 74–87 (2005)

Nenadic, Z., Burdick, J.: A control algorithm for autonomous optimization of extracellular recordings. IEEE Transactions on Biomedical Engineering 53(5), 941–955 (2006)

Olson, B., Si, J., Hu, J., He, J.: Closed-loop cortical control of direction using support vector machines. IEEE Transactions on Neural Systems and Rehabilitation Engineering 13(1), 72–80 (2005)

Oweiss, K., Anderson, D.: A multiresolution generalized maximum likelihood approach for the detection of unknown transient multichannel signals in colored noise with unknown covariance. In: Proceedings of IEEE International Conference on Acoustics, Speech, and Signal Processing, ICASSP 2002, vol. 3, pp. 2993–2996 (2002)

Oweiss, K., Anderson, D.: A unified framework for advancing array signal processing technology of multichannel microprobe neural recording devices. In: 2nd Annual International IEEE-EMB Special Topic Conference on Microtechnologies in Medicine and Biology, pp. 245–250 (2002)

Paxinos, G., Watson, C.: The Rat Brain in Stereotaxic Coordinates (2007)

Quiroga, R.Q., Nadasdy, Z., Ben-Shaul, Y.: Unsupervised spike detection and sorting with wavelets and superparamagnetic clustering. Neural Computation 16(8), 1661–1687 (2004)

Sadler, B.M., Swami, A.: Analysis of multiscale products for step detection and estimation. IEEE Transactions on Information Theory 45(3), 1043–1051 (1999)

Santaniello, S., Fiengo, G., Glielmo, L., Catapano, G.: A biophysically inspired microelectrode recording-based model for the subthalamic nucleus activity in parkinson's disease. Biomedical Signal Processing and Control 3(3), 203–211 (2008)

Shima, K., Tanji, J.: Neuronal activity in the supplementary and presupplementary motor areas for temporal organization of multiple movements. J. Neurophysiol. 84, 2148–2160 (2000)

Smith, L.: Noisy spike generator, matlab software. University of Stirling, Department of Computing Science and Mathematics (2006), http://www.cs.stir.ac.uk/~lss/noisyspikes/

Song, M.J., Wang, H.: A spike sorting framework using nonparametric detection and incremental clustering. Neurocomputing 69(10-12), 1380 (2006)

Thakur, P.H., Lu, H., Hsiao, S.S., Johnson, K.O.: Automated optimal detection and classification of neural action potentials in extra-cellular recordings. Journal of Neuroscience Methods 162(1-2), 364–376 (2007)

Volgushev, M., Malyshev, A., Balaban, P., Chistiakova, M., Volgushev, S., Wolf, F.: Onset Dynamics of Action Potentials in Rat Neocortical Neurons and Identified Snail Neurons: Quantification of the Difference. PLoS ONE 3(4), e1962 (2008)

Wood, F., Black, M., Vargas-Irwin, C., Fellows, M., Donoghue, J.: On the variability of manual spike sorting. IEEE Transactions on Biomedical Engineering 51(6), 912–918 (2004)

Wu, G., Hallin, R.G., Ekedahl, R.: Multiple action potential waveforms of single units in man as signs of variability in conductivity of their myelinated fibres. Brain Research 742(1-2), 225–238 (1996)

Xu, Y., Weaver, J., Healy, D., Lu, J.: Wavelet transform domain filters: a spatially selective noise filtration technique. IEEE Transactions on Image Processing 3(6), 747–758 (1994)

Yang, C., Olson, B., Si, J.: A multiscale correlation of wavelet coefficients approach to spike detection. Neural Computation 23, 215–250 (2011)

Yang, X., Shamma, S.: A totally automated system for the detection and classification of neural spikes. IEEE Transactions on Biomedical Engineering 35(10), 806–816 (1988)

Author Index

Printed in the United States
By Bookmasters